A Dictionary of GEOLOGY

W.G. MORRISON

CBS PUBLISHERS & DISTRIBUTORS PVT. LTD.

New Delhi • Bengaluru • Chennai • Kochi • Kolkata • Mumbai • Pune

ISBN: 81-239-0281-6

First Edition: 1994
Reprint: 2000, 2002, 2004

Copyright © by Author & Publisher

All rights reserved. No part of this book may be reproduced or transmitted in any form or by any means, electronic or mechanical, including photocopying, recording, or any information storage and retrieval system without permission, in writing, from the publisher.

Published by:
Satish Kumar Jain for CBS Publishers & Distributors Pvt. Ltd.,
4819/XI Prahlad Street, 24 Ansari Road, Daryaganj, New Delhi - 110002
delhi@cbspd.com, cbspubs@airtelmail.in • www.cbspd.com
Ph.: 23289259, 23266861, 23266867 • Fax: 011-23243014

Corporate Office: 204 FIE, Industrial Area, Patparganj, Delhi - 110 092
Ph: 49344934 • Fax: 011-49344935
E-mail: publishing@cbspd.com • publicity@cbspd.com

Branches:
- ***Bengaluru:*** 2975, 17th Cross, K.R. Road, Bansankari 2nd Stage, Bengaluru - 70 • Ph: +91-80-26771678/79 • Fax: +91-80-26771680 E-mail: cbsbng@gmail.com, bangalore@cbspd.com
- ***Chennai:*** No. 7, Subbaraya Street, Shenoy Nagar, Chennai - 600030 Ph: +91-44-26681266, 26680620 • Fax: +91-44-42032115 E-mail: chennai@cbspd.com
- ***Kochi:*** Ashana House, 39/1904, A.M. Thomas Road, Valanjambalam, Ernakulum, Kochi • Ph: +91-484-4059061-65 Fax: +91-484-4059065 • E-mail: cochin@cbspd.com
- ***Kolkata:*** 6-B, Ground Floor, Rameshwar Shaw Road, Kolkata - 700014 Ph: +91-33-22891126/7/8 • E-mail: kolkata@cbspd.com
- ***Mumbai:*** 83-C, Dr. E. Moses Road, Worli, Mumbai - 400018 Ph: +91-9833017933, 022-24902340/41 • E-mail: mumbai@cbspd.com
- ***Pune:*** Bhuruk Prestige, Sr. No. 52/12/2+1+3/2, Narhe, Haveli (Near Katraj-Dehu Road Bypass), Pune - 411041 Ph: +91-20-64704058/59, 32342277 • E-mail: pune@cbspd.com

Representatives:
- Hyderabad: 0-9885175004
- Nagpur: 0-9021734563
- Patna: 0-9334159340
- Vijayawada: 0-9000660880

Printed at:
J.S. Offset Printers, Delhi

PREFACE

The aim in preparing the Dictionary of Geology is to provide the nonspecialist with accurate definitions of the working vocabulary of earth sciences. While compiling the dictionary, the author has concentrated on terms which seem to him to be in the widest use ; special and local terms have been excluded. A number of obsolete and popular terms have been included since the former may be encountered in older works frequently still in use, and the latter in some of the books and articles which purport to explain geological terms, stratigraphical divisions (other than periods) are restricted to stage names. No local stratigraphic names are included. Definitions of terms in related sciences and various branches of geology such as geomorphology, geophysics, geochemistry, mineralogy, hydrology, petrology have also been explained with their specific geologic meanings. Each entry dealing with a particular mineral is constructed according to a common discipline to provide all the important information including composition, crystal formation, colour, streak, lustre, physical properties and geographical occurrence.

In addition, every term used in the book is not perfectly standard English but is defined and explained in an easily comprehensible way. Even the amateurs will have no difficulty in using the book while its completeness and authority will make it essential for the student and teachers as well. The dictionary is fully cross referenced to cover alternative names.

PREFACE

The aim in preparing the Dictionary of Geology is to provide the nonspecialist with accurate definitions of the working vocabulary of earth sciences. While compiling the dictionary, the author has concentrated on terms which seem to him to be in the widest use; special and local terms have been excluded. A number of obsolete and popular terms have been included since the former may be encountered in older works frequently still in use, and the latter in some of the books and articles which purport to explain geological terms. Stratigraphical divisions (other than periods) are restricted to stage names. No local stratigraphic names are included. Definitions of terms in related sciences and various branches of geology such as geomorphology, geophysics, geochemistry, mineralogy, hydrology, petrology have also been explained with their specific geologic meanings. Each entry dealing with a particular mineral is constructed according to a common discipline to provide all the important information including composition, crystal formation, colour, streak, lustre, physical properties and geographical occurrence.

In addition, every term used in the book, if not ordinarily standard English, is defined and explained in an easily comprehensible way. Even the amateurs will have no difficulty in using the book, while its completeness and authority will make it essential for the student and teachers as well. The dictionary is fully cross-referenced to cover alternative names.

aa. A Hawaiian term for black lava flow, which has solidified into a blocky mass; *e.g.*, Mt. Etna in Italy.

a-axis. (1) One of the crystallographic axes used as reference in crystal description. It is oriented horizontally, front to back. (2) The axis of a reciprocal crystal lattice which is perpendicular to (100). (3) The direction of movement or transport in a tectonite.

abandon. To stop efforts to obtain oil/gas from a well.

abandoned cliff. A sea cliff that is no longer under the influence of sea wave due to a relative drop of sea level, or progr- adation at the cliff base.

abc soil. A soil with well developed profile, including A, B, and C horizons.

abc system. A procedure for correcting seismic data for the effect of irregular weathering thickness.

ablation. (1) The process of wearing away of solid ice from a glacier. It includes surface, internal and basal welting; sublimation, and calving of icebergs. (2) Removal of molten surface layers of meteorites and tektites by direct vaporization during the passage through the atmosphere.

ablybite. A clay-mineral that resembles halloysite in its dehydration but differs from it in its thermal and X-ray diffraction characteristics; mainly consists of an aluminosilicate of calcium, magnesium and potassium.

a-b plane. The surface along which differential movement takes place.

abrasion. The wearing away or rounding of surface by friction or rubbing as from action of glaciers, wind, and water-borne sand on rock fragments and minerals. It is most effective when the impact of particles on bedrock is vigrous, and the particles themselves are coarse, hard, and angular.

abstraction. The process by which a stream, by the lateral extension of its valley, takes over the catchments of neighbouring streams, thus leading to an overall reduction in number of streams. It usually occurs at the upper end of a drainage line.

abtragung. The degradation not resulting directly from stream erosion.

abyssal. Pertaining to the zones of greatest depth at which water has more or less uniform temperature, also pertaining to the organisms of that environment.

abyssal hill. A common low-relief feature of the ocean floor, usually found seaward of abyssal plains and in basins isolated by ridges, rises, or trenches.

abyssal plain. Flat or nearly level area in deepest portions of ocean based, usually fond at the base of a continental rise; formed by the deposition of turbidity-current and pelagic sediments that obscure the preexisting topography.

abyssopelagic. Pertaining to that portion of the ocean which lies below depths of 3,700 meters, and to ocean-derived sediments found at that depth.

accessory minerals. The minerals which may be present or absent in a crystalline rock without prejudice to the identification of that rock, and are thus contrasted with the essential minerals, on which identification is based.

accordant drainage. A drainage pattern in which the streams are mainly guided by the underlying geological structure.

accordant summits. A series or hilltops and plateau summits rising to approximately same height.
accrelion vein. A type of vein formed by the repeated filling of channels followed by their opening because of the development of fractures in the zone undergoing mineralization.
accumulation area. The part of a glacier of snowfield in which during a year's time, the mass balance is positive i.e., accumulation exceeds ablation.
ac-fracture. In deformed rocks, a tension fracture that parallels the *ac*-plane and is normal to *b*.
Acheulian. Stratigraphic stage name for European Lower Pleistocene; does not contain the chondrules that are found in chrondites.
achroite. A white variety of tourmaline ; rich in potassium.
acicular. A tendency due to which crystals are shaped needlelike.
acid clay. A variety of clay that give off hydrogen ions when dissolved or suspended in water.
acid rock. An igneous rock containing 10% or more free quartz, e.g., granite and ghyolite; former defination; a rock containing more than 66% silica.
aclinal. Horizontal, without dip.
acme-zone. A biozone consisting for a body of strata in which a particular species of genus of an organism occurs with maximum frequency.
acmite. $NaFe(SiO_3)_2$. A brown or green mineral of the clinopyroxene gruop found in certain alkali rich igneous rocks.
acritarch. A group of organic microfossils of unknown biological affinity, generally characterized by a smooth or spiny texture. The presence of the remains of these organisms in sedimentary rocks of marine origin indicates they were plankatonic. Acritarchs range from the Precambrian to the present, but are most abundant in the Precambrian and early Paleozoic.
acrobatholitic. A stage in the batholitic erosion where summits of cupolas and stocks are exposed without any exposure of the surface separating the barren interior of the batholith from the mineralized upper part.
actinolite. $(Ca,Na)_{2\text{-}3}$ $(Mg,Fe,Al)_5$, $(Al,Si)_8$ $O_{22}(OH)_2$. A green monoclinic rock-forming amphibole; occurs in needle-like crystals and columnar or fibrous frame.
active fault. Any type of fault in which there is recurrent movement in the form of continuous creep, small, abrupt displacements, or seismic activity.
active layer. The part of the soil which is within the suprapermafrost layer and which usually freezes in winter and thaws in summer. Also known as frost zone.
active permafrost. Permanently frozen surface which after thawing by artificial means, reverts to permafrost under normal climatic conditions.
active volcano. A volcano capable of venting lava, pyroclastic material, and/or gases.
adamite. $Zn_2(AsO_4)(OH)$. A white or light yellow mineral containing basic zinc arsenate; crystallizing in the orthorhombic system.
adaptive radiation. Process of natural selection whereby lineages seprates physically and genetically within a short interval of geologic time.
adit. A horizontal passage made from the surface of the earth, usually to intersect a mineral vein or coal seam. Although frequently called a tunnel, an adit, unlike a tunnel is closed at one end.

adularia. A colourless variety of orthoclase feldspar, with appearance ranging from transparent to translucent.

advance. (1) Gradual seaward progression of a shoreline as a result of accumulation of emergence; also the net seaward progression during a specific time interval. (2) Forward movement of a glacier front.

advection. (1) Horizontal transport of air or atmospheric properties within the earth's atmosphere. (2) The horizontal or vertical movement of sea water as a current. (3) Lateral mass movement of material in the earth's mantle.

adventive cone. A volcanic cone which is on the flank of and subsidiary to a larger volcano. Also knwon as parasitic cone.

aegirine. $NaFe(SiO_3)_2$. A brown or green colour mineral occurring in alkali-rich igneous rocks. Also known as acmite; aegirite.

aegirite. *See* aegirine.

aeolian deposit. Sediments deposited as a result of transport by wind.

aerogeology. The geologic study of the earth features by means of aerial observations including aerial photography.

aerohydrous mineral. A mineral containing water entrapped in small cavities.

aerolite. A class of meteorite comprising of silicate material. Also known as stony meteorite.

aerosiderite. A meteorite composed chiefly of iron minerals.

affine deformation. A type of deformation involving slipping of very thin layers against each other in such a way that each layer moves equally with respect to its neighbour layer; generally does not lead to folding.

aftershock. A smaller earthquake felt after a large earthquake. It originates at or near the larger earthquake's epicentre.

agalmatolite. A soft and waxy mineral (or stone) or various colours; used for carving images. Also knwon as lardite; pagodite.

age determination. Estimation of the gelogic age of samples of biological and geological importance using radiometric techniques.

age ratio. The ratio of the amount of daughter to parent isotope in a mineral which is radiometrically dated.

agglomerate. Pyroclastic rock of consolidated volcanic fragments, rounded and angular, in a finer-grained matrix in which the rounded fragments predominate. Agglomerates are not well sorted and are common in volcanic necks.

agglutinate. A welded pyroclastic deposit in which the pyroclasts are held together by glassy material.

aggradation. A gradational process that contributes to the general leveling of the earth's surface by means of deposition that builds up. Some possible agents of this process are running water, wind, waves, and glaciers.

aggrading stream. A stream that progressively builds up its channel or flood plain by receiving more load than it can transport. A stream can be caused to aggrade rather than deepen its valley by any change in the adjacent land resulting in alluviation that causes the stream's gradient to become too low for transport of its load; *e.g.*, sinking of the surrounding land.

aggregate. (1) Mass of rock paricles, mineral grains, or combination of both. (2) Any hard, inert material (*i.e.*, gravel, sand, blast furnace slag, clinkers) used for mixing with cement, gypsum or other binding material, to form concrete. The aggregate gives volume, stability, and resistance to erosion, to the finished product.

agnatha. A class of the vertebrata, jawless fish, the fossil forms of which are found in rocks of Devonian age.

agonic line. The line through all points on the earth's surface at which the magnetic declination is zero; it is the locus of all points at which true north and magnetic north coincide.

agpaite. A group of igneous rocks containing feldspathoids; includes naujaite, lujavrite, and kekortokite.

agricere. A resinous or waxy coating on the soil particles.

aiguille. The needle-top of the summit of certain glaciated mountains.

alkinite. $PbCuBiS_3$. A mineral occurring in grey coloured needle shaped crystals; crystallizes in the orthorhombic system.

alabaster. Fine-grained, massive gypsum, normally white and often translucent. It is used for statuary, vessels, *etc*.

alaskite. Plutonic rock consisting mainly of oligoclase, microcline, and quartz; mafic constituents are very few or absent; a commercial source of feldspar.

albertite. A mineral chiefly containing hydrocarbons.

albite. $NaAlSi_3O_8$. A variety of plagioclase feldspar. Also known as sodaclase.

albitite. A porhyritric dike rock consisting almost entirely of feldspar. Common accessory minerals are muscovite, garnet, apatite, quartz and opaque oxides.

alee basin. A basin formed in the deep sea by turbidity currents aggrading courses where the currents were deflected around a submarine ridge.

Alexandrian. The Lower Silurian of North America.

alexandrite. Transparent variety of chrysoberyl used as a gemstone. Its colour is green in daylight and deep red in artificial light.

alfisol. An order of soils with grey to brown surface horizons, a medium- to-high base supply, and horizons of clay accumulation.

algal pit. An ablation depression that is small and contains algae.

algal stucture. A calcareous deposit with banding, possessing irregular concentric structures, crusts, and pseudopisolites or pseudoconcretionary forms resulting from organic, colonial secretion and precipitation.

Algoman. A stratigraphic division of the Canadian Precambrian, Pre- Algonkian.

Algonkian. The youngest staratigraphic division of the Canadian Precambrian.

alkali. (1) Carbonate of sodium or potassium, or any bitter-tasting salt, occurring at near the surface in arid and semi-arid regions. (2) A strong base (hydroxide): *e.g.*, NaOH or KOH. (3) Minerals rich in sodium or potassium; *e.g.*, alkali feldspar.

alkali-calcic seris. The series of igneous rocks with weight percentage of silica in 51-55 range, and weight percentages of CaO and K_2O+Na_2O equal.

alkalic igneous rocks. The igneous rocks in which the sodium and potassium content exceeds the average for the group of rocks to which they belong or exceeds the amount required to form feldspar with the available silica. Also, those igneous rocks having an alkali-lime index below 51, and those belonging to the Atlantic suite.

alkali feldspar. A feldspar composed

of potassium feldspar and sodium feldspar, such as albite, microcline, orthoclase, and anorthoclase; all are considered to be rich in alkali.

alkali flat. A level lake-like plain formed by the evaporation of water in a depression and deposition of its fine sediment and dissolved minerals.

alkali gabbro and alkali dolerite. Igneous rocks which, though gabbroic and doleritic in general character, contain alkali feldspar and/ or feldspathoids in addition to normal minerals. The rock names found in this group are used indiscriminately for doleritic and gabbroic types although in many instances the finer grain size could be distinguished by using the prefix 'micro'. The typical minerals found are titanium-bearing augite and labradorite together with analcite and/ or nepheline, with or without an alkali feldapar, and sometimes with olivine. Common accessory minerals are ilmenite and apatite. Aegirine or aegirine-augite may sometimes be present to the exclusion of titanaugite.

alkali-lime index. The percentage of silica (by weight) in a sequence of igneous rocks as shown on a variation diagram, where the weight percentages of CaO and those of $K_2O + Na_2O$ are equal *i.e.,* the intersection point of the curves for CaO and ($K_2O + Na_2O$).

allalinite. An altered gabbro with original texture and euhedral pseudomorphs.

allanite. $(Ca,La,Ce,Y)_2(Al,Fe)_3(SiO)_4(OH)$. Monoclinic mineral containing high rare–earth content. Also known as bucklandite ; cerine; orthite; treanorite.

Alleghenian. Lower Middle Pennsylvanian geologic time.

allivalite. A form of gabbro composed of anorthite and olivine; accessories are auginite apatite, and opaque oxides of iron.

allochem. Sediment formed by chemical or biochemical precipitation within a depsoitional basin; includes intraclasts, fossils, pellets, oolites.

allochemical metamorphism. Metamorphism accompanied by addition or removal of material so that the bulk chemical composition of the rock is altered.

allochetite. A porphyritic igneous rock composed of phenocrysts of labradorite, orthoclase titanaugite, nepheline, magnetite and apatite in a groundmass of augite, biotite, magnetite, hornblende, nepheline, and orthoclase.

allochthonous. A term applied to the material forming rocks which have been transported to the site of deposition *i.e.,* whose primary constituents have not been formed in site.

allogene. Allogenic mineral or rock constituent such as a pebble in a conglomerate or xenolith in an igneous rock.

allogenic. (1) Minerals and rock constituents derived from pre-existing rocks transported from their original location; also, a stream fed by water from distant terrain. (2) Successive ecologic conditions or occurrences that originate outside the natural community and alter its habitat.

allotrio-. A prefix denoting often foreign.

allotriomorphic. A term applied to mineral grains showing no development of crystal form whatsoever.

alluvial. (1) Composed of or pertaining to alluvium, or deposited by running water. (2) Applied to a placer formed by running water, or to the mineral associated with it.

alluvial cone. A sharply inclined alluvial deposit formed where a stream emerges onto a lowland after its descent from a steep upland area. These are a continuous series of landforms with intervening gradations: talus sheet, talus cone, alluvial cone, alluvial fan. Talus forms are deposited largely by the action of gravity, and alluvial forms by alluvial processes. Alluvial cones are steeper than alluvial fans and show a greater average particle size. Fans are characterized by a much greater degree of stratification and by a higher incidence of mudflows.

alluvial fan . Outspread mass of alluvium deposited by flowing water where it debouches from a steep, narrow canyon onto a plain or valley floor. The abrupt change of the gradient eventually reduces the transport of sediment by the issuing stream. The top-view of the deposits is generally fan-shaped: they are especially prominent in arid regions, but can occur anywhere.

alluvial plain. General name for a plain produced by the deposition of alluvium from the action of rivers; *e.g.*, flood plain; delta plain; alluvial fan.

alluviation. Deposition of alluvium along streamways; aggradation. Also, the filling of a depression of covering of a surface with alluvium.

alluvium. General term for detrital deposits made by river or streams or found on alluvial fans, flood plains, etc. Alluvium consists of gravel, sand, silt, and clay, any often contains organic matter that makes it a fertile soil. It does not include the subaqueous sediments of lakes and seas.

alnoite. A variety of biotite lamprophyres characterized by lepidomelane phenocrysts ; it is free from feldspar but contains meletite, pervoskite, olivine, and carbonate in the matrix.

alpides. The long east-west orogenic belt that includes the European Alps and the Himalayas.

alpine. (1) Petaining to the European Alps. (2) Term applied to topographic features that resemble the form of the European Alps; *e.g.*, the alpine topography of parts of New Zealand.

alpine glacier. *See* glacier.

alpine orogeny. A series of diastrophic movements in southern Europe and the Mediterranean regions associated with the collision of Africa with Europe. The Alpine orogeny perhaps began late in the Triassic and continued well into the Cenozoice Era.

alsbaschite. An uncommon porphyritic variety of aplite; a plutonic rock of sodic plagio clase, quartz, and subordiante orthoclase.

alstonite. $CaBa(CO_3)_2$. An orthorhombic carbonate mineral composed of calcium barium carbonate.

altiplanation. The development of terrace-like surface or flattened summits of solifluction and related mass movements. Altiplanation is most common at high elevations and at latitudes in which periglacial conditions or processes predominate.

altiplano. A tableland plateau at high elevation; specifically, the altiplano of western Bolivia and southwestern Peru, which is a series of intermontane basins.

altithermal. (1) A period of elevated temperature, especially the postglacial thermal optimum (2) Pertaining to a climate characterized by high or rising temperatures.

altitude. (1) Vertical distance measured between a point and some reference surface, usually mean sea level. (2) Vertical angle measured between the plane of the horizon and a reference line at some higher point such as a summit peak.

alum. One of a group of minerals existing as double salt of general formula $(M(III))_2(SO_4)_3.(X(I))_2SO_4.24H_2O$ in which M may be aluminium, iron, or chromium and X may be either sodium, poassium, lithium, rubidium, ammonium, thallium or silver. Alums crystallize as octahedrals or as cubes. Ordinary alum has the formla: $Al_2(SO_4)_3$. K_2SO_4. $24H_2O$.

alum coal. Argillaceous brown coal rich in pyrite in which alum is formed due to weathering.

alumina. Al_2O_3. The native form of the oxide of aluminium, occurs as corundum or in the dydrated forms, as a crystalline or powdery substance.

aluminite. $Al_2(SO_4)(OH)_4.7H_2O$. Monoclinic hydrated form of aluminium sulphate. Also known as websterite.

aluminium silicates. Silicates containing aluminium as the only cation. The four main members of the group, having the formula Al_2SiO_5, are all found in metamorphic rocks, and the study of their phase relationships has given important information about the temperatures and pressures which can be assigned to various grades of metamorphism. Some of them are: andalusite, sillimanite, kyanite, mullite, topaz and staurolite.

alum shale. A shale containing pyrite which is decomposed by weathering to form sulphuric acid, which acts on potash and alumina consituents to form alum.

alunite. $KAl_3(SO_4)_2(OH)_6$. A rombohedral white or pink fibrous mineral containing basic potassium aluminium sulphate.

ambatoarinite. A mineral consisting of a mixture of the carbonates of cerium metals and strontium.

amber. Fossil tree resin that has achieved a stable state after ground burial, through chemical change and the loss of volatile constituents.

amblygonite. (Li,Na)Al(PO_4) (F,OH). A triclinic mineral, found in pegmatites as white or greenish masses; used as a source of lithium.

amethyst. SiO_2. A transparent purple or blue-viloet variety of quartz, valued as a semiprecious gem.

amherstite. A syenodiorite containing andesine and antiperhite.

amianthus. A fine and silky variety of asbestos, such as chrysotile.

Ammanian. A stratigraphic stage name for the British Lower Westphalian.

ammonoid. The extint subclass of the Cephalopoda, that divide up the shell are less simple in shape than those of the Nautiloidea and the siphuncle is at the outer edge of the shell, unlike that of a nautiloid. The ammonites are important as fossils.

amorphous. Term applied to material having no regular arrangement of atoms.

amphibia. A class of vertebrates living on land and in water. Fossil amphibia are found in rocks of Upper Devonian to Recent age.

amphiboles. A large group of rock-forming metasilicate minerals. They have a structure of silicate tetrahedra linked to form double endless chains, in contrast to the single chains of the pyroxenes, to which they are closely related. They are present in many igneous and metamorphic rocks. The amphiboles show a wide range of compositional variation but conform to the general formula: $X_{2-3}Y_5Z_8O_{22}(OH)_2$, where X = Ca, Na, K, Mg, or Fe^{2+}; Y = Mg. Fe^{2+}, Fe^{3+}, Al, Ti, or Mn; and Z = Si or Al. The hydroxyl ions may be replaced by F, Cl or O. Most amphiboles are monoclinic, including cummingtonite, $(Mg, Fe^{2+})_7(Si_8O_{22})(OH)_2$; tremolite, $Ca_2Mg_5(Si_8O_{22})(OH_4F)_2$ actinolite, $Ca_2(Mg,Fe^{2+})S\ (Si_8O_{22})\ (OH,F)_2$; *hornblende, $NaCa_2(Mg\,Fe^{2+},Fe^{3+},Al)_5(Si,\ Al)_8O_{22})\ (OH,F)_2$; edenite, $NaCa_2(Mg.Fe^{2+})_5\ (Si_7AlO_{22})\ (OH,F)_2$,and riebeckite, $Na_2Fe_3^{3+}(Si_8O_{22})(OH,F)_2$, Anthophyllite, $(Mg,\ Fe^{2+})_7\ (Si_8O_{22})(OH,F)_2$ and gedrite, $(Mg,Fe^{2+})_6Al(Si,Al)_8O_{22}(OH,F)_2$, are orthorhombic amphiboles.

amphibolite. A metamorphic rock consisting mainly of amphibole and plagioclase, little or no quartz, and having crystallablastic texture. Amphibolite grades into hornblende plagioclase gneiss as the content of quartz increases.

amphibolite facies. One of the major divisions of metamorphic mineral assemblages, the rocks of which form under conditions of moderate to high temperature and pressure. Less intense temperature and pressure conditions form rocks of the epidote-amphibolite facies, and more intense conditions form rocks of the granulite facies.

ampholyte. A substance that can act as either an acid, in the presence of a strong base, or a base, when in the presence of a strong acid.

amphoteric. Describing a compound that can act as both an acid and a base (in the traditional sense of the term). For instance, aluminium hydroxide is amphoteric : as a base $Al(OH)_3$ it reacts with acids to form aluminium salts: as an acid H_3AlO_3 it reacts with alkalis to give aluminates. Oxides of metals are typically basic and oxides of nonmetals tend to be acidic.

amphoterite. A stony meteorite containing bronzite and olivine with some oligoclase and nickel rich iron.

amygdaloid (1) Describing the structure of rock containing amygdules. (2) Igneous rock in which rounded cavities have become filled with mineral deposits.

amygdule. Deposits of secondary minerals in elongated rounded, or alomond-shaped vesicles of igneous rocks, especially basalt. They range in size form 1 mm to 30 cm. The vesicles, bubbles formed as a result of dissolved gas, are generally larger in basaltic lava flows than in other extrusions. Their size appears to be a function of the lower viscosity of the melt and higher temperatures of the basaltic extrusions. Among the secondary minerals that may fill these vesicles are quartz, carbonates, and zeolites, all of which are precipitates from percolating ground water.

anaerobic. Organisms that are anaerobic *i.e.,* do not require free atmospheric or dissolved oxygen.

anaerobic sediment. Sediment that occurs in basins where the limited circulation of water results in the absence or near absence of oxygen at the sediment surface.

analcime. $NaAlSi_2O_6.H_2O$. A zeolite mineral; common in diabase granite and basalt.

analcimite. An extrusive or hypabyssal rock that consists of pyroxene and analeine.

analcimization. The replacement in igneous rock of feldspar or feldspathoids by analcime.

analcite. $NaAlSi_2O_6$. H_2O. A white or light coloured isometric zeolite found in dibase and in alkali rich basalts. Also known as analcime.

anamorphism. A deep-seated constructive metamorphic process by which new complex minerals of greater density are formed from pre-existing simpler minerals of low density through silication, decarbonization, deoxidation, and dehydration.

anastomosing. (1) A pattern of branching and recombining, as in a braided stream, which is referred to as an anastomosing stream. (2) Interveined, as in leaves showing netlike vein patterns.

anatase. TiO_2. A mineral, trimorphous with rutile and brooklite.

anatexis. The partial melting of pre-existing rock, which generates variable magmas that can subsequently be further diversified by mixing, contamination, and magmatic differentiation.

anatexite. Rock formed by anatexis; also spelled anatectite.

anauxite. A clay composed of kaolin and amorphous silica.

anchi–. A prefix denoting almost.

andalusite. Al_2SiO_5. A brown, red, yellow, or green neosilicate mineral crystallizing in orthorhombic system ; generally occurs in metamorphic rocks.

andesine. A plagioclase feldspar of composition ranging from $X_{70}Y_{30}$ to $X_{50}Y_{50}$ where, $X \equiv NaAlSi_3O_8$ and $Y \equiv CaAl_2Si_2O_8$. It is a primary consitituent of intermediate igneous rocks such as andesites.

andesinite. A diorite containing 5% or less of ferro-magnesian minerals, the remainder being andesine.

andesite. A fine-grained, dark-coloured volcanic rock, frequently porphyritic; the extrusive equivalent of diorite. Phenocrysts of porphyritic andesite consist chiefly of sodic plagioclase (andesine) and one or more ferromagnesian minerals such as biotite or pyroxene; the groundmass is composed generally of the same mineral as the phenocrysts. The andesite rock family is subdivided into quartz-bearing andesites (the dacites), hornblende and biotite-andesites, and pyroxene andesites (the most common). It is generally thought that andesitic magma originates from basaltic magma, either by crystal-liquid fractionation or by contamination with sialic material.

andesite line. A boundary separating the basaltic rocks of the Atlantic suite from the chiefly andesitic rocks of the Pacific suite; *i.e.*, the island are and continental regions of andesitic volcanic rock from the typically basaltic volcanic rock of the ocean regions.

andradite. The calcium-iron end member of the garnet group.

anemoclast. A clastic rock that was fragmented and rounded by wind.

angaralite. $Mg_2(Al,Fe)_{10}Si_6O_{29}$. A mineral of the chlorite group occurring in thin black plates.

anglesite. $PbSO_4$. A mineral occurring as white or grey tabular or prismatic orthorhombic crystals as compact masses. Also known as lead spar.

Angoumian. A stratigraphic stage name for the European Upper Turonian (Cretaceous System).

angrite. An achondrite stony meteorite composed mainly of angite with little olivine and troilite.

anhedral. Referring to mineral crystals not bounded by their typical crystal faces (rational facies).

anhydrite. $CaSO_4$. An important rock-forming anhydrous mineral form of calcium sulphate. It is chemically similar to gypsum but is harder and heavier and crystallizes in the rhombic form (gypsum is monoclinic). Under natural conditions anhydrite slowly hydrates to form gypsum. It generally occurs in white and greyish granular masses and is often found in the caprock of certain saltdomes. It is used as a raw material in the chemical industry and in the manufacture of cement and fertilizers.

anhydrous. Denoting a chemical compound lacking water: applied particularly to salts lacking their water of crystallization.

anion. A negatively charged ion *i.e.* an ion that is attracted to the anode in electrolysis.

anisotropic. Denoting a medium in which certain physical properties are different in different directions. Wood for instance, is an anisotropic material : its strength along the grain differs from that perpendicular to the grain. Single crystals that are not cubic are anisotropic with respect to some physical properties, such as the transmission of electromagnetic radiation.

ankaramite. An augite-rich olivine mela-basalt.

ankaratrite. An olivine nephelinite.

ankaerite. (Ca,Mg,Fe)CO_3. An hexagonal carbonate mineral; composed of carbonate of calcium, magnesium, and iron.

annelid. Any wormlike invertebrate belonging to the phylum Annelida, characterized by a segmented body with a distinct head and appendages.

annual layer. Any sedimentary or precipitated layer of material of variable composition and thickness, presumed to have been deposited during the course of one year. Annual layers are most frequently utilized in the study of varved deposits laid down in glacially fed lakes.

anorogenic. (1) Said of features not related to tectonic disturbance. (2) Said of a region that is crustally inactive.

anorthite. $CaAl_2Si_2O_8$. A feldspar mineral, found as white or greyish triclinic crystals in basic igneous rocks. It is the most calcic member of the plagioclase series; also, the pure calcium-feldspar end members of the plagioclase series.

anorthoclase.(Na,K)$AlSi_3O_8$.. A mineral of the alkali feldspar group, It is a sodium rich feldspar with constituents $Or_{40}Ab_{60}$to $Or_{10}Ab_{90}$ (where, Or ≡ Orthoclase, Ab ≡ Albite).

anorthosite. An intrusive igneous rock composed almost entirely of plagioclase feldspar. Although less abundant than basalt or granite, anorthosite often occurs in massive complexes, usually lopolithic or laccolithic.

antecedent stream. A stream established before the onset of some remolding event, such as uplift, and maintaining its course unaffected by that event.

anthoinite. $AlWO_3(OH)_3$. A white mineral consisting of hydrous basic aluminium tungstate.

anthophyllite. $(Mg,Fe)_7Si_8\ O_{22}(OH)_2$. An orthorhombic mineral of the amphibole group; generally produced by regional metamerphism of ultrabasic rocks.

anthozoan. Any marine benthonic coelenterate of the class Anthozoa, characterized by an external skeletons having a stonelike or leathery texture. Sea anemones and the reef-building corals are both included in Anthozoa, but only the corals gave geological importance.

anthracite. A high grade metamorphic coal having semimetallic lusture, high fixed carbon content 92%–98%, and high density; burns with a short blue flame and very little smoke or odour.

anthraxolite. A hard black asphaltite found in sedimentary rocks, especially in association with oil shales, but not restricted to such locations.

anthraxylon. Composite term for a coal component (maceral) that has a vitreous lustre. In banded coal it forms bands interlayered with opaque or dull attritus.

anticline. A fold, generally of inverted U-shape, which, unless overturned; contains stratigraphically older rocks toward the centre of curvature.

antidune. A transient accumulation of sediment on a stream bed, comparable to a sand dune, that moves upstream. Also, any marking or bed feature, whether moving or stationary, produced by unidirectional flow, and in phase with surface water waves.

antiform. Inverted U-shaped fold in strata of unknown stratigraphic sequence.

antigorite. $(Mg,Fe)_3Si_2O_5(OH)_4$. A laminar variety of mineral serpentine of brownish-green colour.

antiperthite. Natural intergrowth of feldspars formed by separation of sodium feldspar and potassium feldspar during slow cooling of molten mixtures; the potassium rich phase is evolved in a plagioclase host; exactly the inverse of perthite.

antistress mineral. A mineral whose formation in metamorphosed rocks is aided by conditions not influenced by shearing stress but rather by thermal action and hydrostatic pressure; *e.g.*, feldspar and pyroxene.

antithetic fault. A minor normal fault whose orientation is opposite to that of the major fault with which it is associated.

apatite. $Ca_5(PO_4,CO_3)_3(F,Cl,OH)$. A mineral found in igneous rocks, especially pegmatites, and in metamorphosed limestones; varieties form bones and teeth. It is the main source of phosphates.

aphanite. A rock with aphanitic texture.

aphanitic. Textural term used to describe igneous rock with crystalline aggregates of such fine grain size that constituent minerals are not visible to the naked eye. These aggregates form by rapid crystallization, or by the devitrification of glasses. A rock having such fine texture is designated as aphanitic, regardless of the presence of phenocrysts.

aphotic. A term for ocean regions that do not allow the penetration of sunlight. These regions are normally found at depth of 200 metres and greater.

aphyric. Said of the texture of fine-grained igneous rocks, showing two generations of the same mineral but without phenocrysts.

aplite. A term applied to an igneous rock of granitic affinities, but syenite, diorite, and gabbro aplites occur. Granite aplites are light-coloured fine-grained rocks, having the same mineralogy as granites although biotite is rarely present. They often contain pneumatolytic minerals such as tourmaline, topaz, fluorite, etc., and concentrations of rare elements such as beryllium and lithium. This suggests a relationship to the pegmatites, and in the field aplites and pegmatites are commonly found in association, mainly as dykes or veins. Aplites have a characteristic sugary texture; the porphyritic variety alsbachite is uncommon. Gabbro aplite may be termed beerbachite.

apo. A prefix denoting derived from.

Apophysis. In geology a tongue or other direct offshoot of a larger intrusive body, such as a vein or dike.

Appalachian orogeny. A far-reaching distrubance in the southern Appalachian orogenic belt of North America. It probably began in the Late Devonian and continued to the end of the Permian. Frequently called Appalachian revolution, it corresponds to the Variscan orogeny of Europe.

apparent dip. The angle of inclination of a fault plane or bedding surface exposed on the face of an outcrop. It is an angular measurement made on a dipping surface other than on a plane perpendicular to the strike. The result is always less than the true dip of the planar feature.

apparent movement of a fault. The movement observed on any randomly viewed section across a fault. It is a function of the disrupted strata, attitude of the fault, particular site of observation, and actual slip of the fault.

apron. A broadly extended deposit of unconsolidated material at the base of a mountain or in front of a glacier.

aquamarine. A transparent, pale blue-green gem variety of beryl. It occurs in pegmatite, in which it may form very large crystals.

aqueous. (1) Describing a solution in water. (2) Said of rock that have been formed of matter deposited by or in water.

aqueous ripple mark. A ripple mark in sand, soil, etc. made by waves or water currents, in distinction to one made by wind.

aqueous rock. A sedimentary rock deposited by or in water. Also known as hydrogenic rock.

aquiclude. A porous formation that absorbs water slowly but will not transmit it fast enough to furnish an appreciable supply for a well or spring.

aquifer. A body of rock that contains water and releases it in significant quantities for use. The rock contains water-filled pore spaces that are sufficiently connected to allow the water to flow through the rock matrix to wells and springs.

aquifuge. A rock body or rock layer having no interconnected openings or interstices, and which, thus, neither absorbs nor conducts water.

aquitard. A confining bed that retards but does not completely stop the flow of water to or from an adjacent aquifer. Although it does not readily release water to wells or springs, it may function as a storage chamber for groundwater.

aragonite. $CaCO_3$. An orthorhombic carbonate mineral, found in sedimentary rocks.

arapahite. A dark coloured, porous, fine grained basic basalt consisting of magnetite, bytownite, and augite.

arcanite. K_2SO_4. A colourless, vitreous orthorhombic potassium sulphate mineral. Also known as glaserite.

archean. Said of the rocks of the Archeozoic, the earlier part of precambrian time.

archeozoic. The earlier part of Precambrian time.

archipelago. Any large body of water interspersed with a group of islands or with many islands. The term is also applied to the island group itself. For example, the West Indies and the islands of Japan are archipelagos.

arcuate delta. A curved delta with the convex margin facing the water-body.

ardealite. $Ca_2(HPO)_4(SO_4).4H_2O$. A white or light yellow mineral composed of hydrated calcium hydrogen phosphate-sulphate.

Ardennian orogeny. A short-lived orogeny during the Ludlovian stage of Silurian period of geologic time.

areal eruption. Volcanic eruption resulting from collapse of the roof of batholith; the volcanic rocks grade into parent plutonic rocks.

arenaceous. Said of sediment or sedimentary rocks that have been derived from sand or contains sand.

arenaceous rock. Any of the group of detrital sedimentary rocks, typically sandstones in which the particles range in size from 1/6 mm to 2 mm. The term psammitic, although precisely equivalent to arenaceous, is now used almost entirely to describe metamorphosed arenaceous rocks. The term arenite is a convenient one for all arenaceous rocks regardless of their individual characteristics.

arenicolite. A hole, groove, or other mark in a sedimentary rock, generally sandstone, interpreted as a burrow made by an arenicolous marine worm or a trail of a mollusk or crustacean.

arenite. Arenaceious rock of any composition.

argillation. Formation of clay minerals by weathering of aluminium silicates.

argillaceous rock. Any of a group of detrital sedimentary rock, commonly clays, shales, mudstones, and marls. Two grades of particle size are recognised, the silt grade, in which the particles range in size from 1/16 - 1/256 mm, and clay grade, with particles of less than 1/256 mm. The term pelitic, although precisely equivalent to argillaceous, is now used almost entirely in relation to metamorphosed argillaceous rocks. In addition to the clay minerals, argillaceous rocks may contain colloidal substances very finely divided quartz, carbonate dust, carbon and iron pyrites.

aridisol. A soil suborder and a major soil of deserts characterized by an ochric epipedon and other pedogenic horzions.

arkite. A porphyritic plutonic foidite consisting largely of pseudoleucite, and nepheline, subordinate melanite and pyroxene, and accessory orthoclase, apatite, and sphene.

arkose. A coarse grained sedimentary rock composed of sand-size fragments that contain a high percentage of feldspar in addition to quartz and other detrital minerals.

arkosite. A quartzite with a high proportion of feldspar.

armenite. $BaCa_2Al_6Si_8O_{28}.2H_2O$. An hydrous mineral of barium-calcium aluminosilicate.

armored clay ball. A spheroidal mass of clay of silt (5 to 10 cm in diameter) in which gravel and sand become embedded as it rolls downstream.

arsenolite. As_2O_3. A mineral crystallizing in the isometric system and usually occurring as a white bloom or crust.

arsenopyrite. FeAsS. An iron sulpharsenide mineral, that forms metallic-white to steel-grey orthorhombic crystals. It is the principal ore of arsenic and a common mineral with lead and tin ores in ore veins, and in pegmatites, probably having been deposited by the action of both hydrothermal solutions and vapours. Also known as mispickel.

arterite. A migmatite formed as a result of regional contact metamorphism during which residual magmas were injected into the host rock.

artesian. Said of ground water or other confined fluid, under a pressure greater than that corresponding to the depth of the parial ; *i.e.*, greater than hydrostatic. Thus, the fluid will flow from an artesian well without pumping specifically, the piezometric level for the fluid is above the local ground surface. Artesian conditions depend upon certain requirements: *(i)* An inclined aquifer whose lower end is buried and upper end exposed. *(ii)* Impervious layers above and below the aquifer that prevent leakage and allow hydrostatic pressure to develop (an aquifer so specified is an artesian aquifer). *(iii)* Precipitation that infiltrates the aquifer at its exposed end. *(iv)* A spring or well that allows the water of the aquifier to discharge.

arthodiran. One of an extinct order of armored, jawed fishes, found in fresh water and marine deposits of the Devonian. They reached their peak during the Late Devonian, when some genera attained lengths of 9 m.

arthrophycus. A trace fossil originally described as a fossil seaweed, assigned to the genus Arthrophycus. It is a curving, branching, sand-filled form, variously regarded as a feeding burrow or trail produced by an annelid, mollusk, or arthropod.

arthropoda. A phylum of invertebrate animals comprising over one million species- the largest in the animal kingdom. Arthropods inhabits marine, freshwater, and terrestrial habitats worldwide. Characteristically, they possess and outer body layer- the cuticle - that functions as a rigid protective exoskeleton ; growth is thus possible only by periodic moults The arthropod body is composed of segments usually forming distinct specialized body regions, *e.g.*, head, thorax, and abdomen. These segments may possess hardened jointed appendages, modified variously as mouthparts, limbs, wings, reproductive organs, or sense organs. The main body cavity, containing the internal organs, is a continuous blood-filled haemcoel, within which lies the heart. The origins and relationships of the various groups of arthropods remain uncertain. The major classes are: Arachnida Insecta, Myriapoda, and the wormlike Onychophora.

arzrunite. A bliush-green mineral consisting of basic copper sulphate along with chlorides of copper and lead.

asbestine. A material with the properties of asbestos.

asbestos. Any one of a group of fibrous amphibole minerals (amosite, crocidolite (blue asbestos), tremolite, anthophyllite, and actinolite) or the fibrous serpentine mineral chrysotile. Asbestos has widespread commercial uses because of its resistance to heat, chemical inertness, and high electrical resistance. The fibres may be spun and woven into fireproof cloth for use in protective clothing curtains, brake linings, *etc.*, or moulded into blocks.

asbolite. A black mineral aggregate of hydrated oxides of manganese and cobalt.

aschistic. Pertaining to rock of minor igneous intrusions that have not been differentiated into light and dark portions but that have essentially the same composition as the larger intrucions with which they are associated.

aseismic ridge. A submarine ridge that is a fragment of continental crust. It is non-seismic and is thus distinguished from the seismically active mid-oceanic ridge.
ashcroftine. $KNaCaY_2Si_6O_{12}(OH)_{10}.4H_2O$. A pink mineral of zeolite family.
ash rock. The material of arenaceous texture produced by volcanic explosions.
ash stone. A rock composed of fine volcanic ash of particles that are less than 0.66 mm in diameter.
aso lava. A type of indurated pyroclastic deposit produced during the explosive eruptions that formed the Aso Caldera of Hyushu, Japan.
asphalt. A dark brown to black semi-solid sticky substance composed of bitumen with mineral matter. It consists mainly of complex hydrocarbons; occurs naturally in asphalt lakes or in deposits mixed with sandstone and lime stone ; made artificially by adding mineral matter to bitumen.
asphaltic sand. A naturally occurring mixture of asphalt and sand.
asphaltite. Any naturally occurring solid black bitumen that is insoluble in petroleum naphthas.
assimilation. A phenomenon by which material is incorporated into an igneous rock by a process of melting and solution. If this proceeds to any large extent, the final composition of the rock may well be significantly altered.
assyntite. A plutonic rock, largely composed of orthoclase and pyroxene, with lesser amounts of sodalite and nepheline, and accessory biotite, sphene, apatite, and opaque oxides.
asthenolith. A body of magma locally melted anywhere at any time within any solid portion of the earth.
asthenosphere. The layer or shell of the earth directly below the lithosphere. Its upper boundary begins about 100 km below the surface and extends to a depth of about 300 km. The upper limit is marked by an abrupt decrease in the velocity of seismic waves as they enter this region. The temperature and pressure in this zone reduce the strength of rocks, so that they flow plastically.
astrobleme. Eroded remains on the earth's surface of an ancient impact structure produced by a cosmic body ; in particular, a meteor or comet. Astroblemes are usually characterized by a circular to elliptical outline and highly disturbed rocks.
atacamite. $CuCl_2.3Cu(OH)_2$. A Copper ore mineral, found in the oxidised zones of copper lodes; composed of cupric chloride and hydroxide.
ataxic. Pertaining to unstratified ore deposits.
ataxite. An iron meteorite that lacks the structure of either hexahedrite or octahedrite and contains more than 10% nickel.
atmophile. A term used in geochemistry for the elements which are concentrated in the atmosphere.
Atokan. A stratigraphic stage name for the Lower Pennsylvanian in the U.S.A.
atoll. A ring-shaped coral reef composed of closely spaced coral islets that enclose a shallow lagoon; may be circular to elliptical in plan view.
attrition. The act of wearing and smoothing of rock surface by water flow charged with sand and gravel, by the passage of snad drifts, or by the movement of glaciers.
auganite. An olivine-free basalt or an angite-bearing andesite.
augelite. $Al_2(OH)_3PO_4$. Natural, basic aluminum phosphate.

augen. Large lenticular mineral grains or aggregates in foliated metamorphic rock. feldspar, quartz, and gamet are common as augen, which are generally associated with the schistose and gneissic varieties of metamorphic rock. Such rocks are usually referred to as augen schist and augen gneiss, and are said to show augen structure.

augite. $(Ca,Na)(Mg,Fe(II)_2Al)_2O_6$. Dark-coloured pyroxene mineral commonly occurs as dark crystals in many basic igeous rocks.

aureole. A ring-shaped contact zone surrounding and igneous intrusion. Also known as contact aureole; contact zone; metamorphic zone.

authigenic. A term applied to mineral constituents formed in a rock at the site where the rock is found. It is applied chiefly to sedimentary material.

autobrecciation. The process whereby portions of the first consolidated crust of a lava flow are incorporated into the still-fluid portion.

autoclasic. A term applied to rocks that have been fragmented in situ by mechanical processes such as faulting, or by shrinkage following desiccation.

autochthon. A massive rock body that has remained essentially at its place of origin, and which may show moderate to extensive deformation. Although continental drift could have moved the entire framework containing the rock body into a different geographical location, the autochthon has not been displaced by "local" tectonic processes, such as overthrust faulting, that transport the allochthon.

autogenetic. (1) Term applied to landforms that have developed within the boundaries of local conditions, without the influence of orogenic movement. (2) Said of a certain drainage system that is determined solely by the land surface over which streams flow.

autogenic. 1. Said of an ecologic sequence resulting from factors that originated within the natural biotic community and altered its habitat.

autointrusion. The injection of the residual liquid fraction of a magma that is differentiating into fissures formed in the crystallized fraction at a late stage by some deformation.

autolith. (1) An inclusion in an igneous rock to which it is genetically related. It is an aggregate of early-formed crystals precipitated from the magma itself. (2) Fe-Mg minerals of uncertain origin that are found in a granitoid rock, and which may have the form of round, or elongate clots.

autometamorphism. (1) Pervasive mineralogical and fabric changes that occur in an igneous rock body under conditions of decreasing temperature, and which are attributed to the action of the volatiles of the rock body itself. (2) Alteration of an igneous rock body that proceeds more or less automatically during cooling of the body in the presence of its own aqueous fluids. This process is called called deuteric alteration.

autometasomatism. A deuteric effect in which newly crystallized igneous rock is altered by its own residual fluids that are held within the rock.

automorphic. (1) A term applied to the holocrystalline texture of an igneous of metamorphic rock that is characterized by crystals bounded by their typical crystal faces. Also applied to a rock with such texture. The texture is fairly rare, since its development requires nearly simultaneous crystallization of all minerals and minimum interference among the grains during the process. Automorphic texture is well developed among lamprophyres.

autopneumatolysis . The occurrence of metamorphic changes at the pneumatolytic stage of a cooling magma in the approximate 975–1175 K temperature range.

auxiliary fault. A minor fault branching from a major one of abutting it. Also known as branch fault

auxiliary minerals. As defined in the Johannsen classification of igneous rocks, those light-coloured, relatively rare minerals such as apatite, corundum, fluorite, muscovite, and topaz.

aven. A vertical shaft leading upward form a cave passage, and sometimes connecting with passages higher up.

aventurine. A mineral containing sparkling gold-coloured particles, usually cupric or/and chromic oxide.

aviolite. A type of hornfels containing mica-cordierite.

avogadrite. $(K,Cs)BF_4$. A fluoborate mineral occurring in small orthorhombic crystals.

awaruite. Native nickel-iron alloy containing about 57.7% nickel.

axes, crystallographic. A set of imaginary lines used for measuring and describing the various forms of crystals. Axes may be divided into two groups; the orthogonal, in which the three directions chosen are mutually at rights angles and the non-orthogonal, in which one or more directions is are/not at right angles to the other directions. The cubic, tetragonal, and orthorhombic systems are referred to orthogonal axes and the monoclinic and triclinic systems to non-orthogonal axes. The hexagonal and the trigonal systems are referred to a special set of axes : three horizontal axes, arranged mutually at 120°, at right angles to a vertical axis.

axial angle. The acute angle between the two optic axes of a biaxial crystal.

axial culmination. Distortion of the fold axis upward in a form similar to an anticline.

axial element. The lengths, length ratios, and angles which define the unit cell of a crystal.

axial plane. (1) A crystallographic plane in which two of the crystal axes are included. (2) The plane in which lie the optic axes of an optically biaxial crystal. (3) A planar surface connecting the hinge lines of the strata in a fold.

azoic. Said of an environment that is devoid of life.

azonal soil. Any group of soils without well developed profile characteristics, owing to their youth, conditions of parent material, or relief that prevents development of normal soilprofile characteristics.

azulite. A translucent pale-blue variety of smithsonite found in large masses in Arizona.

azurite. $2CuCO_3.Cu(OH)_2$. A blue monoclinic mineral consisting of basic carbonate of copper.

azurmalachite. A mixture of azurite and malachite, usually occurring massive with concentric banding ; used as an ornamental stone.

B

bache. A U.K. term for the valley of a small stream.
back bay. A small, shallow body into which coastal streams drain and which is connected to the sea through the pass between barrier islands.
backfolding. A process in mountain forming in which the folds are overturned toward the interior of an orogenic belt.
backset bed. A cross-bed that inclines against the flow direction of a depository wind current.
backset eddy. A small current swirling in the direction opposite to those of the major oceanic eddies; often observed between the main current and the coastline.
backshore. Upper zone of a beach of shore, bounded by the high-water line of mean spring tides and the upper limit of shore-zone processes. Only unusually big rides or severe storms affect this area.
back slope. (1) Dip slope. (2) The gentle slope of a fault block of questa; it is not necessarily related to the dip of the under lying rocks.
back thrusting. Thrust faulting that occurs toward the interior of an orogenic belt; its direction of displacement is opposite to the general direction of tectonic transport.
bacteria. A diverse group of ubiquitous microorganisms all of which consist of only a single cell that lacks a distinct nuclear membrane and has a cell wall of a unique composition. A bacterial cell may be spherical, rodlike, spiral, comma-shaped, cordscrew-shaped, or filamentous, resembling a fungal cell. The majority of bacteria range in size from 0.5 to 5 pm. Many are motile, bearing flagela, possess an outer slimy capsule, and produce resistant spores.
baculite. A crystallite that appears as a dark rod.
baddeleyite. ZrO_2. A colourless, yellow, or brownish black monoclinic oxide of zirconium; used as heat and corrosion resistant material.
bahamite. A variety of limestone formed of sediment similar to a type currently found accumulating in the Bahamas.
baikerite. A waxlike mineral which is about 60% ozocerite with other tarry, waxy, and resinous hydrocarbons.
bajada. An alluvial plain found in arid or semi-arid regions, formed by deposition of debris in fanshaped spreads, commonly as a result of sheet floods.
ballas. An aggregate of spherical masses consisting of small diamond crystals.
ball clay. An extremely plastic clay, commonly containing organic matter, used as a bonding substance of ceramic wares. Its unfired colours range from light buff to various shades of grey.
ball coal. A variety of coal occurring in spheroidal masses. Also known as peble coal.
banakite. One of the varieties of saturated or oversaturated trachybasalts. It consists of plagioclase, sanidine, and biotite, with small quantities of analcime, augite, and olivine.
bandaite. A dacite type of extrusive rock composed of hypersthene and labradorite.
banded. Term applied to a vein, sediment, or other deposit showing alternating layers differing in texture or colour, and possibly in mineral composition.
banded coal. A variety of bituminous

and sub-bituminous coal made up of a sequence of thin lenses of highly lustrous coalified wood or bark interspersed with layers of more or less striated bright or dull coal.

banded ore. Ore made up of layered bands composed either of the same minerals that differ from band to band in colour or textrues or proportion, or of different minerals.

banded peat. Peat formed of alternate layers of vegetables debris.

banded structure. A feature of igneous and metamorphic rock, resulting from the alternation of layers or streaks differing clearly in texture and/or mineral composition.

bar. 1. Any of various types of submerged or emergent elongated embankments of sand and gravel built by waves and currents. 2. A river bar, channel-mouth bar or mid-bay bar. 3. A unit of pressure equal to 10^6 dynes/cm^2 equivalent to 750.076 mm at 0°C (mercury barometer). It is equal to the mean atmospheric pressure at about 100 m above sea level.

barbados earth. A deposit of earth fossil radiolarians.

barchan. A dune shaped in plain like a crescent. The ends of the dume point in the direction in which the wind generaly blows. Barchans may be upto 400 metres wide and 30 metres high.

barite. $BaSO_4$. A colourless white, or yellow orthorhombic mineral consisting of barium sulphate; often occurs as vein mineral. Also known as baryte; barytine; heavy spar.

barite rose. A flower–like group of tabular, sand-filled crystals of barite, arranged in radially symmetrical form. Barite roses usually form in sandstone and are generally reddish brown in colour.

barkevicite. A monoclinic mineral of the amphibole group, looks like basaltic hornblende but differs from it in its iron content.

barrens. Level or gently rolling land, usually with a sandy soil and sparse tree growth, and relatively infertile.

barrier island. An elongate coastal island representing a widened barrier beach; a sand ridge rising above high tide level and roughly paralleling the coast but separated from it by a marsh or a lagoon.

barrier reef. A long, narrow reef running roughly parallel to a shore and separated from it by a body of water, *e.g.*, a lagoon, usually too deep to permit coral growth. It may include a volcanic island, either completely or in part.

Barrovian metamorphism. A progressive regional metamorphism that can be noted in the mineralogical and textural changes of a sequence of zones, each one of which is named after an index mineral.

basal. Pertaining to or located at the base.

basal arkose. Partially rewoked feldspathic residuum in the lower section of a sandstone that overlies granitic rock.

basal conglomerate. A well-sorted homogeneous conglomerate at the base of a sedimentary sequence resting on a surface of erosion. It forms the initial stratigraphic unit in a marine series, and marks an unconformity.

basalt. A crystalline, sometimes glassy, basic igneous rock of volcanic origin. The essential minerals are a calcic plagioclase and pyroxene (usually augite), with or without olivine. Magnetite is an important accessory, while quartz, hornblende, and hypersthene are sometimes present in significant amounts. Glassy basalt is called tachylite; basalt glass containing olivine and augite phenocrysts is called limburgite, but analysis suggests that limburgites are commonly undersaturated. Basalts are characterised by low SiO_2 content; (45-50%), tholeiites having the higher silica content have high FeO, MgO, and CaO , while Na_2O and K_2O are low, especially in the olivine basalts.

basalt glass. *See* basalt; jaspoid.

base level. The theoretical limit toward which the vertical erosion of the earth's surface progresses but rarely, if ever, reaches. The general base level (ultimate base level) is the lowest possible base level; for land surface, it is about 60 m below sea level, because of the reducing effect of wave action over a long period of time; for a stream, it is the plane surface of the sea projected inland as an imaginary surface beneath the stream. A region is at the temporarty base level whenever it is graded toward some level, other than sea level, below which the land area cannot be reduced for the time being by erosion. For example a level locally controlled by a resistant stratum in a stream bed.

base line. Any line established with precision instruments for the purpose of making additional azimuthal measurements.

basic. Said of igneous rock, having low silica content (generally less than 54%) and usually rich in iron, calcium; or magnesium, Basic is one of four subdivisions (acid, intermediate, basic, ultrabasic) of a commonly used system for classifying igneous rocks by their silica content. 2. Generally said of any igneous rock compound composed mainly of dark coloured minerals.

basidiomycetes. A class of fungi in which sexual reproduction is by means of basidiospores (spores produced externally on a club-shaped or clyindrical cell, the basidium). Basidia are often grouped together forming fruiting structures, such as mushrooms, puffballs, and bracket fungi. Exceptions are the rusts and smuts, which do not produce obvious fruiting bodies.

basification. Formation of a more basic rock, usually with more hornblende, biotite, and oligoclase, by contamination of a granitic magma in the assimilation of country magma.

basimesostasis. A process of the partial or entire enclosure of plagioclase crystals in a dibase by augite.

basset. The out cropping edge of a layer of rock exposed to the surface.

batholite. An older massive protrusion of magma that solidifies as coarse crystalline rock in the deep horizons of the earth's crust.

batholith. Typically a large, cross-cutting plutonic mass with more than 100 km^2 of surface exposure and with no visible or clearly inferred floor, hence showing no direct evidence of emplacement by simple injection into the earth's crust. Deep-earth temperature and density measurements suggest that the floors of some batholiths are as much as 30 km beneath earth's present surface. Although the origin and mechanism of emplacement of batholiths are still controversial, their formation is generally believed to involve magmatic processes. The most common of these plutons are granitic or granodioritic in composition and are usually, but not always, associated with orogenic belts.

bathyal zone. The marine ecological zone that is deeper than the continental shelf but shallower than the deep ocean floor. This range is usually discribed as 200 to 3000 m. Fluctuations in temperature, oxygen concentration, and sedimentation in the bathyal zone result in conditions that are much more variable than in the abyssal egion. The bathyal zone is almost entirely aphotic, with general temperatures between 5° and 15° C. Its salinity may range from 34 to 36%, and current movement in this region is essentialy geostrophic. *See* also continental shelf.

Baume gravity scale. A scale originally developed for the purpose of hydrometer calibration given by the relationship: Baume° = (degrees Baume) = 140/sp. gr. at 60°F) 130. Although still used in parts of Europe, the Baume scale has been largely replaced by the API gravity scale.

bauxite. A whitish to reddish brown mineral rock, formed under very hot and wet conditions. It contains hydrous aluminium oxide and aluminium hydroxide along with impurities such as free silica, hydroxides of iron, and clay minerals.

bay. A narrow neck of water arising from the sea between two headlands.

bay delta. A usually triangular alluvial deposit formed at the point where the mouth of the stream enters the head of a drowned valley.

bayou. (1) A name given (chiefly in the southern states of North America) to marshy offshoots and overflows of rivers and lakes. (2) General term for a stagnant inlet or outlet of a bay or lake.

beach. The zone of unconsolidated material that extends from the low-water line to where there is marked change in material or physiographic form or to the line of permanent vegetation; the shore where it is formed of sand or pebbles.

beach cycle. Periodic retreat and out-building of beaches resulting from waves and tides.

beach profile. Intersection of a beach's ground surface with a vertical plane perpendicular to the shoreline.

beach ridge. Near-parallel ridges of sand, shell, or pebbles varying in height from a few centimetres to several metres. Depressions between them are called swales.

beach rock. A friable to well cemented rock made of calcareous skeletal debris that is cemented together by calcium carbonate.

beck test. A microscopic test in which indices of refraction are compared for minerals. When the illumination of a thin section of rock being viewed under a microscope is reduced by means of the substage iris diaphragm, a bright line can be seen at the boundaries of grains which differ in refractive index from the adjoining material (either another grain or the mounting medium) If the microscope is defocused by racking up slightly, the bright line (Becke′ line) is observed to move into the material with the higher refractive index. Conversely, on racking down, the line moves into the material with the lower refractive index. This test is widely used in identifying minerals in thin section, and grains immersed in liquids of varying refractive index.

biohermite. A term used for limestone composed of debris broken from a bioherm and forming pocket fillings or talus slopes associated with reefs.

biohydrology. The study of interactions of water, plants, and animals, including both the effects of water on biota and plysical and chemical changes in water or its environment caused by biota.

biolite. (1) An old term for a concretion formed through the action of living organisms. (2) A group name for minerals formed by biologic action.

biolith. A rock of organic origin composed of organic remains.

biolithite. A limestone constructed by organisms (faunal or floral) that grew and remained in place; characterized by a rigid frame work of carbonate material that binds allochem grains and speletal elements. It is typical of reef cores.

biologic artifact. An organic compound whose chemical structure demonstrates its derivation from living matter.

biomass. The amount of living material in a particular area stated in terms of the weight or volume of organisms per unit area or the volume of the environment.

biome. A large naturally occurring assemblage of animal and plant species that are of the same general type, being adopted to the particular conditons in which they occur, *e.g.*, tundra and tropical rain forests.

biomicrite. A limestone consisting of a variable proportion of skeletal debris and carbonate mud; specifically a limestone containing less than 25% interclasts and less than 25% ooliths, with a volume ratio of fossils and fossil fragments to pellets greater than 3 to 1, and the carbonate-mud matrix more abundant than the sparry-calcite cement.

biomicrosparite. A biomicrite in which the carbonate mud matrix has recrystallized to microspar.

biomicrudite. A biomicrite containing fossils or fossil fragments that are greater than 1mm in diameter.

biomorphic. Pertaining to or incorporating the forms of organisms *e.g.*, biomorphic sedimets containing fossil forms.

biopelite. An organic pelite such as a black shale.

biopelmicrite. A limestone intermediate in content between biomicrite and pelmicrite.

biopelsparite. A limestone intermediate in composition between biosparite and pelsparite.

biophile. (1) Said of those elements that are concentrated in and by living plants and animals. (2) Said of those elements that are the most typical in organisms and organic material.

biopyribole. A mnemonic term used in classification of rocks to indicate the pressure of an otherwise unidentified biotite, pyroxene, or amphibole.

biosparite. A limestone consisting of a variable propartion of sketelal debris and clear calcite (spar); especially, a limestone containing less than 25% intraclasts, and less than 25% ooliths with a volume ratio of fossils and fossil fragments to pellets greater than 3 to 1, and the spary-calcite cement more abundent than the carbonate mud matrix.

bioseleology. The scientific study of the organisms that live in caves.

biosphere. (1) All living organisms of the earth and its atmosphere. (2) All the area occupied or favourable occupation by living organisms. It includes parts of the lithosphere, hydrosphere and atmosphere.

biostary. Maximum development of organisms at the time of tectonic repose when residual soils form extensively on land and deposition of calcium carbonate is widespread in the sea.

biostratigraphic classification. The organization of strata into units based on their fossil content.

biostratigraphic unit. A stetum of body of strata that is unified by its fossil content or paleontological character and thus differentiated from adjacent strata.

biostramal. Pertaining to a biostrome.

biostrone. A distinctly bedded and widely extensive or broadly lenticular, blanket-like mass of rock built by and composed mainly of the remains of sedentary organisms, and not swelling into a mound-like or lens-like form.

biota. All living organisms of an area; the flora and famuna considered as one unit.

biotic province. A geographic area that supports one or more ecologic associations which are distinct from those of adjacent provinees.

biotite. A black brown or dark green abundant species of rock forming mineral of the micagroup having variable composition: $K_2[Fe(II), Mg]_{6\text{-}4}[Fe(III), Al, Ti]_{0\text{-}2}[Si_{6\text{-}5}, Al_{2\text{-}3}]O_{20\text{-}22}(OH,F)_{4\text{-}2}$

biotope. (1) An area characterized by uniform ecology and organic adaptation. (2) An envoronment in which an assemblage of flora and fauna lives or has lived.

bioturbation. The stirring of a sediment by organisms, often destroying its structure. For example, the feeding by burrowing of many sea-floor organisms through freshly deposited sediment destroys bedding and lamination.

biozone. A general term for any type of biostratigraphic unit. The fundamental unit in biostratigraphic classification, and usually the smallest such unit on which intercontinental correlation can be established. In oceanography, a depth division within the ocean, characterized by particular conditions and organisms.

bisectrix. A line that is the bisector of the angles between the optic axes of a biaxial crystal.

bismite. Bi_2O_3. A monoclinic mineral; it is straw yellow and earthy or powdery, polymorphous with sillinite.

bismoclite. BiOCl. A pale-greyish or white mineral; isomorphous with daubreeite.

bismutite. $(BiO)_2CO_3$. A mineral of bismuth, usually dull with or grey in colour; earthy and amosphous.

bitumen. Any of various solid and semisolid substances consisting mainly of hydrocarbons; *e.g.*, petroleum, asphalt.

bituminous coal. A dark down to black variety of coal which has a high proportion of carbonaceous matter. Also known as soft coal.

bivalve. An animal having a shell composed of two distinct parts that open and shut; a mollusk of the class Bivalvia (Pelecypoda). Clams, oysters, scallops, and mussels are examples.

black alkali. A deposit of sodium carbonate that has formed on or near the surface in arid to semiarid areas.

black band. (1) A thin layer of black band interbedded with clays or shales in black band ironstone. (2) A dark earthy variety of the mineral siderite, occurring mixed with clay, sand and considerable carbonaceous matter, and frequently associated with coal.

black chalcedony. The correct designation for most so called black onyx.

black chalk. A bluish-black carbonaceous clay, slate, or shale; used as a pigment or crayon.

black granite. A commercial granite that when polished is dark grey to black. It may be a diabase, diarite, or gabbro.

black ice. Dark glacier ice formed by freezing of silt laden water.

black jack. (1) A thin stratum of coal interbedded with layers of slate; a slaty coal with a high ash content. (2) A term used in Arkansas for a soft, black, carbonaceous clay or earth associated with coal.

black manganese. A term applied to dark-coloured manganese minerals, such as pyrolusite, hausmannite, and psilomelane.

black metal. A black shale associated with coal measures.

black sand. An alluvial or beach sand consisting predoninantly of grains of heavy, dark minerals or rock (*e.g.*, magmetite, rutile, garnet, or basaltic glass), concentrated chiefly by wave, current, or surf action. It may yield valuable minerals.

black shale. Very thin bedded shale rich in sulphides such as pyrite and organic material deposited under barred basin conditions so that there was an anaerobic accumulation.

blanket deposit. (1) A sedimentary deposit of relatively uniform thickness covering a large area. (2) A flat deposit of ore, relatively thin as compared with its length and breadth.

blanket sand. A blanket deposit of sand or sandstone covering as unusually great area; typically deposited by a transgressive sea. Also known as sheet sand; sand sheet.

blastoid. A class of echinoderms of Middle Ordivician to Late Permian occurrence, possibly evolved from cystoids. They were attached to the sea floor by a stalk of circular plates, surmounted by a calyx or cup, and commonly had fivefold radial symmetry. Some blastoids are useful as index fossils.

bleb. An informal term for a small, rounded inclusion of one mineral in another; *e.g.*, a bleb or olivine poikilitically enclosed in pyroxene.

block. Angular rock fragment with a diameter greater than 256 mm. It shows little alteration as a result of transporting agents.

block caving. A method of mining in which blocks of ore are caused to cave in or collapse by undercutting and blasting away the supporting pillars.

block diagram. A plane figure illustrating the three-dimensional geologic structure of an area. A rectangular block of the earth's crust is drawn in three-dimensional perspective to show a top surface and at least two vertical cross-sections. If an orthorhombic projection is drawn, each of the three sides of a rectangular block is foreshortened equally. In isometric block diagrams, equal parts of the three faces of a cube are seen.

block faulting. Type of normal or gravity faulting in which the crust is separated into structural units (fault blocks) of different orientations and elevations. Fault-block mountains are formed by this process.

block field. Accumulation of coarse detritus on level mountain top surfaces or gently sloping mountain sides; also termed as felsenmeer or blockmeer.

block lava. Lava flows which occur as a tumultuous assemblage of angular blocks.

block mountain. A mountain formed by the combined processes of faulting, uplifting, and tilting.

bloedite. $MgSO_4.Na_2SO_4.4H_2O$. A

colourless or white mineral consisting of hydrous sulphates of magnesium and sodium. Also known as astrochanite; astrakanite.

blonde'au method. A seismic method of determining vertical time to a predetermined depth based on first-break data and the assumption that the instantaneous velocity is proportional to a power of the depth.

bloodstone. A semi-translucent green variety of chaldony speckled with red or brownish-red spots of jaspar resembling drops of blood.

blowhole. (1) A vertical, cylindrical fissure in a sea cliff, extending from the bottom of the inner end of a sea cave upward to the surface. A blowhole is formed if a joint, which is a line of weakness, extends from the tunnel end to the top of the cliff. A fountainlike effect is produced as waves and tides force water and air into it. (2) An opening through a snow bridge into a crevasse, the presence of which is usually indicated by a current of air. (3) A small vent on the surface of a lava flow.

blowing cave. A cave that has a movement of air through its entrance.

blowing well. A water well that has a movement of air through its entrance.

blown sand. Sand that has been transported by the wind; sand consisting of wind-borne particles, eolian sand.

blowout dune. A dune consisting of large accumulation of sand derived from the formulation of blowout.

blue amber. A variety of osseous amber with a bluish tinge that is probably due to the presence of calcium carbonate.

blue ground. Unoxidised slate-blue or bluish green kimberlite, usually a breccia, that is found below the surficial oxidised zone of yellow ground.

blue hole. A term used on the Bahama Banks for a drowned sinkhole.

blue ice. An ablation area created by wind erosion on the Antractic Ice Sheet; characterized by bare glacier ice showing at the surface.

blue malachite. A misnomer for azurite.

blueschist. A schistose metamorphic rock with a blue colour occuring is the presence of sodic amplibole, glaucophane or crossite and commonly mottled bluish-grey lawsonite.

bluestone. Commercial name for a bluish or blue-grey argillaceous sandstone used as building or paving stone.

bluff. A headland, hill, or cliff having a broad, steep face.

boehmite. AlO(OH). An orthorhombic mineral, a dimorph of diaspore; its colours may be grey, brown, or reddish, yellow. It is an important constituent of some varities of bauxite.

bog. (1) Waterlogged, spongy ground, consisting primarily of mosses, containing acidic, decaying vegetation that may develop into peat. (2) The vegetation characteristic of this environment, especially sphagnim, sedges and heaths. The term is often used synomymously with peat beg.

bog burst. The bursting of a bog under the pressure of its swelling, due to water retention by a marginal dam of growing vegetation. The escaping water produces muddy peat flows over the surrounding area.

bog coal. An earthy type of brown coal.

bog flow. The outflow from a bog burst.

boghead coal. A sparopelic coal resembling cannel coal in its physical properties but consisting mainly of algal matter rather than spores.

boghedite. An old synonym of torbanite.

bog iron ore. A soft, spongy, porous deposit of iron hydroxide deposited in swamps and lakes, probably by bacterial action. Such deposits are of some commercial importance, as they may have a low content of phosphorus and sulphur and can yield very pure iron.

bog soil. A great soil group in the 1938 classification system, an intrazonal, hydromorphic group of soils having a mucky or peaty surface horizon and an underlying peat horizon. These soils developed in swamps or marshes in humid or subhumid climates. Bog soils are now classified as histosols.

bogue. A term used in U.S. for the mouth or outlet of a stream, or for the stream itself.

bogusite. An intrusive rock of the same general composition as teschenite, but of lighter colour.

bohdanowiezite. $AgBiSe_2$. A mineral.

Bohemian garnet. A yellowish-red to dark, intense red gem veriety of pyrope obtained from Bohemia.

Bohemian ruby. A red variety of crystalline quartz; specifically rose quartz cut as gem.

Bohemian topaz. Yellow quartz when cut as a gem.

bojite. A gabbro in which primery hornblende substitutes for most of the pyroxene, although some augite and biotite may be present.

bole. (1) A fossil laterite interbedded with lava flows, usually basalts. (2) Any of various red, yellow, or brown earthy clays consisting chiefly of hydrous aluminium silicates.

bolson. A term applied in the desert areas of the southwestern U.S. and nothern Mexico to an extensive basin more or less rimmed by mountains from which drainage flows into the depression. As a result of centripetal drainage, which is characteristic of a bolson, there is usually at or near its near its centre, which is the lowest part, a level plain or playa.

bone beds. Any sedimentary stratum, usually a thin bed of limestone, sandstone, or gravel, in which fossil bones, teeth, or scales are abundant. Bone, beds suggest unusual features of deposition, such as mass extinction due to salinity changes, earthquakes, or outbursts (blooms) of certain organisms.

bone breccia. An accumulation of bones or bone fragments, after mixed with earth and sand and cemented with calcium carbonate.

bone cave. A cave that has served as a trap for fossil vertebrates.

bone chert. A weathered residual chert that appears chalky and porous with a white colour but may sometimes acquire red or some other colour.

boninite. An andesitic rock that contains much glass and abundant phenocrysts of bronzite and less of olivine and augite.

book structure. In ore deposits, a formation showing an alternation or inter leaving with gangue.

boomer. A marine seismic energy source in which capacitors are charged to high voltage and then discharged in a body of water through a transducer, a component of which consists of two metal plates. Abrupt separation of the plates produces a low-pressure region between them, into which water rushes, thus generating a pressure wave.

boracite. $Mg_3B_7O_{13}Cl$. A white, yellow, blue, or green orthorhombic borate of magnesium, occurring in isometric crystalline form.

borax. $Na_2B_4O_7.10H_2O$ A mineral mined as an ore of boron found in arid regions, where it forms as an evaporite in playa deposits and as a surface efflorescence. In a dry environment its clear, transparent crystals lose water and become chalky white tincalconite.

bore. (1) Abrupt and violent wall-like wave of water produced as an incoming tide rushes up a shallow, narrowing bay or estuary. (2) Submarine sand ridge in very shallow water that may reach intertidal level.

boreal. Pertaining to the north; northern.

borehole. Any exploratory hole, usually a deep hole of small diameter, made by drilling into the earth. The purpose can range from oil, mineral, or water prospecting to determining emplacement sites for dynamite.

bornhardt. A large inselberg. Bornhardts are dome-shaped monoliths, examples of which are found on every continent except Antarctica. They are best developed in the tropics.

bornite. Cu_2FeS_4. Copper-iron sulfide, an ore of copper, occasionally occurring with chalcopyrite and chalcocite. Bomite is a fairly common mineral, but crystals of it (isometric) are rare. Freshly exposed surfaces of the mineral show a bronze-like colour that quickly tarnishes in air to a purplish iridescence, giving rise to "peacock ore."

borolanite. A hypabyssal rock that is essentially orthoclase and melanite with subordinate nepheline, biotite, and pyroxene.

borovskite. Pd_3SbTe_4. A cubic mineral.

bort. (1) A granular to very finely crystalline aggregate consisting of imperfectly crystallized diamonds or of fragments produced in cutting diamonds. It often occurs in spherical forms with no distinct cleavage and having a radial fibrous structure. (2) A diamond of the lowest quality; an industrial diamond. 3. A term formerly used as synonym of carbonate.

biochung. A synonym of gravity slope.

boss. (1) A mass of plutonic igneous rock that occurs at the surface of the earth, and which may be knob-like, circular, or elliptical. Its contacts are vertical or steeply inclined. (2) A knob-like excrescence or protuberance on some organ of a plant or animal.

bostonite. A light-coloured hypabyssal rock, characterized by bostonitic texture and composed chiefly of alkali feldspar; a fine-grained trachyte with few or no mafic components.

bostonitic. Said of the texture of bostonite in which microlites of rough irregular feldspar tend to form clusters of divergent laths within a trachytoid groundmass.

botallackite. $Cu_2(OH)_3Cl.3H_2O$. A bluish-green mineral containing copper.

botryogen. $MgFe(SO_4)_2(OH).7H_2O$. A deep red to deep yellow mineral; usually botryoidal.

botryoidal. Having the form of a bunch of grapes; said of certain minerals, *e.g.*, smithsonite, having the form of a bunch of grapes.

botryolite. A radiated, columnar variety of datalite with a botryoidal surface.

bottleneck bay. A bay with a narrow entrance which is guarded with the waves by features other than barrier islands.

bottle spring. A fresh water spring that issues through the floor of a saline lake or pool.

bottom. (1) A lowland usually fertile, along a stream; an alluvial plain. (2) The bed of a body of water. (3) The lowest and usually richest part of an alluvial placer; also, the lower limit of an ore body.

bottom-hole pressure. The pressure produced in a well bore at or near the depth of a reservation formation. It may be ineasured as a flowing bottom hole pressure" or as shut in pressure to record the rate of pressure build up during the survey period.

bottomset bed. One of the horizontal or gently inclined layers of sediment deposited in front of the advancing foreset beds of delta.

boudin. One of a series of sausage shaped segments occurring in boudinage structure.

boudinage. A structure found in greatly deformed sedimentary and metamorphic rocks, in which an originally continuous rigid (competent) band or layer between relatively plastic (less competent) layers is stretched and thinned until rupture occurs; the rigid band, thus separated into long cylindrical pieces, resembles a string of sausages. In the pinched parts, or points of separation, quartz or calcite is usually crystallized.

bouguer anomaly. A gravity value calculated after corrections for latitude, elevation, and terrain.

bouguer plate. An imaginary layer, having infinite length and a thickness equal to the height of the observation point above the reference surface.

boulangerite. $Pb_5Sb_4S_{11}$. A bluish–grey to lead–grey mineral occurring in plumose masses.

boulder. Rock fragments with diameter exceeding 256 mm, being somewhat rounded or otherwise distinctively shaped by abrasion in the course of transport.

boulder barricade. An accumulation of many large boulders visible along a coast between lowtide and half-tide.

boulder barrier. A short ridge created by great pressure from floating ice under the influence of strong winds and gentle slope of the shore and measuring over 6 m in height and 800 m in length.

boulder clay. A term used for glacier deposits containing undecomposed fragments of parent rock owing to absence of chemical decomposition. Its constituents are widely dependent on rock over which the ice has travelled.

boulder conglomerate. A consolidated rock consisting mainly of boulders.

boulderet. A term for a rounded, coarse fragment of glacial drift having a diameter range of 0.15 m to 0.38 m.

boulder facet. One of the small plane surfaces on a faceted boulder.

boulder fan. A fan shaped assemblage of clasts (mainly boulders) diverging from their bedrock source in the direction of the movement of the glacier by which they were transported and deposited.

boulder flat. A level tract covered with boulders.

boulder gravel. An unconsolidated deposit consisting mainly of boulders.

boulder pavement. (1) A surface of till consisting largely of boulders, and worn down by glacier movement. (2) An accumulation of boulders, once part of a glacial moraine, remaining after finer materials have been removed by wind, waves, and currents. (3) A desert pavement composed of boulders.

boulder rampart. A type of beach ridge system composed of boulders thrown up along the seaward side of a reef.

boulder quarry. A quarry in which weathering has produced so much jointing in the stones that is is not possible to mine large blocks from it.

bouldef stone. A consolidated sedimentary rock consisting of boulders-size particles.

boulder train. A line or series of glacial boulders and small clasts extending from the same bedrock source, ofter for many kilometres, in the direction of the movement of the glacier by which they were transported and deposited.

boulder wall. A boulder-built glacial moraine.

boundary monument. A pile of stones or other objects placed at or near a boundary line to identify the location of the line.

boundary stratotype. That point in a particular sequence of rock strata that is used as the standard for recognition and definition of a stratigraphic boundary.

boundary waves. A mode of wave propagation along the interface between media of different properties. Where the interface of boundary is a free surface, such as an earth-to-air surface, the mode is a surface wave.

barnonite. $PbCuSbS_3$. An orthorhombic mineral containing large, intergrown crystals producing radiating effect; shows sub conchoidal to uneven fracture and good cleavage in one direction.

bowenite. An apple green or translucent greenish white variety of serpentine; used for carving and jewellery.

bowlingite. A synonmym of saponite.

bowralite. An igneous rock of the syenite clan; a coarse-grained holocrystalline hypabyssal rock containing automorphic sanidine tablets, subalkaline amplibole, and some aegirine.

boxwork. A skeletal framework of intersecting plates composed of a dydrated iron oxide (any of several are possible) that has been deposited along fractures and in cavities from which intervening material has been dissolved by groundwater related processes. Boxwork is common in the oxidized zone of sulphide ores and on the ceilings of caves. In such sites of occurrence, several different boxwork structures may exist separately or become joined together in a complex manner.

braccianite. A variety of cecilite containing no melilite.

bracewellite. CrO(OH). An orthorhombic mineral of chromium; trimorphous with grimaedite and guyanaite.

brachy–. A prefix denoting short.

brachone. The short arm of an anomoclon or ennomoclone of a sponge, or the different fourth ray of a trider of a sponge.

brackebuschite. $Pb_2(Mn,Fe)(VO_4)_2.H_2O$. A black to reddish mineral.

brackish. Term applied to water with salinity intermediate between that of normal fresh water and normal sea water.

bradytely. A slowing down in the development of a group of organisms, possibly causing particular members to lag behind the normal rate of progress in certain or all of their characteristics.

braggite. $(Pt,Pd,Ni)S_x$. A steel-grey coloured mineral of variable composition.

braid. (1) To branch and rejoin

repeatedly to form an intricate pattern or network of small interlacing streams channels. (2) A reach of a braided stream, characterized by relatively stable branch islands and hence two or more separte channels.

braid bar. Any exposed sand or gravel bar that divides flow and causes a braided pattern in a stream.

braided stream. A stream that divides into numerous channelways that branch, separate, and rejoin, thus becoming a tangle of channels, islands,and sandbars resembling a complex braid. Braided streams occur under conditions of abundant bed load and highly variable discharge. These conditions are provided where a river in a semi-arid region receives most of its discharge from preciptious headwaters, Braiding also occurs where glaciers are melting and the debris supply is consistently large whereas the discharge varies as the ice metlts and refreezes. Examples of this type of braiding are found wherever melting glaciers occur. Information concerning ancient environments is often based on recognition of braided stream deposits in older sediments.

brait. A term, sometimes used for rough diamond.

braitschite. $(Ca,Na_2)_7(Ce,La)_2$-$B_{22}O_{43}$. $7H_2O$. A rare mineral of variable composition.

brammallite. A micaceous clay mineral, representing the sodium analogue of illite.

branch. (1) One of two or more reflecting events that may be observed at a given location from a reflector that is concave upward or discontinuous. (2) A small stream that flows into another, usually larger stream; a tributary. (3) A stream flowing out of the main channel of another stream and not rejoining it, as on a delta or alluvial fan; a distributary. (4) A stream flowing out of another stream and rejoining it, such as anabrach; a by channel . (5) A fork of a tidal river.

branching bay. A bay having a dendritic pattern, produced by drowning or flooding of a river valley by the sea.

branching fault. A fault that splits in two or more parts or branches.

branchwork cave. A cave in which passages interect as tributaries.

brandtite. $Ca_2Mn(AsO_4)_2$. $2H_2O$. A mineral isomorphons with roselite.

brannerite. $(Ca,Ce,U)(Ti,Fe)_2O_6$. A mineral containing rare element uranium.

brash ice. An accumulation of floating fragments not more than 2m across, and representing the wreckage of other forms of ice.

braunite. $3Mn_2O_3.MnSiO_3$. A mineral of manganese occurring in tetragonal form; generally of steel-grey or brown-black colour.

bravais indices. A modification of the Miller indices; frequently used for hexagonal and trigonal crystalline systems; they refer to four axis: the *c*-axis and three others at 120° angles in the basal plane.

bravioit. $(Ni,Fe)S_2$. A yellow suphide mineral; related to pyrite.

brazilite. A mixture of baddeleyite, zircon, and altered zircon.

brea. A rarely used term for a viscous asphalt formed by evaporation of volatile components from oil in seepages.

breach. The formation of a deep opening in a land form, expecially by erosion.

breached anticline. An anticline where erosion has been concentrated along the fold axis, to give an alongated valley bounded by infacing escrapments. Its formation represents an early stage in inversion of relief.

bread crust surface. A surface, resembling the crust of bread, characterizing certain concretions formed where abundant salts are being precipitated by evaporating water in a semiarid climate.

break of slope. A clearly visible sharp change of steepness in a slope profile or river long profile Breaks of slope results from geological influences and mechanisms of slope recession.

break-point bar. An elongated sand bar parallel to beach, deposited as a result of breaking waves.

break thrust. An overthrust developed during the formation of an anticline at that point at which folding becomes fracturing, and strata are overthrust along the fault surface.

breccia. (1) A Type of rock consisting of angular fraqments which have been joined together by a fine-grained cementing material such as clay. The rock fragements are often of deasert origin, but some come from volcanic eruptions Breccias commonly result from weathering proceses, or from tectonic activity. (2) A common lunar rock with clasts produced by meteoroid impact.

breccia, fault. Breccias produced in a fault zone where rocks graind against each other, breaking off angular clips.

breccia, reef. Angular fragments of coral and limy algae broken from reefs and deposited on their slopes.

breccia, residual. Fragments produced by weathering and cemented by iron compounds or calcite.

breccia vein. A fissure containing numerous wall-rock fragments, with mineral deposits in the interstices.

breciola. A well-graded, intrainformational breccia consisting of small, angular limestone fregments in well-defined beds separated by dark shale.

brewsterite. $(Ca,Sr,Ba)Al_2Si_6O_{16}.5H_2O$ A calcium containing mineral of zeolite group.

bright-banded coal. Banded coal conecting mainly of vitrain and clarain, with little quantities of durain and fusain.

brimstone. The commercial name for sulphur, especially native or fine sulphur.

briolette. A pear-shaped gemstone with a circular cross section having the entire surface cut in triangular sometimes rectangular facets.

brochantite. $Cu_4(SO_4)(OH)_6$. A green coloured mineral found in the oxidation zones of copper sulphide deposits.

brockite. $(Ca,Ce,Th)PO_4.H_2O$. A red or yellow coloured mineral contiaing phosphates of cerium, thorium and calcium.

broken belt. The transition zone between open water and consolidated pack ice.

bronzite. An iron-bearing variety of enstatite having bronze like metallic lusture; an orthopyroxene intermediate in composition between enstatite and hypersthene.

brookite. TiO_2 . A brown or black cofoured mineral consisting of titanium dioxide; trimorphous with rutile and anatase.

brown coal. A variety of coal containing upto 50% carbon; intermediate in rank between peat and lignite. It burns with a much smoke and leaves a lot of ash.

brown petroleum. Solid a or semisolid products produced by the action of air in asphalt.

brown soil. A group of zonal soils having a brown surface and a light coloured subsurface.

brown stone. Ferruginuous sandstone where grains are coated with iron oxide.

butte. A flat-tapped, steep-walled hill, usually a remnant of horizontal beds, smaller and narrower than a mesa; characteristic of arid or semi-arid regions.

buttres sand. A sandstone that interacts an underlying surface of unconformily, as on the flank of a burried kill. It often forms a trap for oil.

bystromite. $MgSb_2O_6$. A pale yellow or blue, tetragonal mineral of antimony.

bytownite. A mineral of plagioclase feldspar group with compostion ranging from $Ab_{30}An_{70}$ to $Ab_{10}An_{90}$ Where $Ab=NaAlSi_3O_8$ and $An = CaAl_2Si_2O_8$; occurs in basic and ultrabasic igneous rocks.

bytownitite. An igneous rock of the gabbro clan; an anorthosite, a leucocratic, coarse grained plutonic rock.

caballing. The mixture of two water masses to produce a blend that sinks because it is denser than either of the parent components. The phenomenon occurs when two water masses have same density but different salinities and temperatures.

cable break. An event on a well geophone record caused by elastic waves being transmitted down the supporting cable to the well geophone.

cable-tool drilling. The standard and earlier method of drilling for oil. A chise-like bit is suspended by a cable to a lever at the surface, and the lever alternately lifts and drops the bit, causing it to chip away the rock, which is removed by a bailer. Gushers are a frequent occurrence with this type of drilling.

cabochon. A gem cut in convex form; highly polished but not faceted.

cabocle. A compact rolled pebble resembling red jespar; supposed to be a hydrated phosphate of calcium and aluminium.

cacoxenite. $Fe_4(PO_4)_3(OH)_3$. $12H_2O$. A mineral containing basic hy drated ferric iron phosphate; forms hexagonal tiny, acicular crystals in tufted or radial aggregates or as fibrous coating. Its colour varies from yellow to brownish, to reddish or golden-yellow; rarely green; shows silky lustre and no fracture.

cactolith. An irragular intrusive igneous body of obscurely cactus like form, more or less confined to a horizantal zone and appering to consist of irregularly related and possibly distorted bracling of and anaestomosing dikes that fed a lacolith.

cadwaladerite. $Al(OH)_2Cl$. $4H_2O$. A mineral consisting of a hydrous basic aluminum chloride.

cahnite. $Ca_2B(OH)_4(AsO_4)$. A mineral consisting of borate and arsenate of calcium; occuring in white sphenoidal.

cairngorm stone. A variety of quartz; gemstone. Its colour varies from smoky, yellow or brown often transparent varying to brownish black and nearly opaque in thick crystals. The colour probably results due to radiation bombardment from surrounding rocks.

caking coal. A variety of coal that agglomerates and softens upon heating. A hard, grey cellular mass of coke is left behind after volatile material has been expelled at high temperature. Also known as binding coal.

cal. A term used in U.K. for iron tungstate.

cala. (1) A short narrow ria formed in a limestone coast; a small semicircular shallow bay along a calacoast. (2) A term applied to a creek corresponding to a lateral stream of a main drainage.

calamine. (1) A term used for hemimorphite. (2) A former term for smithsonite. (3) A commercial name for oxidized ores of zinc, as distinguished from the sulphide ores.

calamite. Giant scouring rush (sphenopsid) with a jointed, ribbed trunk and leaf whorls at the joints.

calaverite. $2[(AuTe)_2]$. An ore containing gold ditelluride; sometimes argentiferous with ratio of silver to gold about 6:1. It possesses monoclinic, bladed crystals lathlike or short to slender prisms. It has strongly striated faces; also massive granular to indistinct crystalline. Its colour varies from brass yellow to silver-white and possesses metallic lustre. It shows yellowish to greenish-grey streaky and subconchoidal to uneven fracture with no cleavage.

calc-A prefix indicating that the substance contains calcium carbonate.

calc-alkalic series. A rock suite comprised of those igneous rocks in which the percentage of silica by weight is between 53 and 61% when the weight percentages of CaO and of K_2O+Na_2O are equal.

calcarenite. A variety of limestone or dolomite composed of coral or shell sands or of sand formed by erosion of older limestones, with particle size ranging from 0.05 to 2.0 mm.

calcareous. Said of a substance containing calcium carbonate. When used with a rock name, it generally implies that as much as 50% of the rock is calcium carbonate.

calcareous algae. Certain algae that precipitate calcium carbonate to build up large globular or irregular masses of organic limestone. Such algae have formed characteristic limy deposits in all ages since early in the Precambrian.

calcareous ooze. A fine-grained pelagic sediment containing undissolved sand or slit-sized calcareous skeletal remains of small morine organisms mixed with amorphous clay-sized material.

calcareous rock. A sedimentary rock containing an appreciable amount of calcium carbonate, such as limestone, chalk, tuffa or shelly sandstone.

calcareous sandtone. A sandstone cemented with calcite.

calcareous schist. A coarse-grained metamorphic rock derived from impure calcareous sediment.

calcareous soil. A soil rich in calcium and magnesium carbonate.

calc-finta. A fine grained cale-silicate rock of flinty appearance formed by thermal metamorphism of a calcareous mudstone, possibly with some accompanying pneumatolytic action.

calcian dolmite. A dolomite mineral that contains atleat 8% calcium in excess of the ideal composition of dolomite.

calcic. A term meaning derived from or containing calcium.

calciclasite. An igneous rock of the gabbro clan; a leucocratic granular plutonic rock; contains largely anorthite.

calciclasitite. Synonym of calciclasite.

calcification. Any process of soil formation in which the soil colloids are saturated to a high degree with exchangeable calcium, thus rendering them relatively immobile and almost neutral in reaction.

calcify. (1) To make hard or strong by the deposition of calcium salts. (2) To become hard or stonelike because of the deposition of calcium salts.

calcikersanitite. An igneous rock of the gabbro clan. A hypabssal rock; a variety of kersantite; differs from kersanlite in that plagioclase in bytownite or labradorite, usually containing considerable augite.

calci-lutite. Clastic limestones of grain-size less than 0.62 mm.

calcimicrite. A limestone in which the micrite component is greater than the allochem component, and particles are less than 20 microns in diameter. *See also* micrite.

calcination. (1) The result of prolonged heating of a material at fairly high temperature. (2) The heating of ores to expel water and carbon dioxide.

calciocelsian. $4[Ba,Ca)Al_2Si_2O_8]$. A mineral with 4 percent calcium oxide; variety of celsian; shows scarcely observable cleavage.

calci-rudite. Clastic limestones of grain-size 2 mm and above.

calcisiltite. A sedimentary rock composed mostly of detrital, sit-sized calcite particles that have been consolidated.

calcite. $CaCO_3$. A widely distributed rock-forming mineral the chief constituent of limestone and most marble, and common in the shells of invertebrates. Its crystals, which have perfect rhombohedral cleavage, are usually white or grey; its transparent variety is iceland spar. It is polymorphous with aragonite, vaterite, and some other exotic forms. Calcite is the stable form of calcium carbonate at all those temperatures and pressures obtaining at or near earth's surface. Because of this, it is possible that all other forms convert to it over geologic time. It shows conchoidal fracture and its hardness is 3.0 on cleavage but varies on different surfaces. It dissolves with effervescence in cold and dilute hydrochloric acid. Its varieties are anthraconite, aphrite, satinspar, calcareous sinter, lublinite, rock meal, thinolite, verd antique and pisolite.

calcite dolomite. A carbonate rock with a composition of 10-50% calcite and 90–50% dolomite.

calclithite. Limestone with 50% or more fragments or older limestone that was redeposited after being eroded from the land.

calcrete. Superficial gravels cemented by calc tufa. Also known as calcicrete.

calc-schist. A metamorphosed, argillaceous limestone showing a schistose structure.

calc-silicate hornfels. Medium to high grade contact metamorphic rock with variable, texture but generally fine-grained; derived from marls and other calcareous sediments and therefore contains a great variety of minerals, mostly calc-silicates.

calcspar. A synonym of calcite.

caldera. A more or less volcanic depression whose diameter is many times greater than that of the volcanic vent. Usually formed by a combination of the explosion and collapse of the top of a volcanic cone or group of cones. Such subsidence of the roof can be caused by displacement of the underlying body of magma. The collapse is often that of a composite cone that rapidly depleted the underlying magma reservoir by great eruptions of pumice. The caldera may later fill up with water Super depressions larger than calderas, are found in volcanic areas. However, their angular outlines and angular indentations at the edges indicate that their structure is effected by joints and faults in the underlying older rocks. Therefore, these structures are referred to as volcanic-tectonic depressions. The rapid extrusion of large amounts of lava seem at least partly responsible for their collapse.

calderite. (1) A variety of antradite. (2) A manganese iron end member of the garnet group.

Caledonian orogeny. A name commonly used for the Lower Paleozoic deformation in northwest Europe, extending from The Middle Ordovician to Middle Devonian.

caledonides. The orogenic belt formed by the Caledonian orogeny. It extends from southern Wales across Scotland, and northeastward through Scandinavia.

caledonite. $2[Cu_2Pb_3(SO_4)_2CO_3(OH)_5]$. A mineral composed of basic lead copper carbonate sulphate. It contains orthohombic crystals usually small, striated, elongated in divergent groups; occasionally as coatings; rarely massive. Its colours varies from deep verdigis-green to bluish-green and shows resinous lustre, greenish white streak and uneven fracture. It is brittle and shows perfect cleavage in one direction.

caliche. A sedimentary rock, consisting of alluvium cemented together with sodium nitrate, chloride and other soluble salts. It forms high grade saline layers that are nearly free from debris and are sometimes associated with normal type of deposit. The term can be used to describe deposits of calcrete.

californite. $Ca_{10}Al_4(Mg,Fe)_2Si_9O_{34}(OH,F)_4$. A variety of idocrase that resembles jade in its characteristics. It has dark yellowish, olive or grass-green colour and occurs in transluscent to opaque compact or massive form.

Callovian. A stage in uppermost Middle or lowermost Upper Jurassic which marks a return to clayey sedimentation.

calluna peat. Peat that is mainly derived from the common heather Calluna vulgaris.

calomel. $2[Hg_2Cl_2]$. A mineral composed of tetragonal crystals often tablular and prismatic; also massive, earthy; its colour varies from colourless to white, greyish and yellowish-white to ash-grey or brown. It shows adamantine lustre and pale yellowish-white streak and conchoidal fracture with good cleavage in one direction.

caltonite. An igneous rock of the gabbro clan; bluish black, compact or ophitic extrusive rock. It contains olivine, analcite, pyroxene (usually augite), optional accessory magnetite, apatite and palagonite and abundant plagioclase, either labradorite or bytownite.

calving. The act of breaking of ice from its parent glacier, iceberg, or ice shelf.

camber. 1. A terminal, convex shoulder of the continental shelf. 2. A structural feature that is caused by the plastic clay beneath a bed flowing toward a valley so that the bed sags downward and seems to be draped over the sides of the valley.

Cambrian. The earliest period of the paleozoic era, thought to have covered the time span 570 to 500 million years ago; also the corresponding system of rocks.

campanite. An igneous rock of the syenite clan; a microlitic extrusive rock, a type of leucite tephrite. It contains large leuclite crystals.

camptonite. An igneous rock of the diorite clan. A dark grey to black, holocrystalline-porphyritic to saccharoidal hypabyssal rock; it contains phenocrysts of biotite and hornblende in a groundmass of feldspars, amphibole and pyroxene.

camptospessartite. An igneous rock of the diorite clan; a melanocratic, medium-grained equigranular hypabyssal rock; it contains titanaugite.

campylite. $2[Pb_5(As,P)O_4)_3Cl]$. A mineral comprising lead chloroarsenate phosphate; a variety of mimetite; contains small, barrel-shaped crystals of pyromorphite mimetite.

canadian. The Lower Ordovician of North America.

canadite. An igneous rock of the syenite clan; a plutonic rock, a variety of litchfieldtite; It contains pure albite, soda-microcline, nepheline, amphibole and biotite with accessory cancrinite (0-17 per cent), zircon and apatite.

cancrinite.$Na_3CaAl_3Si_3O_{12}CO_3$-$(OH)_2$. A feldspathoid tectosilicate occurring in hexagonal crystals, usually in compact or disseminated masses.

cannel coal. A fine textured volatile bituminous coal distinguished by a greasy lustre and blocky, conchoidal fracture; burns with a steady luminous flame. Also known as cannelite.

cannelite. See cannel coal.

canneloid. (1) Coal intermediate between bituminous and cannel. (2) Cannel coal of anthracite or semi-anthracite variety.

cantalite. An igneous rock of the granite clan; An extrusive rock; transparent variety of pitchstone or rhyolite-pitchstone; sometimes referred to as yellow quartz.

canyon. A deep, steep walled gorge cut by a river or steam, generally into bedrock. Canyons are most frequently found in arid or semi-arid regions where the effect of stream action greatly outweighs weathering. Also spelled as canon.

capacity. (1) The ability of a stream or wind current to transport debris, as gauged by quantity per unit time at a given point. Stream capacity increases with an increase in slope and discharge, and a decrease in width. (2) The yield of a well, reservoir, or pump. (3) Ability of a particular soil to retain water.

cape. A relatively large section of land jutting into the sea or some other large body of water from a larger land mass.

cape ruby. Synonym of pyrope.

capillitite. $2[(Mn,Zn,Fe)CO_3]$. A mineral consisting of manganese zinc ferrous iron carbonate; a variety of rhodochrosite, identical in form to rhodochrosite, but with yellowish to grey colour.

cappelenite. (BA, Ca, Na) $(Y, La)_6$ B_6 Sil_3 $(O, OH)_{27}$. Yttrium barium borosilicate with small amounts of calcium, potassium, sodium and thorium. It forms hexagonal. thick, prismatic crystals of greenish-brown colour. It shows vitreous to greasy lustre, brownish streak, and conchoidal fracture.

cap rock. A hard covering atop the salt covering in a salt dome, consisting mainly of gypsum and anhydrite, some calcite and sulphur in varying amounts. Large sulphur deposits occur in the cap rock of certain salt domes in the Gulf Coast area of the United States.

carat. A unit of weight for gemstones. The metric carat equals 200 mg.

carbonaceous. Relating to or composed of carbon.

carbonaceous rock. Rock with carbonaceous material included in it.

carbonado. A dark coloured, finegrained diamond aggregate with absence of cleavage planes. Also known as black diamond; carbon diamond.

carbonate-apatite. A member of apatite series; conforms to general formula for apatite with basic carbonate substituting for the anion. Formula approaches $Ca_{10}(PO_4)_6(CO_3)$. H_2O in extreme cases. It dissolves with effervescence in dilute hydrochloric acid.

carbonate cycle. The biogeochemical carbonate pathways involving the conversion of carbonates to carbon dioxide and bicarbonates, and the metabolism and regeneration of it in biological systems.

carbonate scapolite. A member of scapolite series with carbonate ion as sole anion; contains considerable sodium. Its colour ranges from to colourless, to yellowish or greenish; shows prismatic cleavage.

carbonate vishnevtite. A member of the solid solution series cancrinite; contains 20-50 per cent cancrinite.

carbonatite. A rock of apparently magmatic origin consisting chiefly of the carbonates of Ca, Mg and Na, with subordinate feldspar, pyroxene, olivine, and other materials.

carbonation. (1) A process of chemical weathering during which minerals containing calcium, iron, magnesium, potassium, and sodium are transformed into carbonates of these metals by carbon dioxide. (2) Saturation of a fluid with carbon dioxide.

carboniferous. The time interval between 345 and 280 million y.b.p. The period is divided into the Lower and Upper Carboniferous, called the Mississippian and Pennsylvanian Periods in North America. The division is based on the presence of coal measures in the upper strata and barren rock beneath.

carbonisation. Decomposition ofan imal or (more often) plant remains to pure carbon as a form of fossil.

carbon ratio. (1) Ratio of the fixed carbon content in a coal to the fixed carbon plus volatile matter. (2) Ratio of ^{12}C. the most common carbon isotope, to either ^{13}C or ^{14}C. If unspecified, the term is usually $^{12}C/^{13}C$.

carbopyrite. Any coal microlithotype containing 5-20% by volume of iron sulphide.

carbuncle. Term describing almandine gemstone garnet cut in hollow cabochon form; old name for all fiery-red precious stones.

carburan. A pitchy hydrocarbon containing uranium.

carex peat. Peat that is mainly derived from carex, a genus of sedges of the family Cyperaceae.

carinate. (1) Said of a fold that is almost isoclinal. (2) Said of an antiform or synform that is confined to incompetent strata, so that adjacent strata are undisturbed.

carinthine. $2[(Ca,Na)_{2\text{-}5}(Mg,Fe_2,Al)_5(Si,Al)_8O_{22}(OH)_2]$ with silicon c7, magnesium c3.5, ferrous c0.5 and ratio of calcium to sodium c2:1, dark green variety of hornblende with properties intermediate between hornblende and glaucophane.

carmeloite. An igneous rock of the gabbro clan; a porphyrite extrusive rock; a type of oilvine basalt with olivine phenocrysts altered to iddingsite.

carminite. $PbFe_2(AsO_4)_2(OH)_2$. A carmine to tile-red mineral consisting of a basic arsenate of iron and lead.

carnalite. $KMgCl_3 \cdot 6H_2O$. A white or red mineral crystallized in orthorhombic fashion; composed of hydrous chloride of potassium and magnesium.

carnelian. A translucent, reddish semi-precious variety of chalcedony; as it becomes brownish, it grades into sard. The colour is due to colloidally dispersed hematite. Also called cornelian.

carnian. Lower Upper Triassic, geologic time. Also spelled Karnian.

carnotite. $2[K_2(UO_2)_2(VO_4)_2 \cdot 3H_2O]$ Hydrated potassium uranium vanadate:

with zeolitic water. It forms powdery or minute orthorhombic crystal plates; possesses dull or earthy lustre, pearly or silky when coarsely crystalline; powdery and crumbling fracture; shows distinct cleavage parallel to basal pinacoid.

carpholite. $MnAl_2Si_2O_6(OH)_4$. Hydrated maganese aluminium silicate with traces of fluorine; monoclinic; radiated and stellated tufts and group of acicular crystals. Its colour varies from pure straw yellow to wax-yellow and possesses silky, glistening lustre.

carphosiderite. $(H_2O)Fe_3(SO_4)_2((OH)_5\ H_2O))$. Basic hydrated ferric iron sulphate; occurs as earthy or minutely scale crusts and aggregates.

carrara marble. A general name for marbles quarried in the vicinity of Carrara, Italy. Several pure varieties are found there, including fine-grade statuary marble. Its colours are mostly white to bluish or blue-veined.

carrollite. $8[CuCo_2S_4]$Member of linnaeite family with copper substituting for cobalt in the lattice; other properties are those of as linnaetite.

caryinite. $(C,Pb,Na)_5(Mg,Mn)_4(AsO_4)_5$. A mineral consisting of arsenates of calcium, lead, magnesium, manganese and sodium.

caryopilite. $(Mg,Mn)_3Si_2O_5(OH)_4.xH_2O$. A mineral of kaolinite-serpentine group.

cascade. (1) A series of shallow, closely spaced waterfalls descending over a steep, rocky surface. (2) A series of descending recumbent folds associated with gravity sliding. They are found when a bed or an inclined surface buckles as it slides downward under the effect of gravity.

cascadian orogeny. A time of regional mountain building and deformation that was especially effective in the Miocene and Pliocene, ending in the late Pleistocene. It was centralized in the Pacific Northwest as a broad upwarp of the rocky mountain regions, which had been peneplained by late Oligocene time.

cascadite. An igneous rock of the syenite clan; a melanocratic, porphyritic hypabyssal rock. It contains phenocrysts of biotite in a nearly aphanitic groundmass of pyroxene, soda orthocl- ase and accessory magnetite apatite and olivine.

cascajo. Reef derived material composed of coral detritas and other sediment, occurring in old deposits.

casing point. The depth in a drill hole to which a given string of casing extends.

cassidyite. $Ca_2(Mg,Ni)(PO_4)_2.\ 2H_2O$. A mineral composed of phosphates of calcium, magnessium and nickel; found in meteorites.

cassiterite. SnO_2. A yellowish or reddish brown mineral consisting of tetragonal, short prismatic crystals and occurs in high temperature hydrothermal veins and in placers. Also known as stannite.

cast. (1) A sedimentary structure representing the infilling of an original mark or depression made on the top of a soft bed and preserved as a solid from the under side of the overlying and more durable stratum. (2) Secondary rock or mineral material that fills a natural mold.

castorite. A transparent variety of petalite.

cata. A prefix denoting deepest, at greatest depth.

cataclasite. A low to high grade dislocation metamorphic rock; formed by ruptural deformation of brittle parent rocks without recrystallisation or chemical reaction; extreme cataclasis with accompanying decrease in grain size and development of banded structure leads to gradation into mylonites.

cataclastic. A high grade dynamic metamorphic rock; brittle minerals within rock have been broken and flattened at right angles to the pressure stress, giving rock a characteristic granular appearance.

cataclastic flow. Flow involving intergranular movement; *i.e.*, mechanical displacement of particles with respect to each other.

cataclastic metamorphism. A type of local metamorphism confined to vicinity of faults and overthrusts, involving purely mechanical forces causing crushing and granulation of the rock material.

cataclinal. Said of a dip stream or of a valley that descends in the same direction as that of the general dip of the underlying strata it traverses.

catacline. Said of the inclination of the cardinal area in either value of a brachiopod lying at right angles to the orthocline position.

cataclysm. Any geological occurrence that effects sudden and widespread changes of the earth's surface; *e.g.*, an earthquake or volcanic eruption of exceptional magnitude.

catapleite. $2[Na_2,Ca)ZrSi_3O_9.\ 2H_2O]$. A mineral containing hydrated zirconium calcium sodium silicate with calcium replacing sodium in the lattice; monolinic, passing to hexagonal at higher temperatures; forms thin, tabular hexagonal prisms of light yellow to yellowish-brown, greyish-blue to violet; colour shows dull lustre, weakly vitreous on fracture surface.

catastrophe. A sudden and violent disturbance of nature ascribed to unusual or supematural agents, and affecting both physical conditions and inhabitants of the earth.

catastrophism. An hypothesis that preffered recurrent, violent, worldwide events as the reason for the sudden disappearance of some species and the abrupt rise of new ones.

catophorite. $Na_2CaFe(II)_4(Fe(III),Al)Si_2AlO_{22}(OH,F)_2$. Basic aluminium calcium sodium and iron silicate with magnesium as possible accessory; a soda-iron amphibole between arfvedsonite and barkevikite in optical properties.

catawberite. A high grade contact metamorphic rock; contains talc and magnetite.

catchment area. (1) The waterproofed area of a storage reservoir. (2) The recharge area and all other areas that contribute water to an aquifer. (3) Drarinage basin.

cat coal. A variety of coal containing pyrite.

catena. A term used to denote a chain. A soil catena is soil series consisting of a group of soils that developed from similar parent material, but show different profiles because of variations in topography and drainage conditions. Inasmuch as each topographic setting affects the nature of the soil profile, a soil catena holds both lithologic and topographic implications.

cat face. (1) A miner's term for glistening balls or nodules of pyrite in the mining face of the coal. (2) Lenticular deposits of pyrite associated with coal.

catoctin. A residual hill, knob, or ridge

of resistant rock material rising above a peneplained surface and maintaining on its summit a remnant of a still older peneplain.

cat's eye. A crystal penetrated by a series of micro-scopoic tubes of fibres running in a direction. Parallel to the vertical axis of the crystal. Cabochon stones cut with canals or needles running parallel to the base show a single ray of light crossing the gem. This term is generally restricted to chrysoberyl cat's eye, but quartz, amphibole, beryl, tourmaline, scapolite, apatite, diopside, kornerupine, syanite, prehnite and orthoclase exhibit chatoyancy.

cat silver. A synonym of mica.

cattierite. 4[CoS_2]. Cobalt disulphide composed of cubic granular; crystals its colour is generally pinkish and it shows cubic cleavage.

caustic metamorphism. The indurating baking burning and frithing effects of lava flows and small dikes on the rocks with which they come in contact.

caustobiolith. A combustible organic rock, formed as a result of direct accumulation of plant materials; includes coal peat.

caustolith. A combustible rock of organic origin but inorganic deposits such as sulphur, asphalt, graphite also occur.

caustophytolith. A caustobiolith formed by the direct accumulation of vegetable matter, *e.g.*, peat, lignite, and coal.

coustozoolith. A rare caustobiolith formed by direct accumulation of animal matter, *e.g.*, some oils.

cavalorite. An igneous rock of the diorite clan. A granular plutonic rock; contains orthoclase in excess of oligoclase, abundant hornblende and accessory quartz, apatite, magnetite and ilmenite.

cave. A natural, hollow chamber or series of chambers and galleries beneath the surface of the earth, or inside of a mountain or hill, with an opening to the surface.

cave breccia. Sharp fragments of limestone debris deposited on the floor of a cave.

cave coral. Coral-shaped cave deposit of calcite.

cave draperies. Sheetlike cave deposits, usually of calcite, formed by water trickling along a slanted ceiling.

cave onyx. A compact, banded cave deposit of calcite or aragonite. It resembles true onyx and can take a high polish.

cave pearls. Concretions of calcium carbonate, which may be calcite or argonite. It shows pearly lustre and possess reasonably soft with strong cleavage; golden-yellow variety has been facted as collector's stones formed by agency of water in limestone caves.

caver. A person who explores caves as a hobby. Usually synonymous with spelunker, although cave explorers tend to avoid the latter term.

cavern. A synonym of cave, especially one of large size. The term may also be used for a cave system.

cavernous. Characterized by caverns, cells, or large pore spaces, as in certain volcanic rocks and limestone.

c-axis. (1) A vertically oriented crystal axis, usually the principal axis; the unique symmetry axis in tetragonal and hexagonal crystals. (2) The reference axis perpendicular to the plane of movement of rock of mineral strata.

cay. A flat mound of sand built up on a reef slightly above the tide.

Cayugan. Upper Silurian of North

America.

cebollite. $Ca_5Al_2Si_3O_{12}(OH)_4$. Hydrated calcium aluminium silicate; generally occurs as orthorhombic, compact, fibrous aggregates; its colour varies form white to greenish or brownish and shows dull lustre.

cecilite. An igneous rock of the gabbro, clan; dark grey, dense texture extrusive rock; it contains nearly 50 per cent leucite with augite, melilite, nepheline, olivine, anorthite, magnetite and apatite.

celadonite. $KMg_3Fe_3Si_9O_{25}(OH)_2 . 9H_2O$. Hydrated iron potassium magnesium silicate; occurs as monoclini, deep olive green, celandine-green or apple-green very softmasses

celestine. $SnSO_4$. An orthorhombic mineral usually existing as elongalied tabular crystals. It shows uneven fracture and vitreous lusture inclining to pearly on cleavages with perfect cleavage in one direction. Its colour varies from colourless to pale-blue; sometimes brownish-green. Celestine occurs chiefly in sedimentary rocks or as primary mineral in hydrothermal veins. Also known as celestite.

celestite. *See* celestine.

cell texture. In mineral deposits, a network pattern formed by solution or replacement or organic structures, *e.g.*, cell walls.

cellular. Said of the texture of a rock, characterized by openings or cavities, which may or may not be connected. The term vesicular is preferred when describing igneous rocks.

celsian. $4[BaAl_2Si_2O_8]$. Barium aluminium silicate occurring as monoclinic, short-prismatic or slender acicular crystals; usually massive. Generally it is colourless but sometimes reddish brown or grey due to included impurities. It shows somewhat greasy lustre on cleavage surface: streak and fracture as for orthoclase, perfect cleavage in one direction. It is an uncommon member of the feldspar family. Its varieties are calciocelsian and kasoite.

cement. (1) A manufactured, finely ground powder that, when mixed with water, sets and hardens. Widely used in building and engineering construction. *See also* portland cement; concrete. (2) Chemically precipitated mineral matter that is part of the cementation process. (3) Ore minerals that have replaced, or are part of, mineral cement.

cementation. Process of binding together of sedimentary particles by some secondary material, such as that relased from suspension or solution in percolating water; important component of diagenesis.

cement stones. A sedimentary rock; a marly deposit with lime, silica and alumina mixed in such proportions that it can be directly used for manufacture of hydraulic or Portland cement. It is very hard and occurs as nodular masses distributed irregularly. Also known as cement rock.

cenology. An old term for the geology of surficial deposits.

centotypal. Said of a fine-grained porhyritic igneous rock having the appearance of fresh extrusive rocks.

cenozoic. An era of geologic time covering the history of earth during the last 70 million years. It is subdivided into the Tertiary and Quaternary (last 2.5 million years) periods. The early Tertiary includes the Paleocene Eocene, and Oligocene epochs. the Quaternary includes the Pleistocene and Holocene (last 10,000 years) perods. The term "Tertiary" is used for the younger rocks that rest on "Secondary" or Mesozoic strata, and was once used as a synonym for the entire Cenozoic Era.

central vent eruption. Ejection of lava and debris from a ground hole that is fed by a single, deep-lying channel. Ejected material accumulates around the vent and may form a conical volcano or any of several other forms. Central vent eruptions occur along subduction zones, mid-oceanic ridges, and certain rift valleys. The differentiation between central vent and fissure eruption is not always easy. The large volcanoes of the Hawaiian Islands all appear to have central vents, but the summit craters are located on a fissure zone, and most of the lavas are erupted from the fissures.

centre of symmetry. A point within a configuration through which any lines cuts the configuration in pairs of points equidistant in opposite directions.

centroclinal. Refering to geologic strata dipping towards a common centre, as in structural basin.

ceragyrite. Ag(Cl,Br). A silver-bearing mineral with proportion of chlorine greater than bromine; end member of cerargyite-bromyrite series. It usually occurs as cubic crystals and is colourless when pure and fresh, developing greenish-grey line with increasing substitution of bromine and becomes violet or purple when exposed to light; shows white or grey and shiny streak.

cerite. (Ca Fe)$Ce_3Si_3O_{12}$. H_2O. A brown rare-earth silicate of cerium and other metals found in gneiss.

cerolite. A mixture of serpentine and stevensite occurring in yellow or green waxy masses.

cerrusite. $PbCO_3$. A white or yellow mineral of aragonite group; occurring in orthorhombic crystals.

cervantite. Sb_2O_4. A white or yellow mineral of antimony, crystallizing in orthorhombic system.

cesarolite. $PbMn_3O_7.H_2O$. A steel-grey mineral consisting of hydrous lead manganate; occurring in spongy masses.

ceylonite. 8[(Mg, Fe) Al_2O_7]. A very dark green, brown, or black, iron-rich variety of spinel with ratio of magnesium to iron 3:1; sometimes used as gemstone. Also known as candite; pleonaste; zeylanite.

chabazite. $(Ca,Na,K)_7Al_{22}(Al,Si)_2 Si_{26} O_{80}.40H_2O$. Hydrous calcium sodium potassium and aluminium silicate; hexagonal, rhombohedral, cube-like, lenticular crystals; also massive, compact. Its colour varies from colourless, white, pink to yellowish to brownish; it shows vitreous lustre, and white streak, eneven fracture and rhombohedral cleavage. It occurs in basic igneous rock with analcite, stillbite, harmotome, laumontite and heulandite and occasionally in gneiss or schists; member of zeolite family. Its varieties are phacolite, seebachite and haydenite.

chalcanthite. $2[CuSO_4.5H_2O]$. Hydrous copper sulphate occurring as triclinic, short prismatic crystals or cross-fibre grainlets; also stalactitic or reniform; also granular massive. Its colour ranges from sky-blue to a little greenish. It shows vitreous lustre and colourless streak; conchoidal fracture and imperfect cleavage in one direction. It is the secondary mineral found in arid climated or in rapidly oxidising sulphide deposits.

chalcedony. A cryptocrystalline variety of quartz; occurs as crusts with rounded or botryoidal surface and as a major constituent of nodular and bedded chers; varieties include agate, banded agate, carnelian, chrysoprase, dendritic agate, flint, moss agate, onyx, sapphrinine, sard and sardonyx.

chalchihuitl. Green jadeite of the ancient; also synonym of chalchuite.

chalcocite. Cu_2S. A black or dark lead-grey mineral with metallic lustre. It is commonly massive, but also occurs infrequently as orthorhombic crystals. Chalcocite is an important ore of copper, occurring most frequently as the result of secondary enrichment.

chalcomenite. $4[CuSeO_3.\ 2H_2O\ ]$. A mineral occurring transparent as orthorhombic, minute acicular crystals with prism faces often striated or rounded of bright blue colour. It shows vitreous lustre, no fracture and no cleavage it is found with azurite and cerussite as oxidation product of primary selenides of copper in veins.

chalcophanite. $ZnMn_3O_7.3H_2O$. A mineral occurring probably as hexagonal, minute tabular crystals with octahedral habit; commonly drusy botryoidal or stalactitic crusts: also massive, dense, granular, platy or fibrous. Its colour varies from bluish to iron black. It shows metallic lustre, dull steak and no fracture. It shows perfect cleavage in one direction. It is found as secondary mineral associated wtih hydrous oxides of manganese and iron.

chalophile. The term used for the substances have great affinity to comine with sulphur.

chalcophyllite. $Cu_{18}Al_2(AsO_4)_3\ (SO_4)_3\ (OH)_{27}.\ 33H_2O$. A green mineral consisting of arsenates and sulphates of copper and aluminium; occurring as tabular crystals or foliated masses. It occurs as secondary mineral found in oxidised zones of copper deposits. Also known as copper mica.

chalcopyrite. $4[CuFeS_2]$. A brass or golden-yellow mineral occurring in the form of bisphenoids. It shows greenish-black steak, uneven fracture and indistinct cleavage. It occurs in mesothermal or hypothermal deposits. Also known as copper pyrites.

chalcosiderite. $Cu(Al,Fe)_6(PO_4)_4(OH)_8\ .4H_2O$. A green mineral with proportion of iron greater than aluminium; isomorphous with turquoize. It possesses vitreous lusture, white or pale green streak and perfect cleavage in one direction. It occurs as secondary mineral along with dufrenite, and rewiste and goethite.

chalcostibnite. $4[CuSbS_2]$. A mineral occurring as orthorhombic, blade-shaped crystals. It possesses metallic lusture, grey streak and shows sub-conchoidal fracture, and perfect cleavage in one direction. It occurs in lining cavities or veins with pyrite.

chalcotrichite. A variety of cuprite occurring in long, straight, needle-like red crystals.

chalk. A sedimentary rock; a type of limestone Containing 90 per cent calcium carbonate almost entirely as calcite; derived from floating micro-organisms and some bottom-dwelling forms in a matrix of finely crystalline calcite, with detrital quartz and glauconite as main accessories. It is usually fine grained and somewhat friable to gritty texture and is deposited from solution under quiet conditions with no wave action.

chalybeate. Containing or impregnated

with iron salts, *e.g.*, a mineral spring.

chalybite. $FeCO_3$. Iron carbonate with nearly 50 per cent iron and a little manganese, magnesium and calcium often present; occurs as hexagonal trigonal, rhombohedral crystals. Also known as siderite.

Champlainian. Middle Ordovician of North America.

chamosite. $2[Fe^{2+},Fe^{3+}Mg,Al)_6(Si,Al)_7 O_{10}(O,OH)_2]$ with silicon = 2.8-3.1 and ferric iron and hydroxide = 8. Hydrous magnesium iron aluminium silicate occurring as monoclinic crystals; member of chlorite group. Its varieties are berthierine and brunsvigite.

channel. (1) The bed of a stream or waterway. (2) The deeper part of a waterway, especially the part of a body of water deep enough for navigation. (3) A wide strait, as between an island and a continent.

channel capacity. The maximum flow volume that a given channel can sustain without overflowing its banks.

channel-fill deposit. A deposit in a stream channel, consisting largely of bed load materials. The deposit occurs in those places where the transport capacity of the stream is not great enough to remove the material it receives.

channel flow. The type of flow exhibited by surface runoff as it streams through long, narrow, V-shaped gullies or troughs.

channel-mouth bar. A bar built at the place where a stream enters a body of standing water; a decrease in the stream's velocity at this point results in a buildup of sand and gravel.

channel pattern. The configuration of a steam or river course, such as braided, meandering sinuous, or straight. A river may follow a relatively straight course for some distance, and a meandering course at a different location.

cannel sample. A rock sample, usually selected across the face of a rock body or vein to provide an average value.

channel sand. Sand or sandy material depostited in the bed of some channel cut into the underlying rock. Such sand may contain valuable minerals, ore or gas.

charnockite. A high grade contact metamorphic rock. Mineral assemblage approximates granite clan of igneous rocks: A dark, granular plutonic rock; contains mainly hypersthene, with microcline-microperthite, quartz and iron ores.

charnockite series. High grade contact metamorphic rocks; coarse grained to granular, granoblastic to foliated rocks; contain hypersthene with varying proportions of quartz, potashfeldspars, plagioclase and garnet.

chassignite. An achondritic stony meteorite composed mainly of (95%) olivine.

chatoyancy. Ability to shine like a cat's eye; having a changeable lustre or colour with an undulating narrow band or white light.

chattermark. A scar on the surface of bedrock formed by the abrasive action of drift carried at the base of a glacier.

Chautauquan. Uppermost Devonian of North America.

cheleutite. $Co(As,Bi)_2$. A mineral With proportion of bismuth smaller than five per cent; variety of smaltite existing as twisted crystal masses; shows cubical cleavage.

chemical denudation. Washing of land surface by water transport of soluble materials into the sea.

chemical precipitates. A sediment formed from precipitated materials as distinguished from detrital particles that have been transported and deposited.

chemical remanent magnetization. A stable remanent magnetization caused by the slow growth of magnetically ordered mineral grains in the presence of a magnetic field.

chemical rock. A sedimentary rock with less than 50% detrital material formed directly by precipitation from solution or colloidel suspension, or by the deposition of insoluble precipitates; *e.g.*, gypsum, rock salt, or tufa.

chemical unconfirmity. An unconformity or stratigraphic boundary determined by chemical analysis such as in the case of a limestone formation, whose basal part has a higher concentration of impurities due to the presence of inwashed fine clastic and organic material.

chemical weathering. The process of weathering by which rocks and minerals are transformed into new, and stable chemical combinations by reactions such as hydrolysis, oxidation, ion-exchange and disolution.

chemocline. The transition in a meromictic lake between the mixolimnion layer and the monimolimnion layer.

chemogenic. Reffering to rock materials that have been deposited entirely by chemical action.

chemungian. Middle Upper Devonian geologic time, below Cassodagan.

chenier. A wooded beach ridge found on the Mississippi Delta composed of sand or shell debris swept onto the deltaic plain. Although beach ridges formed by similar processes are found elsewhere, they are often given other names.

cheralite. $2[CaTh,Ce_2)(PO_4)_2]$. A mineral with some silica replacing phosphorous and calcium replacing cerium group of rare earths. Calcium, thorium and cerium group of rare earths phosphate; existing as monoclinic, brittle masses up to 50mm across. Its colour varies from dark to pale green, shows resinous to vitreous lustre, white streak and uneven fracture. It is distributed throughout kaolinised pegmatite dike outcropping or in gravel of adjacent streams with tourmaline, chrysoberyl, zircon and considerable smoky quartz; isomorphous with huttonite and monazite.

chert. A sedimentary rock; grey to black, homogeneous with flat fracture. It consists of dense aggregates of microcrystalline quartz and chalcedonic silica mixed in any proportion with or without remains of siliceous and other organisms such as radiolarites, diatomites and sponge spicules.

chertification. A kind of silification, especially by a fine-grained or cryptocrystalline quartz.

chert ironstone. A sedimentary rock having ferruginous deposits of chert; consists of alternate layers of chert and siderite or chert and hematite; some time ellipsoidal granules of greenalite embedded in chert matrix with some carbonate minerals. It has markedly banded structure.

chessylite. A synonym of azurite.

chesterian. Uppermost Mississippian of North America.

chiastolite. A variety of andalusite; the stout crystals have angles and axis of different colours. It occur in metamorphic rocks, often as rounded and knotty projections with kyantite, sillimanite, garnet and tourmaline. Also known as macle.

chiastolite-slate. A high grade contact metamorphic rock; a carbonaceous shale formation; contains conspicuous rystals of chiastolite in generally cryptocrystalline groundmass.

chibinite. An igneous rock of the syenite clan; a coarse grained, hypautomorphic-granular plutonic rock; contains mainly microcline microperthite, sodaclase, nepheline, pyroxene and eudialyte.

chiklite. $2[(Na,Ca,Mn)_3\ (Mn,Fe^{2+},Fe^{3+},Al)_5(Si,Al)_8O_{22}(OH)_3$. A violet coloured mineral showing pearly lustre on cleavage faces. It occurs in pegmatite cutting manganese ore band with quartz and feldspars.

childrenite. $8[(Fe,Mn)AlPO_4\ (OH)_2.H_2O]$. A mineral with proportion of ferrous greater than manganese, containing basic hydrated aluminium phosphate with divalent iron and manganese mutually substituting. Its colour ranges between brown and yellowish brown. It shows vitreous to somewhat resinous lustre, white streak and subconchoidal to unven fracture and poor cleavage in one direction.

chile saltpeter. $NaNO_3$. Sodium nitrate, that occurs in caliche in Chile.

chimney. (1) A columnar portion of headland detached from the shore line. (2) A channelway through which magma reaches the surface. (3) A cylindrical, relatively vertical, ore body. (4) A vertical opening in a cave. (5) A pinnacle-like rock body rising well above its surroundings.

china clay. A synonym of kaolinite.

china stone. A sedimentary rock; semi-decomposed, granitic rock; usually kaolinised, firm, does not curmble readily. Sometimes it contains white mica and fluorite.

chkalovite. $Na_2Be(SiO_3)_2$. Sodium beryllium silicate, existing as orthorhombic, large grains, white semi-transparent. It shows vitreous lustre and cleavage in one direction.

chladnite. A very pure non-ferriferous variety of enstatite occurring in most achondritic meteorites.

chloanthite. $8[NiAs_2]$. Nickel arsenide occurring as cubic crystals; also massive; shows metallic lustre, greyish black streak and poor cleavage.

chlor apatite. $Ca_5(PO_4)_3(F,Cl,OH)$. A mineral with proportion of chlorine greater than that of fluorine or of hydroxide; member of apatite series; rare mineral found often chiefly in veins in gabbroic rocks and in some meteorites.

chlorite. group. Hydrous ferrous iron magnesium aluminosilicate; may also have ferric iron, chromium or manganese. It occurs in monoclinic or pseudohexagonal form silicate sheet structure as the micas. It shows basal cleavage in crystallized forms. The series comprises clinochlore, penninite, ripidolite, corundophilite, amesite, daphnite, cronstedtite, thuringite, stilpnomelane, diabantite, aphrosiderite, delessite and vermiculite. It often occurs as secondary mineral from alteration of mainly pyroxene, amphibole biotite and garnet.

chlorite schist. A metamorphic rock whose composition is dominated by members of the chlorite group.

chloritization. Replacement by alteration into or introduction of chlorite.

chloritoid. $(Mg,Fe)_2Al_4Si_2O_{10}(OH)_4$. A micaceous mineral, that crystallizes in the monclinic or triclinic systems. It occurs in yellow-green to black scaly aggregates in low-grade, regionally metamophosed sedimentary rocks. Also known as struverite.

chlorocalcite. $KCaCl_3$. Potassium calcium chloride, occurring as cube-like crystals with octahedral or dodecahedral-like truncations; also prismatic. The colour is generally white, some times stained violet. It is transparent; shows cube like cleavage, with one direction better than other two.

chloromelanite. A variety of jadeite with considerable amounts of iron oxides; the colour ranges from dark green to black.

chloropal. A synonym of nontronite.

chlorophane. A variety of fluorite; emits bright green fluorescent light on heating.

chloropite. A synonym of delessite.

chlorospinel. $8[Mg(Al,Fe)_2O_4]$ A mineral with ratio of aluminium to ferric iron from 3:1; iron rich variety of spinel.

chondrites. One of the two divisions of stony meteorites characterized by chondrules. They are primitive meteorites, having crystallized about 4700 million y.b.p., and make up over 80% of meteorite fills. They are divided into three groups according to Mg and Fe ratio (2 to greater than 9).

chondrodite. $Mg_5(SiO_4)_2F,OH)_2$.. A monoclinic mineral of humite group of yellow-red colour. It occurs in contact-metamorphosed dolomites.

chondrule. A small globular body of various materials, though mainly olivine and pyroxene; found as an inclusion in certain stony meteorites (chondrites). Chondrules are usually less than 3 mm in diameter, may be porphyritic or glassy in texture, and are clearly visible on polished surfaces.

chonolith. An igneous intrusion that cannot be classified (laccolith, dike, sill, *etc.*) because its form is very irregular.

christophite. A synonym of marmatite.

chrome-augite. A synonym of lavrovite.

chrome-diopside. $4[MgCaSi_2O_6]$. A mineral with small amounts of chromate; variety of diopside; clinopyroxene. The bright green coloured variety is used as gemstone.

chrome-mica. A synonym of fuchsite.

chrome-phengite. A synonym of mariposite.

chrome-spinel. A synonym of picotite; sometimes synonym of magnesiochromite.

chrome-trmolite. $2[Ca_2Mg_5S_8O_{22}H)_2$. A mineral with up to 2 per cent chromate; variety of tiremolite; a clinoamphibole.

chromite. $8[FeCr_2O_4]$. A mineral of the spinel group that occurs as small, black octahedral crystals, usually in compact masses in mafic and ultramafic rocks. It is the principal ore of chromium occurring as accessory mineral in most peridotites and related rocks. Chromite is a member of spinel family. Its varieties are beresophite (of Simpson) alumoberesovite, alumochromite and magnesiochromite.

chromitite. An igneous rock of the ultrabasic clan; a dark coloured medium-grained holocrystalline automorphic hypabyssal rock; essentially chromite with up to 5 per cent biotite or pyribole.

chromohercynite. $8[Fe(Al,Cr)_2O_4]$. A

mineral with ratio of aluminium to chromium at least 3:1; a variety of hercynite with chromium replacing aluminium. It is an isomorphous mixture of chromite and hercynite.

chron. The time unit equivalent to the stratigraphic unit, subseries, and geologic name of a division of geologic time.

chronohorizon. A stratigraphy surface, theoretically without thickness, that is the same age at all points. In practice it is a thin and characteristic interval that serves as a good time correlation or time-reference zone; *e.g.*, coal beds, bentonite beds, and many biohorizons.

chrysocolla. $CuSiO_3.2H_2O$. A botryoidal and massive, enamel-like, mineral of copper; its colour ranges from green to brownish-black. It possesses vitreous, shining, earthy lustre, white steak and conchoidal fracture.

chrysolite. $4[(Mg,Fe)_2SiO_4]$. (1) A variety of olivine having a magnesium to magnesium-iron ratio of 0.9-0.7. (2) A gem characterized by light yellowish-green hues, especially the gem variety of olivine but also including beryl, spinel, and topaz.

chrysoprase. Gemstone variety of chalcedony; coloured due to the presence of nickel oxide as impurity.

chrysophyric. Said of a basalt having olivine phenocrysts.

chrysoprase. $2[Mg_2Si_2O_5(OH)_4]$. An apple-green or pale-yellowish-green variety of chalcedony, occurring as monoclinic fibres; valued as gem.

chrysotile. A fibrous white, grey or green mineral of the serpentine group, the most important type of asbestos. Also known as serpentine asbestos.

chute. An inclined channel, *e.g.*, a tube, trough, or shaft. It may also be used to mean a waterfall or a steep descent or rapids in a river.

chute cutoff. A cutoff made through one of the marsh areas of a point bar, thus isolating part of the bar as an island between two river channels.

cienaga. A term, commonly used in arid regions such as the southwestern U.S., for a marshy area whose wetness is due to springs or seepage.

ciminite. An igneous rock of the syenite clan; a medium grey, compact and somewhat porphyritic, extrusive rock; contains phenocrysts of feldspar, olivine and augite in an aphanitic groundmass of orthoclase, labradorite.

Cincinnatian. A stratigraphic stage name for the North American Upper Ordovician.

cinder. Vesicular, dark-coloured basaltic to andesitic volcanic fragments, ranging in size from 4 to 32 mm. When ejected, cinders are in a semi-rigid state and fall to the ground as solid matter.

cinder cone. A conical elevation formed by the accumulation of volcanic debris around a vent.

cinnabar. HgS. The most important ore mineral of mercury; associated with volcanic activity. It is found is veins and impregnations near necent volcanic rocks and not spring deposits.

CIPW classification. A procedure for recalculating the chemical composition of a rock into a hypothetical group of water-free standard minerals. The merit of the normative composition, or norm, is that the limited number of standard normative minerals facilitates comparisons between rocks. The norm disregards the effects of geological processes and P-T conditions, focusing only on the source of matter comprising the magmatic rock. The initials represent the names of Cross, Iddings, Pirsson, and Washington, the four men who devised the system.

cirque. A semicircular basin or indentation with steep walls originating from an ordinary valley head. It is situated high on a mountain slope, and is associated with the erosive activity of a mountain glacier. Walls of the cirque are cut back by disintegration of lower rock. Resulting rock material, embedded in the glacier, gouges a concave floor that may contain a small lake (tarn) if the glacier disappears. The expansion of neighboring cirques produces aretes, horns, and cols.

cirque glacier. A small glacier occupying a cirque or resting against the headwall of cirque.

cirque lake. A small, deep, commonly circular glacial lake occuping a cirque; it has no prominent inlet being fed by runoff from surrounding slopes and by a lip of bedrock or by a small moraine.

citrine. A yellow to yellowish-brown crystalline quartz, sometimes resembling topaz in colour.

cladogenesis. Phyletic branching or splitting, or progressive specialization through evolution.

clan. (1) A group of igneous rocks having closely related chemical compositions. (2) A small ecologic community having only a single dominant species.

clarain. A coal lithotype appearing as stratifications parallel to the bedding plane and usually having a silky lustre and scattered and diffuse reflection. Also known as clarite.

Clarendonian. A stratigraphic stage name, the North American equivalent of the European Sarmatian (Tertiary System).

clarinite. A heterogeneous, generally translucent material making up the major micropetrological ingredient of clarian.

clarite. *See* clarain.

Clarkeforkian. A stratigraphic stage name for the North American Upper Thanetian (Tertiary System).

clast. An individual grain, fragment, or constituent of detrital sediment or sedimentary rock produced by physical break down of a larger mass.

clastation. The breaking-up of rock masses by physical or chemical means.

clastic. (1) Pertaining to a sediment or rock composed chiefly of fragments derived from pre-existing rocks or minerals; also the texture of such a rock. (2) Pertaining to the fragments (clasts) of which a clastic rock is composed. (3) A clastic rock; usually used in the plural. (4) Said of a bioclastic rock.

clastic dike. A tabular body comprised of clastic materials that cuts across the bedding of a sedimentary formation and is derived from overlying or underlying beds; in particular, a pebble dike or sandstone dike.

clastic ratio. The ratio, in a stratigraphic section, of the amount of clastic material (sandstone, conglomerate) to that of nonclastic material (limestone, dolomite).

clastic rocks. Sedimentary rocks composed of fragmental material derived from pre-existing rocks or from dispersed consolidation products of magmas. Detritus transported mechanically into place of deposition. Sub-divided according to grain size. Rock types include breccia, conglomerate, sandstone and shale.

clastic reservoir. An underground oil or gas trap formed in clastic limestone.

clastic sediment. Deposits of clastic materials transported by mechanical agents. Also known as mechanical sediments.

clay. Sedimentary rock; an argillaceous rock composed of fine-grained clastic sediments; consists of minute, flaky crystals of minerals produced by chemical weathering of crystalline rocks. Its cohesive and plastic properties are derived from thin films of water enveloping flaky particles. When heated and water is driven off, it becomes hard and stone-like. It has adsorption and ion-exchange properties and comprises clay minerals, with varying amounts of quartz, feldspars, other silicates, carbonates and ferruginous or organic matter.

clay ball. A chunk of clay released by erosion of clayey bank and rounded by wave action.

clay band. A highly coloured, argillaceous layer in clay ironstone.

clay crete. Weathered argillaceous material forming a layer immediately overlying bedrock.

clay dune. A dune composed of clay fragments heaped up by the wind-action.

clay gall. A dry, curled clay shaving derived from dried, cracked mud embedded and flattened in a sand stratum.

C layer. The seismic region of the earth between 410 km and 1000 km, equivalent to the transition zone of the upper mantle.

clay loam. Soil containing 27-40% clay, 20-45% sand and rest as silt.

clay marl. A chalky clay of white colour, with a smooth texture.

clay minerals. Finely crystalline, hydrous magnesium or aluminium silicates forming the major constituents of clay; crystal structure of the two-layer or three-layer type. Kandite group (kaolinite, dickite, chamosite, greenalite and related halloysite) have two layer structure and composition $Al_2O_3.2SiO_2$. $2H_2O$ with ferrous iron and magnesium substituting for aluminium. Three-layer types have two layers of silica tetrahedral interleaved with alumina di-and tri-octahedra. Smectite group (montronite and montmorillonite) have lattice expandable by water adsorption. Illite group consists of non expandable minerals with complex structure resembling muscovite. Group includes illite, glauconite and related ferruginous chlorites. Smectite and illite group have three-layer structure. Many clay minerals are of mixed-layer types of randomly or regularly interstratified growths of two or more clay minerals.

clay shale. (1) A shale composed of clayey material which becomes clay again on weathering. (2) Consolidated sediment composed of upto 10% sand and having a silt to clay ratio of less than 1:2.

clay soil. A fine-grained inorganic soil that forms hard lumps when dry and becomes sticky on wetting.

claystone. Indurated clay, consisting mainly of fine material of which a major portion is clay mineral.

clay vein. A clay-body which is similar to an ore vein in form and fills a crevice in a coral scan.

cleat. Vertical breakage planes in coal. Also spelled cleet.

cleavage. (1) The property of some minerals to break along planes related to the molecular structure of the mineral and parallel to actual or possible crystal faces. Cleavage is defined by the quality of the break (excellent, good, etc) and its direction in the mineral, *i.e.*, the name of the form it parallels. Some of the more important types of cleavage are: basal, or parallel with the basal plane, *e.g.*, mica, cubic, or parallel with the faces of a cube, *e.g.*, galena; octahedral, or parallel with the faces of an octahedron, *e.g.*, fluorite, prismatic, or parallel with the faces of a prism, *e.g.*, augite; rhombohedral, or parallel with the faces of a rhombohedron, *e.g.*, calcite. (2) The characteristic or tendency of a rock to split along parallel, closely spaced planar surfaces, *e.g.*, slaty cleavage. It is produced by deformation or metamorphism and is independent of bedding.

cleavage banding. A compositional banding, usually formed from incompetent material such as argillaceous rocks, that is parallel to the cleavage rather than bedding.

cleavage crystal. A crystal fragment bounded by cleavage faces, thus forming a regular form.

cleavage fracture. The manner of breaking a crystalline substance along the cleavage plane.

cleavage plane. A plane along which a crystalline substance may be split.

cleavelandite. A white, leaflike variety of albite, found as a late-stage mineral in pegmatites, replacing other minerals.

cleavings. The partings in a coal seam that separates it into beds.

cliachite. A group name for colloidal mixture of aluminium hydroxides containing bauxite also.

cliff. General term for a high, steep face of a rock formation. Cliffs may be formed by faulting, waves and shore currents, volcanic activity, and glaciation. Sea cliffs are formed by wave action.

cliff glacier. A short glacier that occupies a niche or hollow on a steep slope and does not reach a valley.

cliff overhang. A rock mass hanging out of a slope; especially the upper part of an edge cliff projecting out over the lower, undercut part, as above a wave-cut notch.

Cliftonian. Middle Middle Silurian geologic time.

climbing dune. A dune that develops on the windward side of mountains or hills.

clinker. (1) A mass of fused incombustible matter, commonly found when coal is burned. (2) A jagged fragment of lava.

clinoamphibole. A group of amphiboles that crystallize in the monoclinic system.

clinoaxis. The inclined lateral axis that make on oblique angle with the vertical axis in the monoclinic system. Also known as clinodiagonal.

clinochlore. $(Mg,Fg,Al)_3(Si,Al)_2O_5O)_4$. A mineral of the chlorite group having green appearance; occurring as monoclinic crystals.

clino-chrysotile. Intergrowth of monoclinic form of chrysotile with up to 50 per cent of orthorhombic form; nearly same unit cell dimensions as monoclinic form.

clinoclase. $4[Cu_3AsO_4(OH)_3]$. Basic copper arsenate. having monoclinic; tabular or rhombohedral single crystals or rosettes; also crusts or coatings; it possesses dark greenish-black to greenish-blue; vitreous to pealy lustre on cleavage; bluish-green streak, uneven fracture; it is brittle and shows perfect cleavage in one direction.

clinoform. A subaqueous landform, such as the continental slope of the ocean or the forest bed of a delta.

clinohedrite. $CaZnSiO_3(OH)_2$. A colourless, white, or purplish monoclinic mineral consisting of silicates of calcium and zinc.

clinohumite. $Mg_9(SiO_4)_4$ $(F,OH)_2$. A monoclinic mineral of the humite group.

clinopyroxene. Group name for monoclinic pyroxenes, including monoclinic forms of normally orthorhombic pyroxenes: enstatite, ferrosilite and hypersthene.

clintonite. $Ca(Mg,Al)_3(Al,Si)O_{10}\text{-}(OH)$. A yellowish or reddish brown monoclinic mineral of brittle mica group.

clinton ore. An oolitic iron ore, one of the ironstones of the clinton group, or clinton Formation (Middle Silurian) in the Appalachian Mountain belt. Most ironstones of the group are oolitic hematites, and fossils contained within them are usuallypartly replaced by iron minerals. All of the oolitic ironstones are cemented chiefly by calcite and dolomite.

clod. An artificially formed aggregate of soil particles.

clog snow. A sking term for wet, sticky, new snow.

closed basin. A region that drains into a depression or lake within its environment, and from which water escapes only by evaporation.

closed bay. A bay indirectly connected with the sea through a narrow pass.

closed fold. An old (now rarely used. term for isoclinal fold.

closed form. A crystal form whose faces enclose space.

closed lake. A lake that does not have a rareface effluent and that loses water by evaporation or by seepage.

closed structure. A structure, such as an anticline or syncline, whose map representation shows it to be entirely closed by one or more contour lines.

closed system. A system in which there is no transfer of matter or energy into or out of the system during a particular process.

close fold. A fold with an inter-limb angle between 30° and 70°.

close-grained. Said of a rock whose constituent particles are fine and tightly packed together.

close ice. An obsolete term for sea-ice concentration of 8/10 to 10/10.

close-joints cleavage. An old term for both slip cleavage and fracture cleavage.

closure. On a structure contour map of a subsurface anticline, closure is the vertical distance between the anticline's highest point and its lowest closed structure contour. The information is used to estimate gas or oil reserves.

clot. A group of ferromagnesion minerals in igneous rocks.

clough. A cleft in a hill; a ravine or a narrow valley.

clove. A term used for a narrow and deep valley.

cluse. A narrow, gorge, trench or watergap with steep sides that cuts transversely through an otherwise continuous ridge.

Coahuilan. A North American provincial series in Lower Cretaceous geologic time, above the Upper Jurassic and below the Comanchean.

coal. A redily combustible rock containing more than 50% by weight and more than 70% by volume of carbonaceous material including inherent moisture, formed from compaction and induration of variously altered plant remains similar to those in peat. Coal is non-crystalline, brittle, and dull to brilliant in lustre. With increasing compaction, the colour changes from light brown to black, and the specific gravity increases from 1.0 to 1.7. Coals are classified according to their constituent plant materials (type), ratio of fixed carbon content to volatile matter (rank), and amount of impurities present (grade). Coal occurs in rocks of the early Proterozoic, but did not become widespread until the development of woody land plants in the Devonian. Pre-Devonian coals are composed of algal remains. Sedimentary rocks of the Carboniferous-age hold the greatest abundance of coal-bearing rocks. Large deposits of lignite also occur in rocks of early Tertiary age. Coal beds are commonly found with shales and fine-grained sandstones.

coal ball. A concretion of mineralized plant debris occurring in a coal seam or in adjacent rocks.

coal bark. An exposed seam of coal.

coal barrier. A protective pillar composed of coal.

coal basin. A coalfield with a basinal structure.

coal bed. A seam or stratum of coal parallel to the rock stratification. Also known as coal rake; coal seam.

coal breccia. Angular fragments of coal within a coal bed.

coal field. An area containing coal deposits.

coal gas. A fuel made by the carbonization of a high-volatile bituminous coal. Approximate composition by volume: 50% hydrogen, 30% methane, 6 to 8% carbon monoxide, 2 to 4% olefins and 7 to 8% carbon dioxide, nitrogen, and oxygen.

coal gasification. The conversion of coal, char, or coke to a gaseous product by reaction with air, steam, carbon dioxide, or mixture of these.

coalification. Formation of coal from plant material by the processes of diagenesis and metamorphism. Also known as bituminization; carbonification; incarbonization; incoalation.

coal measures. A succession of sedimentary strata consisting chiefly of clastic rocks with interlayered beds of coal.

coal pebbles. Rounded mass of coal occurring in sedimentary rocks.

coal petrology. The science that deals with the origin, history, structure, chemical composition, and classification of coal.

coal plant. A fossil plant found in a coal bed, or a plant whose altered substance contributed to the formation of coal beds.

coal rake. *See* coal bed.

coal seam. *See* coal bed.

coal type. A classification of coal according to its constituent plant materials (macerals).

coarse aggregate. Said of an aggregate composed of particles greater than 4.76 mm in diameter.

coarse-grained. 1. said of a crystalline rock whose individual minerals have an average diameter greater than 5 mm. 2. A sediment or sedimentary rock in which individual constituents are easily visible to the naked eye. Various limits have been suggested.

coarse gravel. A term for gravel whose

particles have a diameter in the range of 19-76mm.

coarse pebble. A geologic term for a pebble having a diameter in the range of 16-32 mm.

coarse sand. A geologic term for sand composed of grains between 0.5 and 1mm in diameter.

coarse topography. A land surface characterized by large-scale erosional features.

coast. A zone of rather indeterminate width extending landward from the seashore. In geographic terms, it may be that section of a region that is near the coast, and sometimes includes the coastal plain.

coastal plain. A broad plain sloping gradually seaward. It generally represents a strip of emerged sea floor. During the emergence, the shoreline of a coastal plain migrates seaward so that progressively younger rocks are found at decreasing distances from the shore.

coastline. (1) The boundary between coast and shore. (2) The boundary between water (espcially sea or ocean water) and land. (3) A broad term for the overall appearance of a coastal land trait. 4. A broad zone of indefinite width extending both seaward and landward from a shoreline.

cobaltite. 4[CoAsS]. Cobalt sulpharsenide; a mineral occurring as striated, cubic crystals, pyritohedra of combinations; sometimes as granular massive to compact. It possesses red or steel-grey colour with violet tinge or greyish black when much iron is present. It shows greyish-black streak, unvent fracture, and perfect cleavage in one direction.

cobble. A fragment of rock with a particle size between 65mm and 250mm.

cobblestone. A rounded stone used in paving or other construction.

cococolite. A granular, white or green, variety of diopside.

coccolith ooze. A type of ooze.

coccolithophore. A minute marine planktonic photosynthetic organism found only in warm, low-latitude waters. After death, it disintegrates to form coccoliths, which are microscopic calcareous plates of many different shapes, built of calcite or aragonite. Coccoliths are found in chalk and in the deep-sea oozes of tropical and temperate oceans.

cocite. An igneous rock of the ultrabasic clan; a yellowish-green, fine-grained hypabyssal rock; contains phenocrysts of olivine and diopside and smaller amounts of olivine, pyroxene, magnetite, biotite and leucite in groundmass of augite, orthoclase and biotite.

coesite. SiO_2 . A high pressure polymorph of formed in nature only under unique physical condition; usually found in meteor impact craters.

cognate. Pertaining to contemporanceous fractures in a system with regard to time or origin and deformational type.

cognate ejecta. Essential or accessory pyroclasts derived from the magmatic materials of a current volcanic eruption.

cokeite. Natural coke formed by action of magmas on coal or by natural combustion of coal in mines.

colina. A term used for a hillock or other small eminence.

colk. A pothole in a stream bed.

collado. A term used for a saddle, gap or pass.

collapse breccia. Angular rock fragments derived from the collapse of rock overlying a hollow space.

collapse caldera. A type of caldera produced as a result or collapse of the roof of a magma chamber due to removal of magma by voluminous pyroclastic or lava eruptions or by subterran withdrawal of magma.

collapse crater. A larger lunar crater supposed to have been formed by roof subsidence of lava-filled cavities.

collapse fault. A normal, gravity fault on the margin of the salt-withdrawal basin in the gulf coast.

collapse structure. A structure resulting from rock slides under the influence of gravity.

collinsite. $Ca_2(Mg,Fe)(PO_4)_2.2H_2O$. Hydrated calcium magnesium phosphate occurring as triclinic, bladed crystals. It possesses pale brown colour, silky lustre and shows indistinct cleavage.

colloform. A term sometimes used to include all more-or-less spherical forms; *e.g.*, botryoidal, reniform, mammillary.

collophane. A variety of apatite, carbon-enriched intermediate member of hydroxyl-apatite-fluor-apatite series occurring as amorphous or massive, opaline or hornlike with dense, layered or colloform structure; sometimes concretionary, nodular, spherulitic or pulverulent.

colophonite. A variety of andradite; a calcium-iron garnet occurring as coarse, granular crystals of dark reddish or brownish colour with iridescent hues.

colluvium. A general term for unconsolidated material at the bottom of a cliff or slope, generally moved by gravity alone. It lacks stratification and is usually unsorted; its composition depends upon its rock source, and its fragments range greatly in size. Such deposits include cliff debris and talus.

columbite. $(Fe,Mn)(Nb,Ta)_2O_6$. A black mineral, the Nb end member of the columbite-tantalite series, It occurs in granites and related pegmatites.

columbretite. An igneous rock of the syenite clan; a compact extrusive rock; contains sanidine, and altered hornblende in dense groundmass of oligoclase, analcite, small augite prisms and magnetite grains, with interstital filling of sanidine, leucite and plagioclase laths.

columnar jointing. Long, parallel columns, polygonal in cross-section, occurring most frequently in basalts, but also in other extrusive and igneous rocks.

columnar section. A graphic depiction of the sequence of rock units in a particular locality. The lithology of a section is indicated by symbols, and thicknesses are drawn to scale. *See also:* geologic column.

columnar structure. A primary sedimentary structure consisting of columns arranged perpendicular to the bedding.

comagmatic. A term applied to igneous rocks that are regarded as having been derived from a common parent magma, because they share a common group of chemical and mineralogical features.

comanchian. Lower Cretaceous geologic time, of North America.

comendite. An igneous rock of the granite clan; a white porphyritic extrusive rock. It contains phenocrysts of quartz, alkali-feldspar and pyroxene in micrograniitic groundmass of quartz and alkali-feldspar. It may be considered as the leucocratic variety of pantellerite.

comleyan. Lower Cambrian geolgic time.

common granet. A brownish-red translucent of opaque variety of almandine.

competence. The ability of a stream or wind current to carry detritus, as determined by particle size rather than amount. The diameter of the largest particle transported is known as the competence value.

complementary rocks. Rocks which are differentiated from the same magma, and whose average composition is the same as the parent magma.

composite cone. A large volcanic cone constructed of lava and pyrolastic substances in alternate layers.

composite grain. A sedimentary clast formed of two or more types of particles.

composite intrustion. A term applied to bodies of igneous rock such as dikes, sills, or laccoliths, which are composed of two or more intrusions that differ in chemical and mineralogical composition. For example, an acid phase may be injected into a more basic one.

composite sequence. An ideal sequence of cyclic sediments containing all the lithological types in their proper order.

composition plane. A planer composition surface in a crystal uniting two individuals of a contact twin.

composition surface. The surface joining individuals of a crystal twin; may or may not be planer.

compound twins. Individuals of one mineral group joined in accordance with two or more different twin laws.

compound vein. A term used for a vein composed of several minerals; also, a vein comprising many parallel fissures cut across fissures.

compound volcano. A volcano with an associated volcanic dome either in its crater or on its flanks.

compressive stress. A stress that tends to push together on opposite sides of a plane, in a direction perpendicular to that plane. The maximum compressive stress that can be applied to a material under specified conditions before rupture or facture occurs is the compressive strength of the material.

comptonite. A variety of thomsonite with deficiency of sodium.

concentric. Pertaining to relationship between two different sized circular, cylinderical, or spherical shapes with the smaller one exactly centered with larger one.

concentric fold. A fold in which the original thickness of the strata is unchanged during deformation. Also known as parallel fold.

conchoidal. Having a smooth and curved surface; used to describe the fracture surface of a rock mineral.

concordant body. An intrusive igneous body whose contacts are parallel to the bedding of the country rock.

concretion. A hard and compact mass of mineral matter in the pores of sedimentary or fragmental volcanic rock. It represents the concentration of a minor constituent of the enclosing rock or of the cementing material.

confining bed. An impermeable bed adjacent to an aquifer.

confining pressure. The pressure exerted equally on all sides, such as geostatic pressure of hydrostatic pressure.

confluence. The meeting point of two glaciers or two streams.

conformity. (1) A term indicating true

stratigraphic continuity of adjacent sedimentary strata; *i.e.*, they have been deposited in orderly series without evident time lapse. Also known as conformability. (2) A surface separating older strata from younger, along which neither erosion nor non-deposition is evident, and with no significant hiatus.

congelifraction. The act of disintegration of rocks as a result of freezing of the water contained.

congeliturabate. Soil or unconsoildated material which has been moved or distrurbed by frost action.

conglomerate. Cemented, rounded fragments of water worn rock or pebbles, bound by a siliceous or argillaceous material. The fragments can vary in size from large boulders to gravel so fine that is almost a coarse sand.

congressite. An igneous rock of the syenite clan; a pale pink, coarse-grained, usually foliated,plutonic rock. It contains much nepheline with mainly orthoclase, albite, sodalite, muscovite, biotite and magnetite.

Coniacian. Lower Senonian geologic time.

consolidation. Process by which loose, soft, or fluid earth becomes coherent and firm.

contact metamorphism. One of the main local processes of thermal metamorphism that is related to intrusion of magmas; it takes place in rock or near their contact with a body of igneous rocks.

contemporaneous deformation. Deformation that takes place in sediments during their deposition or immediately after. Small folds and faults may develop in soft sediments that slide down gentle slopes.

continent. One of earth's main land masses usually reckoned as seven in number and, at present, constituting about one-thirdi of the earth's surface. Ther term includes continental shelves as well as dry land.

continental. Formed or generated on land, as opposed to in the ocean or sea. Deposits originating in lakes, swamps, or streams, or moved by land winds, are all continental.

continental crust. The crustal rocks that underlie the continents; its thickess ranges from 25 to 60 km., the upper continental crust, consisting of rocks rich in silica and alumina, is equivalent to the sial; the lower crust,the sima, consists largely of silicon and magnesium.

continental deposit. Sedimentary material deposited on land or in bodies of water not directly connected with the ocean, as distinguised from a marine deposit. The term thus includes lacustrine, glacial,fluvial,or eolian deposits originating in a non-marine environment.

contineintal drift. The concept of continent formation by the fragmentation and movement of land masses on the surface of the earth.

continental geosyncline. A geosyncline filled with non-marine sediments.

continental glacier. A sheath of ice covering a large tract of land, such as the ice cap of Antractic.

continental growth. The process contributing to the growth of continents at the expense of ocean basins.

continental margin. Those provinces between the shoreline and the deep-sea bottom; generally consists of the continental broderland, shalf, slope and rise.

continental mass. The continental land rising more or less abruptly from the ocean floor and also the shallow submerged areas surrouding this land.

continental nucleus. A large area of basement rock consisting of basaltic and more mafic oceanic crust and periodotitic mantle from which it is postulated that continents have been formed.

continental rise. A transitional part of the continental margin; a gentle slope with a smooth surface,built up by the shedding of sediments from the continental block , and, located between the continental slope and the abyssal plain.

continental shelf. That part of the continental margin that is between the shoreline and the continental slope; charaterized by a very gentle slope of 0.1°. **continental slope.** The part of the continental margin consisting of the dectivity from the edge of the continental shelf extending down to the continental rise.

continental terrace. The continental shelf and slope together.

continuous deformation. Deformation by process of flow rather than by fracture.

contour. A line on a map or chart that joins points of equal value, also called a contour line.

contour current. An ocean current moving along isopyenic lines which are approximately paralled to bathymetric contours.

contrasted differentiation. Natural sorting of magma into acidic and basic magmas,which are thought to react with each other to produces intermediate rock types.

convection cells. Pattern of mass movement of mantle material in which the outer area is downflowing and the central area is uprising due to heat differences. These cells may be of various sizes and shapes.

convergence. (1) Diminution of the interval between geologic horizons. (2) An occanic condition in which currents of water masses having different densities, temperatures and or salinities.

convergence line. Any horizontal line along which horizontal convergence of the airflow takes place.

converted wave. A seismic wave that has changed from a *P* wave to an *S* wave or vice versa by refraction or reflection at an interface.

convolute bedding. Extremely intricately creased, crumpled, or folded laminae that are contained within an undeformed layer; parallel,uncontorted layers lie above and below; is resulted from deformation coincident with deposition.

cookeite. $4[LiAl_4(Si,Al)_4O_{10}\ (OH)_8]$. (1) Hydrous lithium aluminium silicate; a member of mica family along with as alteration product of tourmaline occurring with tourmàline, lepidolite,saphire or spodmene, often as coatings. (2) Synonym of foresite.

coombe rock. A sedimentary rock; a mass of unrolled flints and fragments of chalk with some earth interstitial matter; considered as result of movement by solifluction during glacial period.

cooperite. 2[PtS]. (1) Platinum sulphide mineral of tetragonal, irregular grains and infrequently distorted crystal fragments. Its occurrs in concentrates from platiniferous norites with sperrylite, braggite and native platinum. (2) Synonym of marmolite.

coppaelite. An igneous rock of the ultrabasic clan, a melanocratic porphyritic extrusive rock. It contains phenocrysts of diopside in holocrystalline groundmass of melilite, diopside and phlogopite with accessory perovskite, apatite and magntite.

copper pyrites. Synonym of chalcopyrite.

copperas. Synonym of melanterite.

coprolite. A fossilized fecal pellet or casting of animal droppings (fish,reptiles,birds,or mammals); generally consisting of 50-60 per cent phosphate, 6-8 per cent calcium carbonate, 4-5 per cent organic matter along with moisture, sand and remains of marine life.

coquina. A coarse-grained, porous,easily crumbled variety of clastic limestone of cemented coarse shell debris, *e.g.*,some of the shelly 'grits'of the English Mid-Jurassic.

coral. Marine animals living on the seafloor with skeletons made of calcium carbonate. They belong to the class Anthozoa. As fossils the corals are important in two ways: as builders of coral reefs and zone fossils.

coral head. A small reef patch of coralline material.

coralline alga. A marine, calcareous red alga commonly associated with coral reefs. It may be a branching form or a massive encrusting type.

coral mud. Sedimentary deposits containing abundant fragments of corals. Coral sand is coarse and found near coral reefs. Coral mud is finer-grained and found further out.

coral reef. A ridge or mass of limestone built up of detrital material deposited around a framework of skeletal remains of mollusks, colonial coral,and massive calcareous algae.

cordierite. $(Mg,Fe)_2(Al_4Si_5O_{18})$. A blue, othorhombic variety of magnesium aluminosilicate; associated with thermally metamorphosed rocks derived from argillaceous sediments. It occurs with quartz, feldspar, sillimanite, hornblende and andalusite.

cordylite.$9[Ba_2(Ce,La)_3(CO_3)_5F_3]$. Barium and cerium group of rare earths fluorcarbonate. It occurs in pegmatite veins in nepheline syenite with aegirine,ancylite and neptunite.

core. The innermist part of the earth; marked by a seismic discontinuity (Gutenberg disontinuity) near a depth of 2900 km, at which *P*-wave velocity decreases from 13.7 to 8 km/s, and below which *S* waves are not transmitted.

corkite. $PbFe_3(PO_4)(SO_4)(OH)_6$.Basic lead ferric iron sulphate-phosphate occurring as rhombohedral, generally pseudocubic crystals of dark green,yellowish,green, or pale-yellow colour.

cornelian. A synonym of carnelian.

corneite. A biotite-hornfels formed during deformation of shale by folding.

cornstone. A sedimentary deposit of micrite and dolomitic micrite formed under semiarid conditions; a concretionary or other limestone-mass occurring as irregular nodules or as lenticular beds in matrix of red sandstone.

corona. A zone of mineralization, often showing radial arrangement, surrounding another mineral. The term has been applied to corrosion rims and reaction rims.

corsite. An igneous rock of the gabbro clan;an ophitic, orbicular plutionic rock,containing oval zonal masses of 25-75 mm diameter showing alternate rings of pyribole and feldspar in groundmass of labradorite buytownite and hornblende with small amounts of hypersthene, quartz, apatite pyrite and ore. Also known as napoleonite.

cortlandite. A peridotite consisting of large crystals of hornblende with poikitically included crystals of olivine. Also known as hudsonite.

corundophilite. A ferroan variety of clinochlore.

corundum. Al_2O_3. Aluminium oxide occurring as hexagonal steep pyramidal barrel-shaped, sometimes flat tabular or rhombohedral crystals. Its colour varies from blue to colourless, yellow, golden, purple green, pink to deep red, brown or black. It is the characteristic mineral of rocks lacking in silicon oxide. Its is generally associated with spinel, garnet, kyanite and high-calcium feldspars in plutonic pegmatite and metamorphic rocks. Some of its varieties are oriental ametyst, oriental emerald, oriental topaz, sapphire, ruby (gemstones) and emery(massive).

cosalite. $CuPb_7Bi_8S_{22}$. Lead copper bismuth sulphide occurring as orthorhombic prismatic crystals, frequently elongated to needle-like and capillary forms. Its occurs in hydrothermal deposits formed at moderate temperatures in contact metamorphic deposits and pegmatites.

coseismal line. A line conneting points on the earth's surface at which earthquake wave has arrived simultaneously.

coset. A term used for a sedimentary unit composed of two or more sets.

cosmic erosion. The wearing away or destruction of rocks on a planetary surface, a cumulative effect of impacts of hypervelocity particles from outer space.

cosmoclastic rock. One of the original rocks of the earth.

cossyrite. A variety of aenigmatite with little or no titanium dioxide.

cosmogeny. The scientific study of the age and origin of the universe; in particular the earth and solar system.

cosmogony. The scientific study of various hypotheses and cultural myths regarding the origin of the universe.

cosmology. The branch of astronomy concerned with the study of both theoretical and observational aspects of the universe as a systematized whole.

cotectic. Said of conditions of temperature pressure and compostition under which two or more solid phases crystallize simultaneously fron s single liquid over a finite interval of decreasing temperature.

couloir. A steep valley or gorge on a mountainside,especially the Alps. The term is also used in mountain-climbing parlance for a passage in cave.

country rock. Rock bodies which enclose and intrusive mass of igneous rock or series of mineral veins.

cove. (1) One of the lesser indentations of a coast, often the result of differentiated marine erosions. (2) A sheltered hollow of recess in a mountain. (3) A sheltered area between hill or woods.

covelline. CuS. Copper sulphide occurring as horizontally striated hexagonal plates of dark blue colour. It shows uneven fracture and perfect cleavage in one direction. It occurs in veins in zones of secondary enrichment associated with copper minerals. Also known as covellite.

covellite. A synonym of covelline.

covite. An igneous rock of syenite clan; composed of sodic orthoclase,hornblende,sodic pyroxene,nepheline, and accessory sphene, apatite, and opaque oxides.

cowpesite. $CaAl_2Si_3O_{10}.5\text{-}6H_2O$. A

mineral of zeolite group.

crack. A partial or incomplete fracture.

crag and tail. A streamlined ridge or hill resulting fron glaciation. It consists of a knob (the "crag")of resistant bedrock with an elongate body (the tail)or more erodible bedrock on its lee side.

craigmontite. An igneous rock of the syenite clan; a meso type coarse-grained plutonic rock; contains repheline, oligoclase,muscovite and corundum. It differs from congressite in containing more nepheline and less plagioclase and corundum.

crater. A saucer-shaped depression on the earth's surface resulting from a meteorite impact or bomb explosion.

crater cone. A cone built around a volcanic vent by lava extruded from the event.

craton. The large and immobile portion of continents consisting of both shield and platform areas.

creedite. $Ca_3\ Al_2(SO_4)(F,OH)_{10}.\ 2H_2O$. A colourless or white variety consisting of monoclinic crystals of hydrous calcium aluminum fluoride along with calcium sulphate.

creep. (1) Slow, imperceptible gravity movement of soil and/or broken rock material from higher to lower levels, or the material itself that has so moved. (2) Slow deformation produced by long-acting stressed below the elastic limit of the material subjected to stress. Part of the deformation is elastic (non-permanent) and part is permanent.

creep limit. The maximum stress that a material can withstand without observable creep.

crest. (1) The highest part of a mountain or hill. (2)A ridge or ridgelike formation. (3) The highest part of an anticline fold. The line connecting the highest points on a given stratum in an infinite number of cross-sections is the crest line. The surface formed by all crest lines is the crest surface or crestal plane. (4) The highest part of a wave.

cretaceous. The interval of geologic time that began about 135 million years and lasted about 71 million years.

crevasse. An open nearly vertical fissure in a glacier or other mass of land ice or the earth, especially after earthquake.

crevice. Any narrow opening cleft of rift in a structure or formation often used in the sense of crevasse.

crichtonite. $[Fe^{2+}{}_{16}Fe^{3+}{}_{14}Ti_{66}\ \ O_{169}]$. Ferric ferrous iron titanate; mineral near like ilmenite,but distinct species containing more titanium than ilmenite.

crinanite. An igneous rock of the gabboro clan; a melanocratic, fine grained hypabyssal rock. It contains olivine augite basic plagioclase and considerable analcite and zeolites.

Crinoid. An echinoderm of the class crinoidea,commonly called"sea lily" Range, Lower Cambrian to present.

crinoidal limestone. A rock composed or mainly crystalline joints of crinoids with foraminiferns corals, and mollusks.

critobalite. SiO_2. A silica mineral that is high temperature form of quartz. stable above 1773K; crystallizes in tetragonal system at low temperatures and in isometric system at high temperatures.

crocidolite. A common variety of asbestos appearing lavender blue, indigo-blue or leek-green; occurs in fibrous, massive and earthy forms. Also known as blue asbestos; krokidolite.

crocoite. $PbCrO_4$. Lead chromate occurring as monoclinic prismatic acicular crystals; massive columnar, granular, as crusts. It occurs as alteration product of galena with galena, quartz, pyrite and vanadinite.

cromaltite. An igneous rock of the ultrabasic clan; a black to dark green, coarse to medium grained hypabyssal rock; contains aegirine, augite, melanite, biotite, ilmenite, apatite and pyrite.

cronstedtite. [$Fe_2^{2+}Fe_2^{3+}SiO_5$ $(OH)_4$]. An iron silicate mineral with small amounts of magnesium occurring as vertically striated pyramidal crystals.

crookesite. [$(Cu,Tl,Ag)_2Se$]. copper thallium selenide with small amount of silver.

Croixan. The name for the North American Upper Cambrian.

cross-correlation. A measure of the similarity of or a method for the comparrison of sequences of data.

crush breccia. A braccia formed basically in place by fragmentation, as a result of faulting or folding.

crush conglomerate. A rock similar to crush breccia but the fragments of which are more rounded. It closely resembles a sedimentary conglomerate.

crush fold. A fold of large dimensions that may involve considerable minor folding and faulting such as would produce a mountain chain or an oceanic deep.

crust. The outer layers of the earth, the lower boundary of which is the Moho. Part of the crust is equivalent to the sial and part to the sima. The entire crust represents less than 0.1% of the earth's total volume, and is non-homogeneous in composition and distribuion.

cryconite. (1) A dark dust transported by wind and deposited on the surface of snow or ice. (2) A mixture of garnet, pyroxene, quartz, sillimanite, zircon, and various other minerals.

cryogenic period. A period of geological history during which large masses of ice at or near poles and climate favoured the formation of continental glaciers.

cryolite. [Na_3AlF_6]. Sodium aluminium fluoride with sodium deficiency and water replacing fluorine. Monoclinic masses occur as modified cuboidal or short prismatic crystals with midified faces. Its colour ranges from Colourless to white ,brownishm reddish or browhish,reddish or brick red; rarely black. It shows virteous to greasy lustre somewhat pearly on basal plane, white streak, uneven fracture and no cleavage.

cromorphology. The subspeciality of geomorphology pertaining to the various processes of cold climates.

cryopedology. A branch of geology that deals with the study of intensive frost action and permanently frozen surfaces.

cryoplanation. Land erosion at high latitudes of elevation due to intensive frost action.

cryosphere. The part of the earth's surface that is frozen throughout the year.

cryptoclactic. Composed of very fine, almost submicroscopic, broken or fragmental particles.

cryptocrystalline. A term used to describe a very finely crystalline aggregate in which the crystals are small as to be indisinguishable except under powerful magnification.

cryptoperthite. An extremely finegrained, intercombined sodic and potassic feldspar on which lamellae are discernible only by meanbs of X-rays or with the use of an electron microscope.

cryptozoic. Term meaning "hidden life." It is that part of geologic time represented by rocks in which evidence of life is slight and only of primitive forms.

crystal. A three-dimensional body, the bounding surfaces of which are arranged symmetrically and, for a given substance with constant angular relationships arising from the inherent regular atomic structure of the substance.

crystal axes. Three imaginary lines in a crystal (four in hexagonal system) that pass through its centre; the structure and symmetry of a particular crystal are described by the relative lengths of these lines and their mutual angular orientation.

crystal class. One of 32 categories of crystals according to inversions and rotations about axis, reflections, and combinations of these which leaves the crystals invarient. Also known as symmetry class.

crystal face. One of the outward planer surfaces which define a crystal and its internal structure.

crystal habit. The general shape of crystals, *e.g.*, cubic, fibrous.

crystal lattice. A lattice from which the structure of a crystal may be obtained by associating with every lattice point an assembly of atoms identical in composition, orientation, and aggangement. Also known as space lattice; lattice.

crystalline limestone. General term for metamorphosed limestones. Mineral composition depends on character of original limestones thermodynamic conditions under which metamorphism was affected and composition of material introduced from external sources. The term includes cipolino, marble,ophicalcite and predazzite.

crystalline rock. (1) Igneous and metamorphic rock, as opposed to sedimentary rocks. (2) Rock made up of minerals in a crystalline state.

crystalloblastic. Said of a crystalline texture formed by metamorphic conditions of high viscosity and regional pressure, as distinct from igneous rock textures that are the result of a sequential crystallization of minerals under conditions of low viscosity and near uniform pressure.

crystal mush. Magma that is partially crystallized.

crystal settling. In a magma the sinking of crystals of greater density, resulting in crystal accumulation.

crystal structure. The arrangement of atoms or ions in a crystalline body.

crystal symmetry. The existence of non-trivial operations, consisting inversions, rotations around an axis reflections and combination of these, which bring a crystal into a position indistinguishable from its original position.

crystal tuff. A variety of volcanic tuff comprising detached crystals formed by explosive disruption of lava in an time of eruption. It contains crystals of biotite; hornblende, quartz and feldspar in glassy interstitial groundmass of fine glass dust iron oxide and secondary chalcedony or hydrous silica.

cubanite. [$CuFe_2S_3$]. Copper iron sulphide composed of orthorhombic, striated, elongated or thick tabular crystals.

cubic packing. The spacing pattern of uniform solid spheres in a clastic sediment or crystal lattice in which the unit cell is a cube.

cubic plane. A plane that is at right angles to any one of the three crystallographic axes of the cubic system.

cubic system. A crystal system divided into five classes. All cubic crystals are referred to three axes mutually at right angles, employing same parameters on all three axes.

cuesta. A low, asymmetrical ridge with a gentle slope on one side, conforming with the dip of the underlying strata, and a steep face on the other.

culm. A local term used in eastern pennsylvania, for fine particles of anthracite coal. In certain regions, tremendous culm piles accumulate and eventually are overgrown with vegetation. The term is also applied to certain rocks consisting of sandstones and shales with some fine bands of crushed coal.

cumberlandite. An undersaturated igneous rock of the ultrabasic clan; a greenish-black hypabyssal rock with a few white plagioclase crystals. The black granular groundmass comprises 50 per cent olivine 40 per cent magnetite and ilmenite and accessory spinel.

culmination. The highest point of some geological structure; *e. g.*, a dome or anticline. Several culminations may be found along a mountain range. The highest topographic point of an eroded-fold mountain may or may not be coincident with the culmination.

cumbraite. An igneous rock of the granite clan, porphyritic extrusive rock containing phenocrysts of bytownite-anorthite in a groundmass of labradorite,enstatite-augite and much glass which is potentially andesite sanidine and quartz.

cummingtonite. $[(Mg,Fe)_7Si_8O_{22}(OH)_2$. A ferrosilicate mineral with ratio of magnesium to iron greater than 1:1. It contains hydrated magnesium ferrous iron silicate and occurs as a clino-amphibole, identical with anthophyllite in composition and therefore an amphibole asbestos. It form series with grunerite.

cumulate. A layered igneous mass commonly contains rocks which apper to have formed by 'accumulation'of 'primary precipitate crystals.'

cumulophyric. A synonym of glomeroporphyritic.

cuprite. $[Cu_2O]$. Cuprous oxide composed of cubic octahedral dodecahedral or cubic crystals. It occurs in oxidised zone of copper deposits with mainly native copper malachite, azurite, chalcocite and chrysocolla. Its varieties are chalcotrichite, ruby copper and tile ore.

current-bedding. Cross-bedding resulting from water or air currents.

cuselite. An igneous rock of the diorite clan; a melanocratic porphyritic to granular hypabyssal rock containing phenocrysts of plagioclase in granular groundmass of plagioclase, orthoclase, biotite, augite, hornblende and diopside.

cusp. One of a series of low crescent-shaped mounds of beach material separated by smoothly curved, shallow troughs spaced at almost regular intervels along and generally perpendicualr to the beach face.

cuspidine. $Ca_4Si_2O_7(F,OH)_2$. Basic calcium fluorsilicate ocurring as monoclinic; minute, spear-shaped crystals showing vitreous lustre.

cutoff. (1) For map making and cross-sections, a cutoff is a boundary normal to the bedding indicating the areal borders of a particualr stratigraphic unit that is not defined by natural features. (2) An impermeable barrier placed within or beneath a dam to block or check seepage.

cutoff grade. The lowest grade of mineralized material *i.e.,* material of lowest assay value considerd as ore in a particular deposit.

cutout. Mass of siltstone, sandstone, or shale occupying the space of an erosional channel cut into a coal seam.

cuvette. A basin of deposition which is not tectonic in origin; *i.e.,* often a lower region between mountain range.

cuyamite. An igneous rock of the gabbro clan; a dark coloured holocrystalline, xenomorphic grangular to ophitic hypabyssal rock; contains apatite, feldspar, magnetite, augite and analcite.

cycle of erosion. A concept in which the various stages in the erosion of a region are regarded as parts of a cyclic process. Ideally the cycle begins with uplift of a land surface followed by initation and developement of relief, and finally degeneration of relief and ultimate plantation. Further uplifts cause the cycles to recommence. It is very likely that complete planation is never achieved, uplift occurring at almost any stage in the cycle. In certain circumstances traces of an earlier cycle may be preserved even after serveral later cycles of erosion have operated.

cycle of sedimentation. (1) An ordered repeated sequence of associated processes and conditions recorded in a sedimentary deposit. (2) The deposition of sediments in a basin between the onset of any two successive marine transgressions.

cyclic evolution. The concept that in lineages of many life forms, evolution had a strong and rapid initial course which was followed by a long phase of slow and moderate changes. Overspecialized or inadaptive forms became extinct during a final brief episode.

cyclosilicate. One of the group of silicates in which the SiO_4 tetrahedron forms rings by linking together. Regardless of the number of such links in a ring (3,4, or 6),all the rings have a 1:3 ratio of silicon to oxygen.

cyclothem. A rock stratigraphic unit applied indiscriminately to the repeat unit of either cyclic sedimentation or rhythmic sedimentation. The cyclothem for a given set of deposits is probably an idealised concept, since detailed examination of the actual units usually show gaps in or repetitions of the cyclothem.

cylindrite. $Pb_3Sn_4Sb_2S_{14}$. A blackish-grey mineral consiting of sulphides of lead,tin,and antimony; occurs in cylindrical forms.

cymatolite. A mixture of albite and muscovite possessing fibrous to wavy structure.

cymrite. $Ba_2Al_5Si_5O_{19}(OH)\ 3H_2O$. A mineral of zeolite group consisting of a basic aluminosilicate of barium.

cyprine. A variety of idacrase containing traces of copper. It ocquires pale to sky-blue or greenish-blue colour due to presence of copper.

crytolite. Radioactive variety of zircon. It contains uranium and yttrium. The crystals are identical to zircon, but have dull, convex, pyramidal faces. It possesses reddish-brown colour; used for recovery of rare earths.

dachiardite. $(Ca,K_2,Na_2)_3Al_4Si_{18}O_{45}.14H_2O$. Hydrated calcium potassium sodium and aluminium silicate zeolitic mineral with composition near epistilbite but higher alkali and silica content.

dacite. Very fine crystalline or glassy rock of volcanic origin; usually contains phenocrysts of quartz, plagioclase, pyroxene, hornblende and minor sanidine and biotite in finely crystalline groundmass of alkali-feldspar and silica materials.

dacite glass. A natural glass formed by rapid cooling of dacite lava.

dactylitic. Said of a rock texture, characterized by finger like projections of a mineral that penetrate another mineral.

dahamite. An igneous rock of the granite clan; a porphyritic hypabyssal rock; variety of paisanite charcterised by abundant albite.

dahllite. A synonym of carbonate-apatite.

dalles. A term sometimes used for rapids in a deep, narrow river or stream flowing through a gorge or canyon.

damkjernite. An igneous rock of the ultrabasic clan; a dark-coloured porphyritic hypabyssal rock; contains phenocrysts of biotite and pyroxene in fine-grained groundmass of pyroxene, magnetite and biotite in base of nepheline (altered to muscovite and chlorite) and less microcline.

damourite. A variety of muscovite; a hydromica; derived from alteration of kyanite, topaz or corundum.

danaite. (Fe,Co)AsS. An ore containing up to 12 per cent cobalt in lattice; cobalt-rich variety of arsenopyrite.

dancalite. Igneous rock intermediate in composition between the syenite and diorite clans a porphyritic extrusive rock; -contains phenocrysts of sodic oligoclase, green augite and rare amphibole in trachytic groundmass of plagioclase laths and interstitial analcite.

darcy. A unit for measuring the permeability coefficient of a rock; *e.g.*, in calculation the flow of water, gas, or oil.

darwin glass. A synonym of queenstownite.

dashkesanite. $(Na,K)Ca_2(Fe^{2+},Mg,Fe^{3+})_5(Si,Al)_8O_{22}Cl_2\}$ A mineral with ratio of sodium to potassium 1, ratio of ferrous to magnesium to ferric iron 3:1:1 and ratio of silicon to aluminium 3:1.

dating techniques. Methods of estimating the age of rocks, palaeontological specimens, archaeological sites, etc. Relative dating techniques date specimens in relation to one another; for example, stratigraphy is used to establish the succession of fossils. Absolute (or chronometric) techniques give an absolute estimate of the age and fall into two main groups. The first depends on the existence of something that develops at a seasonally varying rate, as in dendrochronology and varve dating. The other uses some measurable change that occurs at a known rate, as in chemical dating, radioactive(or radiometric) dating (*see* carbon dating; fission-track dating; potassium-argon dating; rubidium-strontium dating; uranium-lead dating), and thermoluminescence.

datum. The top or bottom of a bed of rock on which structure contours are drawn.

daubree. A unit of measure delineating the extent of wear of a sedimentary particle which is equivalent to the removal of 0.1 g from 100 g sphere of quartz.

dauphine law. A law according to which the twinned parts are related by rotation of 180° around the *c*–axis.
davainite. An igneous rock of the ultrabasic clan; a black, granular plutonic rock containing pyroxene, brown hornblende and a little orthopyroxene.
davidsonite. A greenish or greenish-yellow variaty of beryl.
davyne. $(Na,K,Ca)_{6-8}Al_6Si_6O_{24}(SO_4,CO_3,Cl_2)_{1-2}$. A member of cancrinite series containing considerable chlorine; resembles nepheline.
dawsonite. $(NaAlCO_3(OH)_2)$. Basic sodium aluminium carbonate; occurss in thin crustations or rosettes of bladed to acicular crrystals or tufts of fine needles.
dead glacier. A glacier that does not have an accumulation area.
dead sea. A body of water in which there are no normal aquatic organisms, and from which evaporites are being or have been precipitated.
debris. Large fragments arising from disintegration of rocks and strata.
debris avalanche. The sudden and rapid downward movements of incoherent mixtures of rock and soil on deep slopes.
debris fall. The nearly free falling of weathered mineral and rock material from a vertical or overhanging face. Debris falls are very common along the undercut banks of streams.
debris flow. Moving mass of mud, soil, and rock fragments of which more than half the particles are larger than sand size.
debris slide. The rapid sliding or rolling of predominantly unconsolidated earth debris without backward rotation in the movement.
decay. The spontaneous transformation of one radioactive nuclide into a daughter nuclide, which may be radioactive or may not, with the emission of one or more particles or photons. The decay of No. nulides to give *N* nuclides after time t is given by $N=N_0\exp(-Y)$, where Y is called the decay constant or the disintegration constant. The reciprocal of the decay constant is the mean life. The time required for half the original nuclides to decay (i.e. $N=1/2N_o$) is called the half-life of the nuclide. The same terms are applied to elementary particles that spontaneously transform into other particles.
declivity. A slope descending downward from a point of reference.
decomposition. Chemical reaction in which a compound breaks down into simpler compounds or into elements.
deconvolution. A technique designed to restore a wave configuration to the shape it is assumed to have taken prior to filtering. It is applied to seismic reflection and other data for the purpose of improving the visibility and clarity of seismic phenomena.
decrepitation. A crackling noise produced when certain crystals are heated, caused by changes in structure resulting from loss of water of crystallization.
dedolomitization. A process resulting from contact metamorphism, wherein dolomite is broken down into its two components, $CaCO_3$ and $MgCO_3$. The former crystallizes into a coarse calcite, whereas the latter further breaks down into magnesium oxide and carbon dioxide.
deep. An area of great depth in the ocean, representing a depression in the ocean froor; generally implying depths exceeding 5.5 km.
deep coal. Coal that is far enough below the surface to require underground mining.
deep lead. A lead or alluvial placer that

is burried under soil or rock.

deep percolation. Precipitation moving downward, below the plant-root zone, toward storage in subsurface strata.

deep phreatic. Said of cave formation at considerable depth below the top of the water saturated zone.

deflation. The transport of loose surface debris by wind action.

deformation. A general term for processes of rock deformation resulting from various earth forces; they include faulting, folding, shearing, compression, and extension.

deformation plane. Twinning produced by deformation and gliding. A portion of the crystal structure is sheared in such a way as to produce a mirror image of the original crystal.

deglaciation. (1) The stage of retreat of a widespread glaciation. (2) The exposure of an area from beneath a glacier or ice sheet as a result of shrinkage of the ice by melting.

delafossite. $CuFeO_2$. Copper iron oxide occuring as hexagonal, tabular to equant crystals; massive as botryoidal crusts. It shows metallic lustre, black streak, brittle fracture and imperfect cleavage in one direction.

deldoradite. An igneous rock of the syenite ckan; a leucocratic, coarse grained plutonic rock, containing microperthite with cancrinite and accessory biotite, acgirine, apatite, titanite and iron ore.

dellenite. An igneous rock of the granite clan; a fine grained extrusive rock; contains basic plagioclase with orthoclase, occasional quartz and some hornblende; its type is intermediate between rhyolite and dacite.

delta. An alluvial deposit, usually triangular in shape, at the mouth of a river stream, or tidal inlet. It is normally built up only where there is no tidal or current action capable of removing the sediment as fast as it is deposited, and hence the delta builds forward from the coastline. This process of building-up is complex, and leads to the formation of a numbers of separate channels isolated lagoons, levees, marshy ground, and a network of small creeks. Most deltas are complicated and multiple, but in a simple delta three main types of bed may be distinguished: (1) The bottom set beds-forming on the sea-bottom beyond the seaward face of the delta. (2) The forest beds-building outward from the seaward face of the delta over the bottom set beds. (3) The topset beds-which are deposited above the already deposited forest beds.

demantoid. A lustrous, green variety of andradite; used as a gem.

dendrite. A crystal that has branched in growth into two parts. Crystals that grow in this way (dendritic growth) have a branching treelike appearance.

dendritic agate. A variegated variety of chalcedony . It has brown or black dendritic markings due to visible impurities distributed throughout the mass..

dendrochronology. The study and correlation of tree rings for the purpose of dating events in the recent past.

density. The mass of a substance per unit of volume. In SI unitm it is expressed in kg m^{-3}.

density stratification. The process of stratification of water in a lake as a result of density differences. It is usually caused by temperature changes, but also may be caused by differences in the quanitity of dissolved material, *e.g.*, where a less dense freshwater surface layer overlies denser salt water.

denudation. The end resullt of processes

such as weathering and erosion which cause a lowering of the land surface.

depletion. The act of seriously diminishing or exhausting a quantity of something, *e.g.*, natural resources.

deposit. Consolidated or unconsolidated material that has been accumulated by a natural agent or process.

depositional mark. An irragularity formed on the bedding plane of a sediment during deposition.

depression. (1) Any relatively sunken part of the earth's surface; especially any low-lying area surrounded by higher ground and having no natural outlet for surface drainage. (2) A structurally low area in the crust,produced by negative movements that sink or downthrust the rock . The term includes the basins and furrows. (3) The angular distance of an object beneath the horizontal plane that passes through the observer.

depth. The vertical distance from a specified datum to the bottom of a body of water,such as distance between the sea-leval and the sea-floor.

depthhoar. A layer of ice crystals formed between the ground and the snow cover by sublimation. Also known as sugar snow.

depth ice. Small ice particles formed below the surface of the sea agitated by wave activity.

depth zone. A zone within the earth giving rise to different metamorphic assemblages.

dereliction. A recession of water from the sea or other body of water so that the land is left dry.

derivative rock. A rock composed of materials derived from the weathering of older rocks, or a rock formed of a material that has not been in a state of fusion immediately before its accumulation.

dermolith. A synonym of pahoehoe.

desalination. The process of removal of salt from sea water for irrigation of the land or to provide drinking water. The methods employed include evaporation, often under reduced pressure (flash evaporation); freezing (pure ice forms from freezing brine); reverse osmosis; electrodialysis; and ion exchange.

descensional deposit. A sedimentary deposit produced by the disintegration of rock and aggregation; of the resulting particles in beds.

desert. In general a region where precipitation is less than potential evaporation. Although they may be hot or cold, deserts are always characterized by a scarcity of vegetation. The four kinds of deserts are: (1) polar or high altitude deserts marked by perpetual snow cover and intense cold; (2) middle-latitude deserts, characterized by scant rainfall and higher summer temperatures; (3) trade-wind deserts with negligible precipitation; and (4) coastal deserts, as in Per, where there is a cold current on the western coast of a large landmass.

desert armor. A desert pavement surfaced with stony fragments that protect underlying fine-grained material from wind erosion.

desert crust. A hard layer containing binding matter (*e.g.*,calcium carbonate, gypsum), exposed at the surface in a desert area.

desert pavement. A residual layer of wind-polished, closely concentrated rock fragments covering a desert surface due to the removal by wind of loose silt and sand.

desert polish. A slick, polished surface on rocks of desert regions, produced by wind and sand abrasion; sometimes used as a synonym of desert varnish.

desrt roses. Concretions of barytes or sometimes gypsum in sandstone, or clustersofplatycrystalscontainingsand. Barytes acts as cement for sand.

desert varnish. A brown or black stain orcrustofmanganeseoriron oxidechar-acterizingmanyexposedrocksurfacesin the desert.

desiccation breccia. Fragments of a mud-cracked layer of sediment deposited with other sediments.

desiccation crack. A crack formed by shrinking of clay under the influence of heat of the sun.

Desmoinesian. A stratigraphic stage name for the North American Middle Pennsylvanian(CarboniferousSystem).

desorption. The removal of adsorbed atoms,molecules,orionsfromasurface.

destinezite. [$Fe^{3+}{}_2PO_4SO_4OH.\ 5H_2O$] Basic hydrated iron sulphate-phos-phateo; It occurs in gossan of pyritic deposits,asrecentdepositinminework-ingsorformedbyactionofsulphatesolu-tions on secondary phosphate minerals.

detached core. The inner part of a fold that has been detached or pinched from the parent structure by some process, usuallyseverecompressionandfolding. The term core is used even though a fold may be composed entirely of one kind of rock.

detritus. Particles of minerals or rocks which have been derived from pre-exist-ing rock by processed of weathering and/or erosion.

deuteric. A term applied to those reac-tions between primary magmatic minerals and water-saturated solutions thatseparatefromthesamemagmabody at a late phase of its cooling profile.

devitrification. Loss of the amorphous nature of glass as a result of crystal-lization.

Devonian. A time interval of the paleozoic era during which the rocks of Devonian system were formed. It ex-tends from 395 to 345 million years.

devonite. An igneous rock of the gabbro clan; a porphyritic hypabyssal rock. It contains colourless phenocrysts of potash-rich labradorite in dark green groundmass of altered plagioclase and augite.

deweylite. Memberofserpentinegroup; contains mixture of stephanite, chrysotile and lizardite. Also known as gymnite.

diabantite. $(Mg,Fe^{2+},Al)_6(Si,Al)_4O_{10}(OH)_8$. Hydratedaluminiummagnesium ferrous iron silicate; member of chlorite group. It occurs in seams, clefts and amygdaloidal cavities in diabase.

diabase. Anintrusiverock,ophiticintex-ture,consistintg chiefly of pyroxene and plagioclase feldspar, most commonly labradorite. In mineral composition, diabase is a gabbro; its grain size is inter-mediate between basalt and normal gab-bro.

diagenesis. The sum of all changes, physical, chemical, and biological, to which a sediment is subjected after deposition, but before weathering or metamorphism.

diagnostic mineral. A mineral whose presence in a rock indicates whether the rock is oversaturated or undersaturated.

diagonal joint. A joint whose strike is in adirectionthatliesbetweenthestrikeand direction of dip on the associated rocks.

diallage. An igneous rock of the ultrabasicclan;afinegrainedhypabyssal rock;containsmostlydiallagewithmain-ly accessory hornblende, garnet, pyroxene, olivine, feldspar and pyrite.

dialogite. An obsolete synonym for the

mineral rhodochorsite.

diamantine. Consisting of or resembling diamand.

diamictite. A calcareous, terrigenous sedimentary rock that is insorted or poorly sorted and contains particles of many sizes. Also known as mixtite.

diamicton. A nonlithified diamictite.

diamond. Crystalline form of carbon predominantly octahedral, usually much surface curved and often striated; occurs as rounded or irregular grains or pebbles often with internal radial structure. Its colour varies from white to bluish-white, shades of yellow or brown; occasionally orange, pink, mauve, green, blue, red, or black. It shows a damantine to greasy lustre; ash grey streak and conchoidal fracture and perfect octahedral cleavage. It occurs in pipes, alluvial fan, wave-concentrated deposits on shore-lines, in conglomerates and meteorites. Its non-gemstone varieties are bort, hailstone bort, shot bort and stewartite.

diaphthorite. Schistose rocks in which minerals have formed by retrograde metamorphism.

diapir. A vertical columanar plug of less dense rock or magma that is forced through a more dense rock.

diaspore. AlO.OH. Hydrous aluminium oxide occuring as thin platy and elongated crystals, sometimes acicular; massive, stalactic or disseminated. It occurs mainly with corundum, emery, dolomite, margarite, chlorite and magnetite. It is dimorphous with boehmite. Also know as kayserite.

diastem. A theoretical, minor undetected break in a sedimentary sequence due to non-deposition or slight erosion. The concept is based on inferences drawn from the discrepancies between calculated sedimentation rates and those resulting from observational analysis.

diastrophism. A term for all movement of the earth's crust resulting from tectonic processes. This includes the formation of continents, ocean basins, plateaus, and mountain ranges. Diastrophic processes are usually classified as orogenic (mountain building with deformation), and epeirogenic (regional uplift without significant deformation.)

diatomaceous earths. The sedimentary rocks; siliceous organic deposits formed by accumulation of frustules of diatoms; composed of nearly pure silica; friable and earthy. It occurs in oceans, fresh water lakes and swamps.

diatomite. Sedimentary rock; a type of diatomaceous earth; marine deposit. A fairly hard rock containing abundant remains of diatoms lithified by calcite, silica or calcium phosphate.

diatreme. A general term for a pipelike structure formed deep within the earth's crust by the explosive energy of magmatic gases, or by the explosions of hot groundwater.

dichroism. The property of some crystals, such as tourmaline, of selectively absorbing light vibrations in one plane while allowing light vibrations at right angles to this plane to pass through.

dichroite. A synonym of cordierite.

dickite. $Al_2Si_2O_5(OH)_4$. Basic aluminium silicate; member of kaolinite family.

didymite. A variety of muscovite; contains calcium carbonate as impurity and forms greenish or greyish-white fine scales.

diestiean. North European stage: Upper Miocene.

differencial erosion. Erosian occurring at variable rates, caused by differences in resistance and hardness of the surface materials.

differential melting. Partial melting of a rock, resulting from differences in melting points of its constituent minerals.

differential weathering. Weathering that occurs at varying rates because of the intrinsic differences in rock composition and resistance, or differences in the intensity and degree of erosion. It helps to develop and modify many upstanding forms such as columns, pillars and rock pedestals.

differentiation. The process of developing more than one rock type, *in situ*, from a common magma.

differentiation index. The number repressenting the sum of the weight percentages of normative quartz, orthoclase, nepheline, kalsilite, albite, and leucite, a numerical expression of the extent of differentiation of a magma.

digenite. Cu_2S. Copper sulphide, occurring usually with some copper deficiency, as octahedral crystal of blue to black colour; it shows conchoidal fracture and occurs with chalcocite challcopyrite and other copper ores as replacement of bornite.

dihedral. An angle formed by the intersection of two planes (*e.g.*, two faces of a polyhedron). The dihedral angle is the angle formed by taking a point on the line of intersection and drawing two lines from this point, one in each plane, perpendicular to the line of intersection.

dike. (1) A tabular body of igneous rocks that cuts across adjacent rocks or massive rocks. (2) An embankment or wall built to prevent flooding of a lowland area.

dike swarm. A group of parallel, linear, or radially oriented dikes.

dike wall. A topographic feature formed when a dike is more erosion resistant than the rock on either side.

dilatancy. An inelastic increase in the bulk volume of a rock during deformation by squeezing, owing to the opening and extension of small cracks.

dilation. Deformation by change in volume but not configuration. Also spelled dilatation.

dilation vein. A mineral deposit occurring in vein space created by the bulging of a crack, rather than in veins formed by the replacement of wall rocks.

diluvial. Pertaining to, produced by, or resembling a flood.

dimorphism. The existence of a substance in two different crystalline forms.

diopside. $MgCaSi_2O_6$. Magnesium calcium silicate occuring as prismatic, often slender crystals; also lamellar massive to granular and columnar. Its colour varies from white ,yellowish, greyish white to pale green, dark green to nearly black; sometimes transparent and colourless. It shows vitreous lustre, white or grey streak; uneven fracture and perfect prismatic cleavage. Its varieties are chromediopside, coccolite, enstatite-diopside, violane, sahlite, mansjoite, alalite.

diopsidite-jadeite. A synonym of tuxtlite.

diopsidite. An igneous rock of the ultrabasic clan, a fine grained hypabyssal rock containing essentially diopside with accessory iron ore, ceylonite and garnet.

diorite. A dark-colourd, coarse-grained plutonic rock in which plagioclase feldspar is in excess of alkali-feldspar and quartz is absent. Plagioclase is within oligoclase-andesine range associated with mafic minerals (mainly hornblende, biotite, augite and hypersthene) and accessories (mainly sphene, apatite and magnetite). Type-rocks of three subdivisions are diorite (coarse-grained plutonic), micro-diorite (medium-grained hypabyssal) and andesite (fine-grained extrusive).

dip. (1) The angle in degrees between a horizontal plane and an inclined earth feature such as a rock stratum, fault or dike. (2) The angle between a reflecting or refracting seismic wavefront and the horizontal, or the angle between an interface associated with a particular seismic event and the horizontal.

dip fault. A type of fault that strikes parallel with the dip of the strata involved.

dip joint. A joint that strikes at about right angles to the cleavge or the bedding of the constituent rock.

dip log. A log of the dips of formations traversed by boreholes.

diploid. A crystal form in the isometric system having 24 similar quadilateral faces arranged in pairs.

dip separation. The distance between formerly adjacent beds on either side of a fault surface, measured directly along the dip of the fault.

dip shift. The relative displacement of rock units parallel to the fault dip, but beyond the fault zone itself.

dip slip. The component of a fault parallel to the dip of the fault.

dip slope. The slope of the land surface that more or less conforms with the inclination of the underlying rock.

dirt band. A dark layer in a glacier representing a former surface, usually a summer surface, where silt and debris accumulated.

dirt bed. A burried soil containing partially decayed organic material.

discharge. The amount of water flowing down a stream channel; it is measured in cubic metres per second, and data for such measurement is always taken at a specified location along a given stream.

discontinuity. (1) Any interruption in sedimentation that may represent a time interval in which sedimentation has ceased and/or erosion has occurred. (2) In structural geology, abrupt changes in rock type caused by tectonic activity. (3) *S* zone within the earth in which the velocity of seismic waves shows a sharp change; *e.g.*, Conrad discontinuity; Mohorovicic discontinuity.

discordant. (1) Said of intrusive igneous bodies, such as dikes and batholiths, whose margins cut through the bedding or foliation of the country rock. (2) Structurally unconfformable; *i.e.*, said of strata in which parallelism of bedding or structure is absent. (3) Radiometric ages, determined by more than one method for the same specimen, are said to be discordant if they disagree beyond experimental error. (4) Said of topographic features whose elevations are not equal or similar.

dislocation. A line defect in a crystal, the result of slip along a surface of one or more lattice constants.

disordering. The displacement of atoms from their position in a crystal lattice (*e.g.*, as a result of the effect of ionizing radiation) to positions that are not part of the lattice.

dismicrite. Fine grained limestone of obscure origin resembling micrite but containing sparry calcite bodies.

dispersal pattern. Distribution pattern of metals in soil, water, rock, or vegetation.

dispersed elements. Elements which form few or no independent minerals but are present as minor ingredients in minerals of abundant elements.

dissection. The cutting of ravinees, gullies, or valleys by erosion, especially by streams.

disseminated. Spread out, or dispersed, as applied particularly to particles of ore mineral in the substrate rock.

distal. (1) Referring to an ore deposit formed at a significant distance from the source of its constituents. (2) Said of a sedimentary clastic deposit that was laid down far from its source location.

disthene. A synonym of kyanite.

disturbance. A term used in geology for a lesser or minor orogeny such as the Palisades disturbance at the close of the Triassic.

ditroite. An igneous rock of the syenite clan; a coarse to fine-granular plutonic rock. It contains nepheline, microline-microperthite, codalite and cancrinite with biotite or aegirine-augite rimmed with arfvedsonite.

dixenite. $(Mn,Fe,Cu)_{15}As_5^{3+}O_{13}(OH)_7(Si,P,As)O_4)_3$ A mineral consisting of basic hydrate silicate, phosphate and arsenate of manganese, iron, copper and arsenic oxide. It occurs in crevices in dolomite with native lead or in hematite.

dolerine. Metamorphic rock; a variety of talc-schist.; contains feldspar and chlorite as chief minerals.

dogtooth spar. A variety of calcite that occurs as pointed crystals somewhat resembling a dog's teeth.

dolerite. An igneous rock of the gabboro clan; a dark coloured, generally fresh basaltic rock, essentially of labradorite and pyroxene with optional olivine. The term is sometimes expanded to include all rocks of appropriate compoosition regardless of age, texture or mode of occurrence.

doline. A term originally used in the U.S. to mean a solution sink, *i.e.*, a sinkhole that developed downward by solution beneath a soil mantle, without physical disturbance of the ambient rock.

dolomite. $CaMg(CO_3)_2$. Calcium magnesium carbonate occurring as hexagonal, rhombohedral crystals with curved surfaces; it is massive coarsely crystalline, compact, granular, friable; colourless and transparent when pure; grey greenish; yellowish-brown or brown with iron impurities; pink, rose or flesh-red with manganese impurities. It forms group with ankerite and kutnahorite. Its varieties are miemite, pearl spar, rhomb spar, bitter spar and brown spar.

dolomite rocks. Sedimentary rocks of secondary origin. A type of magnesian limestone near dolomite in composition; formed from limestone, coral or marble by the action of solutions containing magnesium. Mostly mixtures of dolomite and calcite. Siliceous types have diopside and other magnesium silicates.

dolomitization. The process by which limestone ($CaCO_3$) becomes dolomite by the substitution of magnesium for the original calcite, either wholly or in part. Dolomitization occurs most commonly very soon after the limestone has formed.

dome. (1) A circular or elliptical almost symmetrical upfold of anticlinal type of structural deformation. (2) A large igneous intrusion whose surface is curved upward.

domite. An igneous rock of the syenite clan; a light coloured, fine grained extrusive rock containing sodic sanidine, oligoclase, biotite and sometimes hornblende with tridymite in pores in the rock.

dopplerite. A naturally occurring gel of humic acids found in peat bags or where an aqueous extract from a low-rank coal can collect.

Dordonian. A synonym of Maestrichtian (cretaceous system)

dorgalite. An igneous rock of the gabbro clan which is generally holocrystalline extrusive rock; contains phenocrysts of olivine and abundant plagioclase in groundmassofmagnetite,olivine,augite and labradorite.

dormant volcano. A volcano that is not actively erupting but that has erupted within historic time and is considered likely to do so again in the future.

double refraction. The property, possessed by certain crystals (notably cancite), of forming two refracted rays from a single incident ray. The ordinary ray oberys the normal laws of refraction. The other refracted ray, called the extraordinary ray, follows different laws. The light in the ordinary ray is polarized at right angles to the light in the extraordinary ray. Along an optic axis the ordinary and extraordinary rays travel with the same speed. Some crystals, such as calcite, quartsz, and tourmaline, have only one optic axis; they are uniaxial crystals. Others, such as mica and selenite, have two optic axes; they are biaxial crystals. The phenomenon is also known as birefringence and the double-refracting crystal as a birefringent crystal. See also polarization.

downthrown. Referring to that side of a fault that appears to have moved downward relative to the other side.

downbuckle. A downfolding of earth's crust due to compression and associated with the formation of oceanic trenches.

Downtonian. A stratigraphic stage name for the top of the European Silurian (basal Devonian by some authers).

drag fold. A minor fold formed in an incompetent bed by movement of a competent bed so as to subject it to couple; the axis lie at right angles to the direction in which the beds slip.

Dragonian. A stratigraphic stage name for the Norh American equivalent of the mid Montain (Tertiary Sysstem).

drainage. The removal of excess meteoric water by rivers and streams.

drainage basin. Region drained by a particular stream or river system, *i.e.* an area that contributes water to a particular river or channel system.

drainage density. A term describing the relative spacing of a stream network; it is expressed as the ratio of the total length of all streams within a drainage basin to the area of the basin. Thus it is a measure of the dissection or, of the topographic texture of the area. It depends on the climate geology and physical characteristics of the basin.

drainage patterns. A particular arrangement or configuration that is collectively formed by the individual stream courses in an area. These patterns reflect the influence of factors such as initial slopes inequalities in rock hardness and structural controls, and the geological and geomorphic history of the drainge basin.

drakonite. An igneous rock of the syenite clan; a light white, yellowish, or grey porphyritic extrusive rock. It contains phenocrysts of sanidine, plagioclase (oligoclase to labradorite), biotiute and/or hornblende in groundmass of alkali-feldspar, diopside and alkali-amphibole.

drewite. Calcareous ooze composed of impalpable calcareous meterial.

driblet cones. Small cones formed by lavas which are very viscous and usually acidic.

drift. A general term for all rock material that has been carried by glaciers and deposited directly from the ice or through the agency of meltwater.

drift theory. One of the theories of coal origin according to which the plant material from which coal developes has been transported from its place of growth to another locality, from where it accumulates and where coalification occurs.

dripstone. General term for calcite or other mineral deposits formed in caves by dripping water. This includes stalactites, stalagmites, and similar deposits formed by flowing water.

dripstone cave. Any cave of calcium carbonate or other mineral formed by the action of dripping water.

drowned coast. A shoreline transformed from a hilly land surface to a archipelago of small islands by inundation by the sea.

drowned valley. A valley that is partly flooded by a sea lake, or reservoir. It may result from a rise of sea level, subsidence of the land or a combination of both.

drumlin. Elongated elliptical hill consisting of unconsolidated material and commonly occurring in swarms. Drumlins are formed under the margin of glaciers or from the erosion of older moraine by reglaciation.

druse. A cavity in rock or mineral often filled or partially filled with large crystals.

drystone. A stalagmite or stalactite formed by dropping water.

dry wash. A wash, arroyo, or coulee whose bed lacks water.

Duchesnian A stratigraphic stage name the North American equivalent of the European Ludian (tertiary system).

ductility. A property especially of metals, of being capable of being drawn out into thin wires.

dull coal. A component of banded coal of grey colour with dull appearance; consisting of small anthraxylon constituents in addition to cuticles and bark like constituents embedded in the attritus.

dumalite. An igneous rock intermediate in composition between the syenite and diorite clans; a fine-grained extrusive rock. It contains postash feldspar, acid plagiocalse, augite and minor ore in glassy interstitial material mostly nepheline.

dune-bedding. A type of largescale cross-bedding.

dune-complex. The totality of topographic forms, especially dunes, comprise the moving landscape.

dungannonite. Undersaturated igneous rock of the diorite clan; a white or grey, coarse grained plutonic rock containing dominant plagiocalse with biotite muscovite, calcite, magnetite and large corundums scapolite and nepheline.

dunite. An igneous rock of the ultrabasic clan a plutonic yellowish green on fresh fractures with greasy to vitreous lustre. Xenomorephic granular rock contains olivine and chromite with accessory magnetite, ilmenite, pyrope and spinel.

durain. Sedimentray rock constituent of bituminous coal consisting of small, resisitant debris of plants, such as fragments of cuticle, spore-cases, bits of resin or remains of allgal colonies and fusinite. It shows close firm granular texture. and rough fracture with dull or matt surface. It occurs in bands or lenticles of varying thickness in bituminous coal; often flecked with hair like streaks of bright coal.

durbachite. An igneous rock of the syenite clan; a porphyritic, hypabyssal rock. It contains phenocrysts of twinned orthoclase in ground orthoclase with accessory plagioclase, hornblende, titanite, zircon and occasionally a little quartz.

duricrust. A general term for the hard crust on the sufrace or within the upper horizons of a soil in a semi-arid climate.

duripan. A soil layer characterized by

cementation by silica. Durians occur mainly in volcanic areas having arid or semi-arid climates.

dyke. Sheet of igneous rock intruding through other strata.

dyke, Neptunean. A sheet of sediment filling a crack in another rock.

dyke-swarm. A group of dykes injected at about the same time and either parallel to each other or in lines meeting at a point.

dynamic allotropy. A kind of allotropy in which the allotropes are in dynamic equilibrium with each other, *e.g.* liquid sulphur.

dynamic equilibrium. (1) A state of equilibrium in which the rate of change of two processes is equal and opposite e.g. a liquid in equilibrium with its saturated vapour. (2) Biochemical processes involving living tissues which are continually being broken down and resynthesised so that their structure remains invariant.

dynamic metamorphism. The process by which strong but local forces (as in thrusting) change the nature of rocks in the crust.

dynamic recrystallization. The formation of new grains in a rock during solid state deformation. It is essentially a recovery process in which simultanous nucltheation and grain growth competes to yield a uniform grain size inversely proportional to the applied differential stress.

dynamic rejuvention. Renewal of the effectiveness of erosion process caused by uplift of the landmass with accompanying tilting and warping.

dysanalyte. $[(Ca,Ce,Na)(Ti,Nb,Ta)O_3]$. A mineral containing calcium cerium sodium tantalalte niobate(columbate) titanate, a variety of perovskite containing niobium (columbium in which proportion of niobium (columbium) is usually greater than tantalum and cerium or sodium may be absent.

dyscrasite. Ag_3Sb. Silver antimonide mineral occurring in silver deposits with antimonian silver; other silver minerals and galena, generally in a calcite gangue.

dyssnite. $(Mn,Fe)_2Si_2O_7$ Trivalent manganese and ferric iron, silicate alteration product of fowlerite.

dystrophiclake A lake in which nutrient matter and oxygen are sparse. It is rich in humic matter, which consists mainly of undecomposed plant fragments. Dystroiphic conditions are often associated with acidic peat bogs.

dzhalindite. $In(OH)_3$. A yellowish brown mineral of indium; an alteration product of indite.

dzhezkazganite. A mineral of lead rhenium sulphide; its composition is not known with certainty.

E

earlandite. $Ca_3(C_6H_5O_7).4H_2O$. A mineral consisting of hydrated calcium citrate; found in sediments in the Weddell sea.

earth. (1) The planet of the solar system lying between venus and mars, at an average distance of 150,000,000 km from the sun. It is almost spherical in shape, being slightly flattend at the poles. It mass=5.976×10^{24}kg; equitorial radius 6378.388 km, and polar radius= 6357 km. (2) Material that can be removed and handled economically. (3) A general term for the solid materials that make up the physical globe. (4) An organic deposit that has remained unconsolidated,although it is no longer in the process of accumalation. (5) Any of the several metallic oxides that cannot be reduced easily.

earth current. Static or alternating electric current flowing through the ground and arising either in natural or artificial electric or magnetic fields.

earth flow. The rapid movement of water laden soil material down a slope.

earth hummock. A low rounded frost mound generally 10 to 20 cm in height, consisting of a central core of fine material surmounted with vegetation.

earth pillar. A tall, conical column of unconsolidated to semiconsolidated earthy marerials, produced as a result of differential erosion in a region of spordic heavy rainfall.

earthquake. A sudden motion or trembling in the earth caused by the abrupt release of slowly accumulated strain. Earthquakes and volcanoes are manifestations of the motion of crustal plates. Shallow earthquakes occur where plates are colliding or separating, and deeper earthquakes occur where the lithosphere plunges into the mantle (subduction zone). Many shallow earthquakes are caused by movement of one crustal block with respect to an adjacent block along a fault planne.

earth science. The science that deals with the study of the earth or any part thereof; includes geology, geography, geophysics and geochemistry(in their widest senses), geodetics, climatology and the meterology, oceanogarphy, and the astronomical aspects of the earth - moon system.

earthquake swarm. The occurrence of a large numner of lesser earthquaks in a particular region over an interval of time perhaps several months, without the event of a major earthquake.

earthquake zone. An area or region in which fault movements occur. Such an area sometimes coincides with a location of volcanic activity.

eastonite. Magnesium sodium potassium aluminosillicate; a silver-white mica similar in composition to a vermiculite.

ebb and flow structure. Rock strata with alternating horizantal and cross-bedded layers, believed to have been produced by ebb and flow or tides.

eclogite. High grade regional metamorphic rock;coarse grained,occasionally fine, texture. It contains mostly omphacite and pyrope-rich garnet with main accessories of rutile, quartz, amphibole muscovite, clinozoisite and diamond; occurs usually as lenses or layers in schist or peridotite; also forms nodules xenoliths or inclusions in kimberlite, tuff or breccia.

eclogite facies. A set of metamorphic mineral assemblages of basaltic composition represented by omphacitic pyroxene and almandine pyrope garnet.

ecological niche. The environment in which an orgainism lives to which it is better suited than any other organism, and which suits it better than any other environment.

ectexis. Magmatization with in site formation of the mobile part.

ectexite. Rock formed by ectexis.

ectinite. Rock formed as result of the essentially isochemical regional metamorphism.

ectogene. Said of external factors that influence the texture of a rock.

eddy. A swirling or circular movement of water that is generally in a different direction from that of the main current. It is generally formed at a point in which a current passes through an obstruction or between two adjacent currents flowing in opposite direction or at the edge of a permanent flow.

edenite. $NaCa_2Mg_5(Si_7AlO_{22})(OH,F)_2$. Basic sodium calcium magnesium and alumunium flurosilicate; an iron free variety of hornblende.

Edingtonite. $2[BaAl_2Si_3O_{10}.\ 4H_2O]$. Hydrated barium aluminium silicate sometimes with appreciable calcium oxide; occurs; associated with thomsonite harmotome, calcite, analcite and prehnite.

edolite. A variety of hornfels consisting essentially of feldspar and mica; sometimes andalusite may be present.

Eemian. North European climatostratigraphic and floral stage; equivalent in time to Riss/Wirm interglacial.

effective stress. The average normal force per unit area transmitted directly from particle to particle of a soild or rock mass.

effective permeability. The ability of earth materials to transmit one fluid such as gas in the presence of other fluids such as water or oil.

effective porosity. The percent of the total volume of a given rock or earth material that consists of interconnecting voids.

effervescence. The escape of small gas bubbles from a liquid, usually as the result of chemical action with the application of heat.

effluent. Flowing out or forth. (1) A stream flowing out of a larger stream or out of a lake. (2) An outflow or an effluence *e.g.*,liquid discharges as waste.

effluent stream. A stream whose channel lies below the water table and into which water flows from the zone of saturation.

effusion. The passage of gases through small apertres under pressure. The relative rates of effusion of different gases under the same conditions are inversely propertional to the square roots of their densities.

effusive. A synonym of extrusive.

ekanite. $(Ca,Fe,Pb)(Th,U)Si_8O_{20}$. Thorium uranium calcium iron and lead silicate; metamict, recrystallising in tetragonal crystals on heating.

eleolite. A synonym of nepheline.

elasticity. The property of a body or material of resuming its original form and dimensions when the forces acting upon it are removed. If the forces are sufficiently large for the deformation to cause a break in the molecular structure of the body or material, it loses its elasticity and the elastic limit is said to have been reached. Hooke's law applies only within the elastic limit.

elastic limit. The limit of stress within which the strain in a material completely disappears when the stress is removed.

elastic wave. A wave propagated by a

medium having inertia and elasticity in which displaced particles transfer momentum to neighouring particles, and are themselves restored to the original position.

elastomer. A material such as synthetic rubber which at room temperature can be stretched under low stress to atleast twice its original length and upon immediate release of the stress will return with force to its aproximate original length. Some of the common elastomers are neoprene, polysulphide rubber; and silicone rubber.

elaterite. A variety of bitumen; it is quite amorphous; shows brown streak, soft and elastic with consistency like India rubber when fresh hard and brittle with conchoidal fracture on exposure.

electric calamine. An American synonym of hemimorphite.

electrometallurgy. The study of electrical processes used in separating a metal from its ore or refining or shaping metals by electrolytic procedures.

electron diffraction. A diffraction phenomenon resulting from the passsage of electrons through matter analogous to the diffraction of visible light or *X*-rays. The phenomenon of electron diffraction is the principal evidence for the existence of waves associated with electrons. The diffraction of electrons when passed through crystals or thin metal foils is used as a method of investigating crystal structure.

electron paramagnetic resonance. A method of spectroscopic analysis similar to nuclear magnetic resonance except that microwave radiation is employed instead of radio frequencies. It is used for studying free radicals, crystal centres, transition elements and structures involving unpaired electrons. Also known as electron spin resonance.

electron probe micronalysis. A method of analysing a very small quantity of a substance by directing a focused electron beam on to it so that an *X*-ray emission is produced characteristic of the elements present in the sample. The diameter of the beam is usulally about 1 μm and quantities as small as 10^{-14} g can be detected by this means. The method may be used quantitatively for elements whose atomic number exceed 11.

electrum. A natural alloy of gold (55%-85%) and siliver.

element. (1) A substance consisting entirely of atoms of the same atomic number. A part of an electron tube, semiconductor device, or antenna array that contributes directly to the electrical performance.

elementary particles. The basic units of which all matter is composed. The stable particles protons, electrons, and neurtinos combine with neutrons to form stable atoms. But many other short-lived particles and resonances have been detected that play an essential part in the structure of matter. For every particle that exists, there is a corresponding anti-particle, which has the same mass and spin but opposite electric charge. Some electrically neutral particles have antiparticles in which some otherproperty is rversed (*e.g.*, strangeness) and some neutral particles are regarded as their own antiparticles.

eleolite. A dark coloured variety of nepheline with a greasy lustre.

elie ruby. A synonym of pyrope.

elkhornite. An igneous rock of the syenite clan; a mottled, medium-grained hypabyssal rock; contains predominantly microcline and sodaorthoclase with small amounts of labradorite and augite, with or without biotite and titanite.

elutriation. A process for separating finely divided particles into sized fractions in accordance with the rate at which they rise or sink in a slowly rising current of water or air of known and controlled velocity. Elutriation is used in the mechanical analysis of a sediment, and in the removal of a substance from a compound.

elvanite. An igneous rock of the granite clan; a porphyritic hypabyssal rock; contains phenocrysts of quartz and orthoclase, often rounded or corroded, in microcrystalline groundmass of the same minerals. Tourmaline occurs as isolated crystals or radiating groups.

eluvial. (1) Pertaining to eluvium. (2) The eluvial phase of a dune cycle is marked by dune degradation, rather than growth. Soil-forming processes, creep, and slope wash predominate. The phase is initiated when vegetation becomes heavy enough to halt deflation. (3) Said of a deposit formed by rock disintegration at the place of origin; no stream transport is involved.

eluviation. The downward movement of materials in suspension or solution that are being carried through the soil by descending soil water.

eluvium. Disintegrated rock material formed and accumulated in situ or moved by the wind alone.

embayed. Characterized by the presence of a bay or bays; *e.g.*, an embayed coast.

embayment. (1) The process by which a bay is formed; also, the bay itself. (2) The incursion of a downwarped region of stratified rocks into other rocks. (3) The penetration of a crystal by another. (4) The corrosion of a crystal or xenolith by the magma in which it has formed.

embrechites. A type of migmatite in which structural features of crystalline shifts are preserved but often partially obliterated by metablastesis.

emerald. $Al_2(Be_3Si_6O_{18})$. Gemquality beryl; it possesses bright green colour owing to the presence of chromium.

emerald nickel. A synonym of zaratite.

emplacement. The intrusion of igneous rock into a particular place or position; also, the development of ore deposits in a particular location.

empirical. Based upon the results of actual experiment and observation rather than theory.

emulsification. The process of dispersing one liquid in another immiscible liquid.

emulsion. A stable mixture of two or more immiscible liquids held in suspension by small percentages of substances called emulsifying agents. All emulsions have a two-phase system in which the dispersed phase consists of minute droplets of other liquid.

enatiomophism. The occurrence of substances in two crystalline forms, one being a mirror image of the other. *See also* optical isomerism.

enantiotropic. Substances that exist in two different physical forms, one being stable below a certain temperature(the transition temperature),the other above it, *e.g.*, sulpher exists as alpha sulphere at all temperatures below 370K above this the stable form is beta-sulphur.

encrinal limestone. A vareity of limestone consisting of more than 10% and less than 50% of fossil crinoidal fragments.

encrinital. A term used mainly as a descriptive for limestones; made up of crinoid debries. Also known as crinoidal.

endellite. $Al_2Si_2O_5(OH)_4.2H_2O$. Hydrate aluminium silicate; hydrated form of halloysite.

enderbite. A metamorphic rock a type of charnockite with dioritic mineral composition; contains andesine in excess of microcline, pyroxene (typically hypersthene), with perthiote and antiper thite.

englacial. Said of something that is contained embedded, or carried within a glacier or ice sheet.*e.g.*, till or drift.

enrichemment. The action of natural agencies that increases the relative amount of one constituent mineral or element contained in a rock. This may be due to the selective removal of other constituents, or to the introduction of increased amounts of other constituents from external sources.

enstatolite. An igneous rock of the ultrabasic clan; a yellowish or greenish grey coarse-grained plutonic rock forming a bladed mass; contions essentially enstatite with chromite as main accessory.

entrainment. The process of picking up and transferring or carrying along such as the collection and movement off sediment by water currents.

eolation Any action of wind on the land.

eolianite. A sedimentary rock consisting of clastic material which has been deposited by wind action.

eolian soil. A variety of soil ranging from sand dunes to loess deposits whose particles are mainly of silt size.

eosite. A variety of quartzite; it is bluish with veins or splotches of brownish-red; contains small crystals of pyrites. It is used for ornamental carvings.

Eötvös unit. A unit of gravitational gradient or curvature, equal to 10^6mgal/cm.

ephemeral stream. A stream that flows(carries water) only during periods of precipittaion and briefly there after. Its channel is dry most of the year and may be referred to as and arroyo, or gully.

epimerism. A type of optical isomerism occurring in carbohydtates and some other organic and complex compounds. It is due to the formation of isomers (epimers) that differ in their molecular arrangements about an asymmetric atom.

epinaatrolite. A synonym of natrolite.

epistilbite. $CaAl_2Si_6O_{16}.5H_2O$. Hydrated calcuim aluminium silicate; member of zeolite family.

epitaxy. Growth of one crystal on the surface of another crystal in which the growth of the deposited crystal is oriented by the lattice structure of the substrate.

epizone. (1) The zone of metamorphism characterized by moderate temperature low hydrostatic pressure and powerful stress. (2) The outer depth zone of metamorphic rock.

epoch. (1) An interval of geologic time longer than an age and shorter than a period. during which the rocks of a particular series were formed. (2) An inexact short space of geologic time, such as a glacial epoch.

epithermal. Said of an hydrothermal mineral deposit formed within about 1 kilometre of the earth's surface in the temperature range 320-475 K.

equant. A term used for the crystal form in which all the faces are of equal dimensions.

equation of state. _Any equation connecting the pressure *p*,volume *V*, and temperature,*T* of the substance. Some equations of state attempt to cover more than one phase of the substance,*e.g.*,Van der Waals'equation of state, and are approximate. Others are intended to be applied to one particualr phase of the substance *e.g.*,the gaseous phase,and that only within certain limits of *P*,*V*, and *T*. With these limitations, these latter equations can represent the actual behaviour of the gases with much greater accuracy.

equatorial axis. The diameter of the earth described between two points on the equator.

equatorial plane. The plane passing through the equator of the earth or of other celestial body perpendicular to the axis of the earth that is equidistant from the two poles.

equilibrium. A state of balance between opposing forces or effects.

equilibrium, chemical. _Many chemical reactions do not go to completion. In such reversible reactions, a condition is attained in which chemical reaction occurs at equal rates in its forward and backward directions so that the concentrations of either the reactants or the products donot change with time, provided any of the substances is not mixture and the conditions such as temperature and pressure remain unaltered.

equipotential surface. A surface along which the gravity is constant at all points and for which the gravity vector is normal to all points on the surface. The geoid is an equipotential surface.

equivalent. Agreeing or corresponding in geological age or position within the stratigraphic profile. Strata in different regions that have been formed at the same time, or contain the same fossil types are said to be equivalent.

era. A geologic time unit during which rocks of the corresponding erathem were formed. It ranks one order of magnitude below eon; *e.g.*, paleozoic, Mesozoic, Cenozoic.

erosion. The process of wearing of a surface by water or wind particularly when containing debris; can also be caused by animals.

erratics. Glacially transported stones and boulders; may be embedded in till or occur on the ground surface.

erubescite. A synonym of bornite.

esythrite. $2[(Co,Ni)_3(AsO_4)_2.\ 8H_2O$. Hydrated cobalt arsenate containing traces of nickel; also occurs in upper parts of veins with smaltite and cobaltite.

esboite. An orbicular diorite in which an disene or oligoclase is the main plagioclase and forms the orbicules.

esbiotic crystillization. The process in which the orbicules of an esboite attain oligoclastic composition.

escarpment. A high more or long, steep slope situated between a lower, more gently inclined surface and a higher surface.

eschynite. $(Ce,Th)(Nb,Ti)_2\ (O,OH)_6$. A mineral consisting of rare earths, titanium, thorium, niobate.

eskebornite. $CuFeSe_2$. The selenium analog of the mineral pyrrhotite.

esker. A term used for a long narrow sinuous or straight ridge formed of stratified glacial meltwater deposits, usually including large amounts of sand and gravel. Most eskers are probably deposited in channels beneath or within slow moving or stagnant ice. Their general orientatinon runs ar right angles to the ice edge. Some eskers originate on the ice and have ice cores.

espichellite. An igneous rock of the gabbro clan; a porphyritic hypabyssal rock. It contains phenocrysts of hornblende augite olivine,magnetite and pyrite in black compact groundmass of magnetite hornblende, augite, mica, and labradorite with orthoclase rims.

essential mineral. A mineral component that is basic to or required in the classification of a rock.

essexite. A rock of igneous origin composed chiefly of plagioclase, hornblende, biotite and titanaugite. Current usage applies the name to a labradorite, titan-augite,olivine, nepheline and/or analcite rock. Some alkli feldspar may be present.

essonite. A synonym of hessonite.

esterellite. An igneous rock of the diorite clan; a bluish holocrystalline hypabyssal rock. It contains rare phenocrysts of quartz with andesine hornblende and accessory magnetite in a groundmass predominantly andesine, quartz and magnitite with accessory pyroxene orthoclase and biotite.

estuarine deposit. A sediment deposited at the heads and floors of estuaries.

ethomolith. A downward tapering funnel-shaped, discordant intrusion of igneous rocks.

etindite. An igneous rock of the syenite clan, a dark grey to green porphyritic extrusive rock. It contains phenocrysts of augite in dense holocrystalline granular groundmass of leucite, nepheline and augite with accessory perovskite, titanite apatite and magnetite.

eu–. A prefix denoting well developed.

euclase. $BeAlSiO_4(OH)$. A mineral occurring as prismatic crystal of blue green or violet colour.

eucolite. A variety of eudialyte possessing prismatic crystals with large or small basal plane. It shows cleavage in one direction and imperfect in two others.

eucrite. An olivine bearing gabbro generally contains calcic; plagioclase; a meteorite.

eucryptite. $3[LiAlSiO_4]$. Lithium aluminium silicate possessing hexagonal symmetrically arranged crystals with basal cleavage.

eucrystalline. A term applied to those igneous rocks which are holocrystalline as well as well-crystallized.

eudnophite. A variety of analcite; possesses six-sided prismatic crystals terminated by a macrodome.

euhedral. Said of mineral crystals that develop fully and show well-formed polyhedral facial planes.

euktolite. A feldspathoid igneous rock; contains phenocrysts of olivine; melilite and mica in roughholocrystalline groundmass of olivine, melilite leucite, phlogopite, and magnetite; shows numerous fractures along flow direction.

euktolith. A synonym or euktolite.

eulite. $16[(Fe,Mg)SiO_3)]$. Iron magnesium silicate with molar proportion of ferrous silicate between 70-90 per cent; a variety of orthoferrosilite.

eulysite. The high grade regional metamorphic rock; a fine grained rock containing manganese-fayalite, diop side iron manganese garnet and amphibole with accessory quartz, biotite apatite and iron ore.

eulytite. A synonym of eulytine.

euphotic zone. An occan zone in which light penetration is sufficient for photosynthesis. It commonly extends from 60-70 m but may be even greater.

eustatic. A term used for worldwide changes in sea level as distinct from local changes.

eustratite. An igneous rock of the

diorite caln; a compact textured hypabyssal rock. It contains rare phenocrysts of olivine and corroded hornblende in groundmass of automorphic augite and titaniferous augite in base of feldspar mica and glass.

eutaxitic. A term used to describe a fabric typical of welded ash-flow tuff in which lapilli' pumice and glass-shards are flattend into a disc-like form more or less parallel to the plane of deposition.

eutectic. A solution or an alloy that has the lowest possible constant melting point.

eutectic crystallization. Simultaneous crystallization of the constituents of an eutectic alloy during cooling of the melt.

eutectic mixture. A solid solition of two or more substances having the lowest freezing point of all the possible mixtures of the components. This is taken advantage of in alloys of low melting point which are generally eutectic mixtures.

eutectic point. Two or more substances cabable of forming solid solutions with each other have the property of lowering each other's freezing point attainable, corresponding to the eutectic mixtures.

eutectic system. The composition and temperature of substances at the eutectic point.

euthalite. A variety of analcite . Its successive layers are; greeinsh or greyish white in colour.

evansite. $Al_3PO_4(OH)_6$. $6H_2O$. Basic hydrated aluminium phosphate with some ferric oxide; a secondary mineral found with limonite and allophane.

evaporite. Deposits of physico–chemical origin mainly involving precipitation and crystallisation from saturated solutions; forms lenticular masses (if of terrestrial origin) or well bedded occasionally, regulary interstratified,with fragmental or organic sediment (if lalid down in lakes or the sea). It deposits mainly from evaporation processes from lakes seas or evaporation of ground water drawn up through soil by capillary action. Chemical deposition from springs leads to formation of calcareous sinters.

event. A more or less catch–all term for any occurrence in which probable tectonic importance is implied by some geologic evidence.

evjite. A gabbro of hornblende in which only light coloured mineral is labradorite or bytownite; hornblende must be primary, not uralitic.

exfoliation. The separationof successive thin, onionlike shells (spalls) from bare surfaces of massive rock, such as granite or basalt. It is found in regions of moderate rainfall.

cxhalation. The process by which radioactive vapour or gases escape from the surface of soil or loose rocks where they are formed by decay of radioactive substances.

exinite. A hydrogen-rich maceral group consisting of spore exines, resins, waxes, and cuticular matter; includes sporinite, cutinite, alginite, and resinite. Also known as liptinite.

exo. A prefix denoting out of, outside.

exogen effect. The effect of an igneous mass on the rock that it invades.

exogenetic. The term used to describe processes or events which take place at or very near the surface of the earth, e.g.,weathering, erosion,etc.

exogeosyncline. A parageosyncline that lies along the creational border and accumulates sediments from highlands in the orthogeosynclinal belt that lies outside the crator.

exomorphism. Changes in country rock produced by the intense heat and other properties of magma or lava in contact with them; contact metamorphical in the usual sense.

exosphere. The outermost layer of the earth's atmosphere, in which the density is such that an air molecule moving directly outwards has a 50% chance of escaping rather than colliding with another molecule. The exosphere lies beyond the ionosphere and starts about 400 kilometres above the earth's surface.

exothermic process. A process in which energy in the form of heat is released.

exotic. (1) Said of an organism that has been introduced into a new area from where it grew naturally. (2) Applied to a rock body that is unrelated to the rocks with which it is associated.

exposure. A place where rocks can be seen in situ, *i.e.,* not as detached boulders or fragments or covered by vegetation.

exsolution. The appearance of two mineral phases in the solid solution when cooled slowly to a certain temperature.

extravasation. The eruption of lava from a vent in the earth.

extrusion. The emission of relatively viscous lava on the earth's surface; also the rock so formed.

extrusive rocks. A term applied to rocks that result from cooling solidification and the formation of igneous materials above the surface of the earth. Such rocks are differentiated from plutonic or intrusive rocks which have solidified from magma below the earth's surface.

eye agate. Variety of chalcedony in which delicate parallel lines of banded colours can be white, wax-like pale and dark brownand black, blue or of some other shade. Lines usually follow wavy, sometimes straight, occasionally concentric, circular courses.

eye coal. Coal characterized by small circular or elliptic structural discs that reflect light and are arranged in parallel planes either in or normal to the bedding. Also known as circular coal.

eye ridge. A raised line or narrow band extending from the forward and inner part of a trilobite eye to the anterior part of the glabella.

ezcurrite. $Na_4B_{10}O_{17}.7H_2O$. A triclinic mineral dimorphous with nasinite.

explosion tuff. A tuff whose pyroclastic fragments are in the place in which they fell, rather than having been washed into place after they landed.

explosive index. The percentage of pyroclastics among the total products of a volcanic eruption.

extension fault. A fault in sedinmentary rocks along which there has been bed-parallel elongation.

extrusive ice. Ice formed subaerially from water that emerges from the ground or below other ice, as in an icing.

exudation basin. A spoon-shaped depression on the ice surface at the head of an outlet glacler.

eyebrow scrap. A fault scrap that crosses a piedmont alluvial fan near its apex and that seldom maintains the dip of the fault surface in the unconsolidated gravels of the fan.

F

fabric. (1) The spatial orientation of the elements of a sedimentary rock. (2) The sum of all the structural and textural features of a rock.

fabric axis. In structural petrology, one of the three orthogonal axes used as references in the orientation of fabric elements and in the description of folding and symmetry of movement in deformed rocks.

fabric diagram. A graphic representation of the data of fabric elements.

fabric element. A surface or line of structural discontinuty in a rock fabric.

face. a single flat surface of a naturally developed crystal,which exists in fixed angular relation to adjoining faces. It may be regarded as one element of a form. **facet.** (1) One of the small polished plane surfaces on a cut gemstone. (2) Nearly planar surface produced on a rock by the action of water or sand. (3) Any plane surface resulting from faultion or erosion and cutting from faulting of erosion,and cutting across the general slope of the land.

faceted pebble. A pebble with three or more faces naturally worn flat and meeting at sharp angles.

faceted spur. A ridge with a flat truncation at its nose.

facies. The sum of features such as sedimentary rock type, mineral content, sedimentary structures, fossil content etc., which characterise a sediment having been deposited in a specified environment.

facies evolution. Gradual lateral or vertical changes in the type of rock or fossils found in contemporaneous sedimentary deposits.

facies fossil. A fossil that represent a particular paleo-environment and therefore, a correspondingly restricted distribution.

facies tract. A system of interconnected sedimentary facies of the same age, including the areas of erosion from which the sediments of the facies originate.

facing. The direction along the axial plane of a fold in which it passes through younger layers.

faheyite. $(Mn,Mg)Fe_2Be_2(PO_4)\ 6H_2O$. A white hexagonal mineral.

fahlband. A band of sulphide impregnation in metamorphic rocks.

fairchildite. $K_2Ca(CO_3)_2$. A hexagonal mineral composed of calcium potassium carbonate; shows good cleavage in one direction.

false-bedding. An inclined bedding produced by currents.

false cleavage. (1) Spaced surfaces about a millimeter apart along which a rock splits. (2) A weak cleavage at an angle to the slaty cleavage.

false topaz. A synonym of citrine.

famatinite. Cu_3SbS_4. A mineral composed of copper sulphide and antimonide; occurs as crusts of minute cubic crystals; usually massive, granular to dense, sometimes reniform. Its hardness on Mohs scale is 3.5.

fan. (1) Alluvial fan. (2) Fan-shaped mass of solidified lava.

fan fold. A fold of strata in which both limbs are overtuned, forming a syncline or anticline.

fanglomerate. Coarse material in an alluvial fan, with the rock fragments being only slightly worn.

faröelite. A silica rich variety of thomsonite having spherical concretions of lamina; possesses pearly lustre on cleavage surfaces.

farrisite. A feld spathoid igneous rock; a chocolate-brown, fine-granular to porphyritic hypabyssal rock. It contains 33 per cent melilite like mineral, barkevikite,pyroxene and rare biotite and altered olivine with accessory magnetite, pyrite and apatite.

fasinite. A feldspathoid igneous rock; a coarse grained to porphyritic under saturated plutonic rock. It contains augite, nepheline and subsidiary olivine and biotite.

fassaite. (Ca,Mg,Fe,Al)(Si,Al,Fe) O_3. Calcium magnesium aluminium and ferrous iron silicate with considerable ferric and aluminium oxides, in which proportion of is greater than 0.5 and proportion of magnesium is greater than ferrous iron; a variety of augite rich in calcium.

faujasite. $Na_2CaAl_4Si_{10}O_{28}.20H_2O$. Hydrated sodium calclium aluminium silicate; member of zeolite family.

fault. (1) A general term for a dislocation in a crystal. (2) A fracture or a zone of fractures along which there has been displacement of the sides relatives to one another paralled to the fracture.

fault basin. A depression separated from the surrounding area by faults.

fault bench. A small fault terrace.

fault block. A crystal unit formed by block faulting ; it is bounded by faults and behave as a unit during block faulting and tectonic activity.

fault breccia. The assembly of angular fragments found frequently along faults. Also known as dislocation breccia.

fault gouge. A fine-grained, clay-like substance formed by the grinding of rock material as a fault developes; it may also be formed by decomposition caused by circulating solutions.

fault line. Intersection of the fault surface with the surface of the earth or any other horizontal surface of reference. Also known as fault trace.

fault plane. The surface along which rock masses on opposite sides of a fault rub against each other. In some instances,it may be a warped and irregular surface. Also known as fault zone; fault surface.

fault rock. A rock often found along a fault plane and made up of fragments formed by the crushing and grinding which accompany a dislocation.

fault scrap. A steep cliff formed by movement along one side of a fault.

fault strike. The angular direction, with respect to north , of the intersection of the fault surface with a horizontal plane.

fault system. Two or more fault sets which interconnect.

fault terrace. A step on the slope formed due to displacement of two parallel faults.

fault trap. Oil or gas reservoir formed by a structural trap limited in one or more directions by subterranean geological faulting.

fault zone. A fault found in a zone consisting of a network of many small faults or by a zone of fault breccia or mylonite. Fault zones may be hundreds of metres wide.

faunal province. A region characterized by an assemblage of specific fossil fauna widely distributed within it.

faustite. $ZnAl_6(PO_4)_4(OH)_8.5H_2O$. An apple-green mineral composed of hydrated bassic zinc aluminium phosphate, having conchoidal to smooth fracture. Its hardness on Mohs scale is 5.5.

fayalite. Fe_2SiO_4. A brown to black silicate mineral of olivine group, found either massive or in crystalline form. Its hardness on Mohs scale is 6.5. It gelatinises in acids and when fused, reduces to black magnetic globule in flame.

feldspar. A group of rock forming minerals that make up about 60% of the earth's crust . Their general formula is $MAl(AlSi)_3O_8$, where M can be K, Na, Ca, Ba, Rb, Sr,or Fe. Feldspars occur in all types of rock, but are essential constituents of most igneous rocks. The kind and amount of feldspar present are used as a means of classifying these rocks. Isomorphous substitution of one alkali metal with another gives rise to varieties of feldspar as when sodium replaces potassium in orthoclase to form soda orthoclase. Plagioclase series consists of mixtures of albite and anorthite in all proportions; replacement of the sodium and silicon with calcium and aluminium results in complete gradation between two end-members. Albite, oligoclase andesine , labradorite, bytownite and anorthite are plagioclase, microcline soda orthoclase and anorthoclase. Alkali feldspars are orthoclase microcline, soda orthoclase and anorthoclase. Barium feldspars are celslian and hyalophane.

feldspathic. A term used for mineral aggregate containing feldspar. The percentage of feldspar present must usually be within certain limits.

feldspathoid. Group minerals similar to feldspars, especially in chemical composition, but deficient in silica. Aluminium, sodium and/or potssium silicates, are generally co-ordinated to a calcium or sodium compound. Series comprises:

Leucite	$KAlSi_2O_6$.
Nepheline	$(NaK)AlSiO_4$
Cancrinite.	$4(NaAlSiO_4)CaCO_3H_2O$.
Sodalite.	$3(NaAlSiO_4)NaCl$.
Hauyne.	$3(NaAlSiO_4)CaSO_4$.
Nosean.	$3(NaAlSiO_4)Na_2SO_4$.
Lazurite.	$3(NaAlSiO_4)Na_2S$.

feldspathic graywacke. Sandstone containing less than 75 per cent quartz and chert and 15%-75% detrital clay matrix, and having feldspar grains in greater abundance than rock fragments.

feldspathization. Formation of feldspar in a rock, generally as a result of metamorphism leading toward granitization.

feldspathoid igneous rock. Type of igneous rock that is undersaturated in silica; contains feldspathoids (or their equivalentes)in place of feldspars and no free quartz. It includes undersaturated equivalents of syenite, diorite and gabbro clans.

felsenmeer. A flat or slightly sloping veneer of angular rock fragments occurring on moderate mountain slopes above the timber line.

felsic. A word derived from feldspar and silica and applied to light coloured silicated minerals such as quartz, feldspar and feldspathoids. This term describes the rocks that contain an abundance of one or all these minerals.

felssite. An igneous rock of the granite clan; a grey or pinkish compact, aphanitic, hypanyssal rock, consists of cryptocrystalline aggregate, essentially quartz and orthoclase. Satellitic to granite masses or may be spherulitic; it includes non porphyritic rhyolites,trachytes, phonolites,latites and andesites.

felsitic. A textural term used for dense, light-coloured igneous rocks that contain microcrystalline or cryptocrystalline aggregates.

felty. A textural term used for dense holocrystalline igneous rocks,or the dense holocrystalline groundmass of porphyritic igneous rocks consisting of tightly and flatly pressed microlites. These are generally feldspar, interwoven in an irregularmanner.

femaghastingsite. A variety of hastingsite containing ratio of ferrous ions to magnesium 0.5-2.0. Chemically it is identical with ferrohactingsite except megnesuim replaces ferrous ions.

fenite. An igneous rock of the syenite clan; a leucocratic, coarse grained plutonic rock. It contains 70-90 per cent alkli-feldspar (usually orthoclase or microcline), 5-25 per cent aegirine or aegirine-augite with subordinate alkali-hornblende and accessory titanite and apatite.

feberite. $2[FeWO_4]$. A monoclinic mineral composed of iron tungstate containing upto 20 per cent manganese tungstate; crystals are elongated somewhat flattened and striated, often wedge-shaped.

fergusonite. $Y_2O_3.(Nb,Ta)_2O_5$. A browinsh-black rare-earth mineral having a tetragonal crystal form.

ferrate. 1. A salt of the hypothetical ferric acid, H_2FeO_4. 2. A multiple iron oxide with another oxide, e.g., Na_2FeO_4.

ferric. Containing trivalent iron. Ferric salts are usually yellow or brown in colour.

ferrierete. Conglomerate formed from cementation of gravels by oxidation of percolating solution of iron compounds.

ferrierite. $(Na,K)_4\ Mg_2Al_{10}Si_{30}O_{73}.20H_2O$. A hydrated sodium postassium magnesium aluminium silicate related to mordenite with magnesium replacing calcium.

ferrite. (1) Alteration product of chrysolite (of Wallerius); iron rich variety of olivine. (2) A salt of the hypothetical 'ferrous acid' that exists in strong alkaline solution only, *e.g.*, $NaFeO_2$.. The name applied to several types of iron ore. Pure alpha iron , or solid solutions of alpha iron is the solvent.

ferrites. A group of ceramic materials that exhibit the property of ferrimagnetism. They consist of iron oxides to which small quantities of other transtion metal oxides *e.g.*, cobalt and nickel oxides) have been added . The spinel ferrites have the formula $MO.Fe_2O_3$ where M is a divalent transition metalion. By suitable combination of metallic oxides ferrites can be made which exhibit ferromagnetism, but as they are electrical insulators and therefore do not suffer from the effects of eddy currents, they can be used as cores in coils and transformers, in electronic equipment at frequencies that would be impossible with ordinary ferromagnetic materials.

ferro–. Prefix denoting iron, especially in name of alloys; *e.g.*, ferromagnese.

ferroaluminium. An alloy of aluminium (up to 80%) and iron.

ferroan. Pertaining to iron-bearing.

ferroan dolomite. A type of ankerite having less them 20 per cent of the manganese positions occupied by iron.

ferro-anthophyllite. $4[Fe_7Si_8O_{22}(OH)_2]$. Hydrated iron silicate; end member of anthophyllite series, or iron rich anthophyllite.

ferroaugite. $8[Ca,Mg,Fe^{2+})SiO_3]$. Calcium magnesium ferrous iron silicate with or without some ferric oxide or aluminium ferric oxide. Molar proportion of calcium silicate varies between 25-45 per cent and proportion of ferrous iron greater than magnesium.

ferroboron. An alloy of iron and boron

that is added to steel to make it hard.

ferrobrucite. [(Mg,Fe)$(OH)_2$]. Magnesium ferrous iron hydroxide with up to 10 per cent ferrous oxide; a variety of brucite.

ferrochrome. An alloy of chromium with 30%-40% iron, obtained by the reduction of cromite with carbon in an electric furnace.

ferroelectric. Dielectric materials that have electrical properties analogus to certain magnetic properties such as hysteresis *e.g.*, barium titanate and potassium sodium tartrate (Rochelle salt). Ferroelectric materials usually also have piezoelectric properties.

ferrogabbro. A gabbro rock in which the pyroxene and olivine constituents have a high iron content.

ferrohastingsite. Na Ca_2 $Fe_4^{2+}(Al,Fe^{3+})$ $(Si_6Al_2O_{22})(OH,F)_2$. Basic sodium calcium aluminium ferric and ferrous iron fluoride and aluminosilicate; iron rich member of hastingsite pargasite series.

ferrohedenbergite. 8[Fe,Ca)SiO_3]. Iron calcium silicate with proportion of iron greater than calium; variety of hedenbergite.

ferrohortonolite. $(Fe,Mg)_2SiO_4$. Iron magnesium silicate with molar proportion of ferric silicate 70-90 per cent; a_variety of horonolite.

ferrojohannsenite.4[Ca(Mn,Fe)Si_2O_6. Calcium manganese ferrous iron silicate; iron rich mineral with magnanese partially replaced by ferrous iron.

ferroselite. Rock forming silicate minerals which contain magnesium and /or iron.

ferroselite. $FeSe_2$. Iron selenide with traces of silver, thorium vanadium; and titanium; formes solid solutions series with pyrite and mixed crystals occur with uranium minerals in sandstone.

ferruginous. Pertaining to iron or containing iron; staining it; *e.g.*, a sandstone cemented with iron oxide or with a cement containing iron oxide. Also said of a red or rustcoloured rocking containing ferric oxide.

festoon. The upfolded or printed of part of a layer in a congeliturbate.

fetch. A term used in wave forecasting for the area of the open ocean over the surface of which the wind blows with constant speed,and thereby creating a wave system.

fiamme. Dark vitric lenses in welded tuffs formed by the collapse of fragments of pumice.

flasconite. An igneous rock of gabbro clan;. a dark grey, compact porhyritic extrusive rock; contains phenocrysts of augite, leucite,anorthite, olivine, biotite magnetite, and apatite.

fibre. A structure resembling a thread.

fibroblastic. Said of a metamorphic rock, having a texture which is homeolblastic as a result of the development of minerals with a fibrous habit during crystallization.

febroferrite. $Fe(SO)_4(OH).5H_2O$. A fibrous mineral consisting of hydrated basic ferric sulphate.

fibrous texture. An array of circular or rodlike crystal found in certain mineral deposits, such as asbestos.

fibrous zeolites. Members of the zeolite family with framework structures of silicate tetrahedra arranged in groups of five, with aluminium generally replacing part of the silicon . These minerals are monoclinic or orthorhombic. The series comprises:

Natrolite.	$Na_2(Al_2Si_3O_{10}).H_2O$.
Mesolite.	$Na_2Ca_2(Al_2Si_3O_{17})_3.\ 8H_2O$.
Scolecite.	$Ca(Al_2Si_3O_{10}).3H_2O$.
Thomsonite.	$NaCa_2\ (Al,Si)_5O_{10}).6H_2O$.
Gonnardite.	$Na_2Ca[(AlSi)_5O_{10}].6H_2O$.
Edingtonite.	$Ba(Al_2Si_3O_{10}).4H_2O$.

fichtelite. $C_{18}H_{32}$. White crystalline hydrocarbon; monoclinic; occurring in small tabular crystals and aggregates .

field classification. Examination or preliminary analysis of fossils or hand specimens of rock or minerals, commonly with the aid of a hand lens.

field geology. That part of geology that is practiced by direct observation in the field.

filiform. A term used for thread like or fibrous materials.

fill. (1) Man-made deposits of soil, rock or various debris materials used for building embankments, filling in soggy ground, extending a shoreline into a lake, or filling unused mine workings. (2) sediment deposited by any agent in such a manner that it fills or partially fills a valley.

finandranite. An igneous rock of the syenite clan; a coarse-grained plutonic rock; contains microcline and amphibole with some biotite, ilmenite and apatite.

fine-grained. (1) Said of the texture of a sedimentary rock in which the average diameter of the particles is less than one sixteenth mm(silt size and smaller). (2) Said of the texture of an igneous rock of less than 1 mm.

fines. Very small particles in a mixture of particles of various sizes. The term is also used for finely crushed ore,mineral, or coal.

fine sand. Sand whose particles are from 0.125 to 0.24 mm m size.

fine Slit. A term used for slit particles having diameter in the range of 8-16 microns.

finger coal. Natural coal occurring as small hexagonal clolumns associated with igneous intrusions.

fingerlakesian. North American stage; lower upper Devonian.

fingertip channel. One of the smaller unbranched stream channels at the head of a drainge network.

fireball. A bright meteo with luminosity that equals or exceeds brightness of planets.

fireclay. A clay that can resist higher temperatures without becoming glassy; it is generally soft, white or grey. Clay rich in hydrated aluminum silicates or silica while deficit in alkali and iron content; consists of dominant kaolinite and illite-chlorite and chlonite-kaolinite.

fire-damp. An explosive mixture of methane (CH_4) and air, formed in coal mines, due to decompostion of coal or other carbonaceous matter.

fire fountain. Rise and fall eruption of incandescent lava either from a central volcanic vent or along a fissure as a jet of molten material in the form of a spray and bombs.

fire opal. Transparent to translucent with red to orange-red body colour which ,may or may not show a play of colour.

firestone. *See* flint.

firm. (1) A geographic term applied to the accumulation area or upper region of a glacier. (2) A material that is transitional between snow and glacier ice being older and denser than snow but not yet transformed into a glacier ice.

firm edge. The boundary on a glacier between glacier ice and firm during the ablation period.

firnification. The process of transformation of snow into firn and then into ice in a glacier.

fissile material. Isotopes that are capable of undergoing nuclear fission. Sometimes the term is restricted to isotopes that are capable of undergoing fission upon impact with a slow neutron

(*e.g.*, U-233, U-235, Pu-239)

fissure. A general term meaning an extensive cleft 'break' or fracture in a rock formation volcano or the earth's surface. It is also used in the sense of crevasse.

fissure eruption. Volcanic eruption from a fissure or series of fissures rather than from a central vent.

fissure vein. A mineral deposit of vein-like shape, characterized by .clearly defined walls as opposed to extensive country-rock replacement.

fixed carbon. A solid, finely divided black elemental carbon whose presence distinguishes coal from peat. The higher the proportion of fixed carbon and the lower the content of volatiles, the higher the rank of coal.

flaser gabbro. A dynamically metamorphesed gabbro which has been crushed and sheared into lenticular masses, separated by wavy ribbons and streaks of finely granulated and recrystallised material, with original pseude-porphyritic igneous texture retained in the rock that has not become schistose. It contains large plagioclases often crushed around peripheries and pyroxenes showing uralitisation around borders.

flaser-geneiss. Low grade Coarse grained dynamic metamorphic rock. Typically gneiss assemblage of minerals. It contains lenticular masses of parts of the original rock in finely crystalline, foliated groundmass.

flasser structure. A metamorphic structure in which small lenses and layers of granular material are surrounded by a matrix of sheared, crushed material, resembling a crude flow structure. The developement of a patchy structure in a granular igneous rock produced by dynamic metamorphism.

finkite. $Mn_3(AsO_4)(OH)_4$. A dark coloured mineral occurring in feathery forms; consists of basic manganese arsenate.

flint. Variety of chalcedony; contains mainly granular chalcedony with small proportion of opal silica and sometimes organic remains. Also known as fire-stone.

flinty crush–rock. Medium to high grade dynamic metamorphic rock; partly fused variety of mylonite with flinty composition and black colour. Usually it is structureless, but occasionally showing traces of crystallisation.

flock. Small masses formed in a liquid as a result of fine suspended particles.

flocculating agent. A substance added to disperseion medium to bring about flocculation of fine particles.

flood basalt. Extremely fluid basaltic lava that erupts as a series of horizontal flows in rapid succession (geologicaly), covering vast areas; generally believed to be the product of fissure eruption.

flood basin. A term applied to a tract of land that is inundated during the highest of floods. It also applies to a flat area between a sloping plain and a natural levee where swampy vegetation often grows.

floodplain. That portion of a river valley, adjacent to the river that is built up of alluvium deposited during the present disposition of the stream flow.

floe. A sheet of floating ice, smaller than an ice field; found especially on the surface of the arctic sea.

florencite. $CeAl_3(PO_4)_2(OH)_6$. Basic cerium earths aluminium phosphate; occurring as hexagonal or rhombohedral crystals.

flos-ferri. A variety pf argonite; forms groups of delicately interlacing and coalescing stems. It is like coral in appearance.

flow. (1) The movement or rate of movement of water, or the moving water itself. (2) Any rock deformation that is not immediately removable without permanent loss of cohesion. (3) The mass movement of unonsolidated material in the semifluid or still plastic state.

flowage line. A contour line at the edge of a body of water, representing the given water level.

flow banding. An igneous rock structure formed as a result of flowing magmas or lavas; characterized by alternation of mineralogically different layers.

flow breccia. A breccia formed with the movements of lava flow while the flow is still in motion.

flow cast. One of the group of bed ding plane structures formed in graywacke.

flow cleavage. Rock cleavage in which solid flow of rock accompanies recrystallization.

flow fold. Folding in beds composed of relatively plastic rock, that assume any shape impressed upon it by the more rigid surrounding rocks or by the general stress pattern of the deformed zone.

flow plane. In igneous and metamorphic rocks the plane along which displacement occurs.

fluid saturation. Measure of the gross void space in a reservoir rock that is occupied by a fluid.

flume. A ravine with a stream flowing through it.

fluoridation. Formation of fluorine-containing minerals in rocks, *e.g.*, topaz; fluorite.

fluorine dating. A method of determining the related ages of Pleistocene or Holocene fossil bones, from the same excavation based on the gradual combination of fluorine in groundwater with the calcium phosphate of buried bone material. *See also* dating methods.

fluorite. CaF_2. A mineral found in pneumatolytic deposists, and pegmatites; forms distinctive interpenetrating cubes which may be coloured, banded and show octahedral cleavage. Octahedra and single cubes of flourite also occur but in veins it is transparent and colourless when pure but often it is yellowish, green, purple, or blue with a vitreous lustre. Its hardness on Mohs scale is 4.0. The mineral can be recognized by its form, cleavage and hardness, which is greater than that of calcite but less than that of quartz. It is a hydrothermal mineral usually found as a gangue in veins with argentite, galena, sphalerite, quartz,and barytes. It also occurs in cavities in limestone and dolomite and as a cement in some sandstones. Also known as fluorspar.

fluorspar. A synonym of fluorite.

fluotaramite. $2(Na,K)_{2\text{-}5}(Mg,Fe^{2+},Fe^{3+}.Ca)_5Si_8O_{22}(OH,F)_2]$. Basic sodium potassium magnesium calcium ferric and ferrous iron fluorsilicate; containing little amount of aluminium occurs as long acicular crystals.

flute. An assymetical, spoon-or scoop shaped groove or depression in a sedimentary bed. It is usually seen as a flute coat. Such impressions are formed by non-laminar, sediment-bearing currents that can scour a soft stream bottom.

fluted. Having a channelled or grooved, structure or texture.

fluvial. A geological term applied to sediments formed by rivers.

fluvial lake. (1) A slowly moving part

of river, generally where its width has been expended. (2) A lake with a perceptible flow of water.

fluviation. The activites engaged in, and the processes employed, by streams.

fluviology. The scientific study of river.

flux. (1) A strem of flowing water. (2) Substance added to assist fusion. (3) The rate of flow of mass or energy per unit area normal to the direction of the flow. (4) The product of the number of particles per unit volume and their average velocity.

fluxing. Making of a liquid phase in a ceramic body under heat treatment by melting low fusion components.

flux mapping. The process of measuring the radiation flux at various points within a nuclear reactor or some other radiation source.

fluxmeter. An instrument for the measurement of magnetic flux.

fluxtone. In metallurgical processes, limesonte, dolomite, or other rock used to lower the fusion temperature of the ore, or to combine with any impurities present, to form a fluid slag.

flyash. Fine, non-combustible particles carried in a gas stream from a furnace. It is a mixture of alumina, silica, unburnt carbon and various metallic oxides.

foid. A collective term used to denote the feldspathoid group of minerals.

foidite. A plutonic or volcanic rock in which the feldspathoids constitute 60-100 percent of the light coloured components.

feld. Rock strata subjected to pressure beyond their elestic limit may yield to show plastic deformation by bending or folding into more or less symmetrical series of folds with series of alternating crusts and troughs. The main types of folded structures are monoclines, anticlines, synclines, basins and domes.

fold, drag. Weak layers thrown into minor indulations. Their symmetry indicates the relative displacement of the adjacent layer.

folded fault. (1) A thrust fault, the hanging wall of which has slightly become folded due to development of step thrusting or step faults beneath it. (2) Any fault that has become deformed by folding.

fold fault. An overfold, the middle limb of which is replaced by a fault surface.

fold mountains. Mountains that have been formed by large-scale and profound folding.

fold system. A set or group of congruent folds that have been produced by the same tectonic episode.

foliated. Having a layered 'onion skin' form; interfoliated structures are interleaved.

foliation. A planer, layered to laminated structure occurring almost exclusively in metamorphic rocks. When caused by parallel oriention of mineral plates and rods, it is called schistosity.

foliose. _Leafy, or resembling a leaf ; said of a corallum with laminar branches..

fondo. A term applied to the environment of sedimentation that ties on the deep floor of a water body.

fondoform . The subqueous land form coasting the main floor of a water body; the site of the fonds environment of deposition.

fondotherm. Rock units formed in the fonds environment of deposition.

font. A term for a stream or a spring, fountain, or source of a stream; used to form part of the place names.

fontainbleau sandstone. A variety of sandstone with a cement of crystalline calcite.

fool's gold. A synonym of pyrite.

forellenstein. A synonym of troctolite.

foresite. A variety of stilbite, forming crystalline crusts, shows pearly lustre on cleavage. It occurs on tourmaline or heulandite lining cavities in granite.

footwall. The rocks that lie below the fault plane of a fault that is not vertical.

foraminifera. Protozoa with shells, which are commonly divided into parts. The shell may be made of calcite or arenaceous material. Most of the foraminifera are marine, and valuable in stratigraphical correlation especially in the Tertiary.

forams. Short from of foramininfera.

force. _An external agency capable of altering the state of rest or motion in a body; measured in newtons (SI units), dynes (c.g.s. units), or poundals (f.p.s units). The force *F*, required to produce an acceleration, *a* in a mass, *m*, is given by $F=ma$. If *m* is in kilograms, *a* in ms^{-2}, *F* is expressed in newtons.

foredeep. An elongate crustal depression that borders an orogenic belt or island are on the convex side.

foreland. An area marginal to an orogenic belt, toward which folds have been overfolded and thrusts have moved. The side from which the surface rocks have moved in the hinterland.

forelimb. The steeper of the two limbs of an asymmetrical anticline.

fore reef. The seaward side of a reef. Parts of it may be slopes covered with reef debris; other areas may be vertical walls built by deposits of marine organisms.

foreslope. The steeply sloping parts of an organic reef, extending from its outer margin to an arbitary depth of about 10 fathoms.

fork. A place where two or more streams join to form a larger waterway; a confluence.

form. The group of faces generated from an initial face by the operation of the symmetry elements of a particular crystal class. All the faces comprising a form are always present together except where the crystal is crudely-developed. An open form is an assemblage of faces which cannot exist in the absence of a second form. A closed form is an assemblage of faces which can enclose a volume of space.

formation. Any assemblage of rocks which have some common character and are mappable as a unit.

forsterite. Mg_2SiO_4. A mineral of the olivine group.

fortification agate. *See* landscape agate.

fortunite. An igneous rock of the diorite clan; a brownish-grey, fine-grained porphyritic extrusive rock; it contains phenocrysts of enstatite and bronzite with orthopyroxene and clinopyroxene, mica, feldspar and some glass.

foshagite. $Ca_5Si_3O_{10}(OH)_2$. $2H_2O$. A orthorhombic mineral consisting of hydrous calcium silicate having silky lustre. Its hardness on Mohs scale is 3.0.

fosse. A long and narrow depression between a glacier and the sides of its valley.

foshallassite. $Ca_3Si_2O_7.2H_2O$. Hydrated calcium silicate; occurs as scaly, sometimes as spheroidal aggregates.

fossil. The organic remains, traces, or imprint of any organism preserved in the curst of the earth since sometime in the geological past.

fossil, derived. A fossil that has been eroded from rocks that were deposited earlier and has been deposited again in a younger bed. A derived fosil is therefore not of the same age as the rocks in which it is found.

fossil dune. An ancient desert dune.

fossil fuel. Any hydrocarbon deposit that can be used as fuel, *e.g.*, coal, natural gas, petroleum.

fossiliferous. Any substance or deposit containing fossils.

fossil man. Ancient man identified from prehistoric skeletal remains which are archeologically earlier than the Neolithic.

fossil ore. An iron bearing sedimentary deposit in which shell fragments have been replaced and cemented together by hematite and carbonate.

fossil peneplain. A peneplain that was burried by younger sediments and was later exposed by their removal.

fossil record. All the remains of past animal and plant life found in rocks.

fourchite. A monchiquite that lacks feldspar and olivine; it contains about 75 per cent pyroxene as phenocrysts of augite in greatly altered holocrystalline groundmass which may have been analcite, together with accessory iron oxide.

fourmarierite. $PbU_4O_{13}.5H_2O$. Hydrated uranium lead oxide occurring as orthorhombic, tabular and usually elongated and striated crystals.

foyaite. A nepheline syenite; mainly composed of potassium feldspar; it contains orthoclase, nepheline and hornblende with accessory pyroxene mica, iron ore and sodalite nosean, and zircon.

fractional crystallization. Separation of a cooling magma into multiple minerals as the different minerals cool and congeal at progressively lower temperatrures. Also known as fractionation.

fractionation. *See* fractional crystallization.

fracture. (1) Breaking behaviour of mineral, not associated with cleavage directions. Its character is sometimes useful in naming a particular mineral. (2) One of the ways in which rocks yield to deforming movements in the earth's crust; *e.g.*, cracks, joints, faults, or other breaks. (3) Deformation resulting from a momentary loss of resistance to differential stress and a release of internal elastic energy.

fracture cleavage. Cleavage that occurs in deformed but only slightly metamorphosed rocks along closely spaced, parallel joints and fractures.

fracture dome. The zone of loose or semi-loose rock which exists in the immediate hanging or footwall of a slope.

fracture system. A stress related group of contemporaneous fractures.

fracture zone. A zone in which faulting has taken place. It runs parallel to the fault line. It can be the site of sesmic and volcanic activity.

francolite. $Ca_2(PO_4,CO_3)_3(F,OH)$. A colourless fluride-bearing carbonate-apatite.

franklinite. $ZnFe_2O_4$. A black mineral of spinel group, which is slightly magnetic in nature. Sometimes, zinc is replaced by iron or manganese. It forms thick beds in crystalline limestone, sometimes admixed with various amounts of zincite, willemite, rhodonite and tephroite; a member of spinel family.

frasnian. A stratigraphical stage name for the base of the European Upper Devonian.

french chalk. A synonym of steatite.
frenzelite. A synonym of guanajuatite.
friable. Capable of readily crumbled to a powdered form.
friedelite. $6[Mn,Fe)_8Si_6O_{15}(OH,Cl)_{10}]$ Hydrous manganese iron chloride-silicate; occurring as hexagonal, tabular, striated crystals.
frohbergite. $FeTe_2$. A mineral consisting of telluride of iron, and is isomorphous with marcasite.
frondelite. $(Mn,Fe(III)_4(PO_4)_3(OH)_5$. A mineral consisting of basic manganese iron phosphate with proportion of manganese greater than ferric iron, crystals are orthorhombic botryoidal masses and crystals with a radial elongated fibrous or fine-columnar structure occurs. Its colour varies from dark green or olive-green to greenish-black in fresh material, becoming brownish-green or dark reddish-brown on oxidation. It possesses vitreous to dull lustre and shows uneven fracture. It is brittle and shows cleavage is one direction, good in second direction and fair in third direction. Its hardness on Mohs scale is 4.5. It is a secondary mineral found in limonite deposits or as alteration product of manganese-iron phosphates in pegmatites.
front. A metamorphic zone of changing mineralization developed outward from an igneous mass.
frontal moraine. A moraine rempart at the front of a former glacier.
front bay. A large, irregular shallow bay connected with the sea through a pass between barrier islands.
front pinacoid. In an orthorho rhombic monoclinic, or triclinic crystal, the {100} pinacoid.
frost. Granular or flaky deposit of ice crystals caused by sublimation of water vapour on a surface whose temperature is, below the freezing point of water.
frost action. The weathering process caused by cycles of freezing and thawing of water in surface pores, cracks, and other openings.
frost boil. An accumulation of mud and water released from ground ice by accelerated spring thawing.
frost line. The lower limit of permafrost.
frustule. Silica-based shell of a diatom.
fuchsite. $Si_3O_{10}(OH)_2]$ $4[K(Al,Cr)_3$. A mineral of the mica group containing upto 5 per cent chromium; a bright-green variety of muscovite.
fucoid. A tunnel like marking on a sedimentary structure identified as a trace fossil but not referred to a described genus.
fulgurite. A type of natural glass. Thin tubes of glassy rock resulting from fusion of sand or loose or compact rocks; caused by the intense heat set up when a lightning flash strikes the sands of a desert or an exposed crag.
fuller's earth. A clay having a high adsorptive capacity, composed mostly of montmorillonite. It is widely used as a bleaching agent, as a filter, and for removing grease from fabrics It mainly contains montmorillonite with other hydrated aluminium silicates and accessory minerals such as quartz, mica and glauconite.
fumarole. A hole, usually found in volcanic areas, from which volatile vapours or gases escape out. It is found at the surface of lava flows, in the calderas and craters of active volcanoes, and in areas where hot intrusive, igneous rock bodies occur.
fussain. The local lithotype strands or patches, characterized by silky lustre, fibrous structure, and black colour. It is a constituent of bituminous coal; char-

coal like substance consisting of fusinite. It is extremely friable and easily reduced to fine powder. It is formed from alteration of wood in oxidising environment.

fusinite. Major constituent of fusain. It consists of wood of which very little is left but woody conducting cells or thick-walled elements so highly carbonised as to contain only traces of ulmins.

fusiform. Spindle-shaped; elongated and tapering at both ends.

fusinization. The process of formation of fusain in coal.

fusite. A coal microlithotype that contains atleast 95 per cent fusinite; a variety of inertite.

fusovitrian. A coal lithotype transitional between vitrain and fusain, but predominantly vitrain.

fusulinid limetone. A foraminiferal limestone composed mainly of fusulinid tests.

G

gabbro. A group of dark-coloured, intrusive igneous rocks with granular texture, composed largely of basic plagioclase more calcic than anorthite associated with clinopyrosene. Orthopyroxene, olivine, apatite and iron ore may or may not be present. Melanocratic. Its silica content vary from 45 to 55 per cent. The term gabbro generally refers to plutonic rocks of coarse grained, grannular texture, but clan definition covers all modes and textures. Some gabbro contain greenish crystals but most are black speckled with white plagioclase. Layering is common and separate layers of dark pyroxene and light plagioclase give gabbro a banded apperance when seen in outcrop though hand specimens are fairly homogeneous. The concentration of dark minerals makes gabbro a relatively dense rock.

gabbro-norite A synonym of hyperite.

gabbrophyrite. An igneous rock of the gabbro clan; a prophy ritic plutonic rock, containing phenocrysts of basic plagioclase and sometimes pyroxene and/or olivine in fine-grained equigranular holocrystalline groundmass of basic plagioclase, pyroxene and magnetite with accessory biotite, hornblende and iron ores. It is also known as gabbro porphyry.

gabbro-porphyry. See gabbrophyrite

gadolinite. $(Fe,Be)_3(Y_2)\ Si_2O_{10}$. A dark coloured mineral containing rare earths. It has vitreous to greasy lustre and its hardness on Mohs scale is 6.5-7.0.

Principally it is found in pegmatitle venis, often associated with allanite.

gageite. $(Mn,Mg,Zn)_8Si_3\ O_{14}\ (2\text{-}3)\ H_2O)$. Hydrated zinc magnesium manganese silicate with 10 per cent magnesium oxide and 4 per cent zinc oxide. It is mainly associated with zincite, willemite and calcite.

gahnite. $ZnAl_2O_4$. A dark colour mineral containing oxides of aluminium and zinc. It is a member of spinel family. Also known as zine spinel. It occurs.

in zinc deposits, crystalline schists, granite, pegmatites or contact (gap) metamorphised limestones.

varieties are dysluite, kreittonite and limaite.

gahnospinel. $8(Mg.Zn)\ Al_2\ O_4)$ Magnesium–zinc–aluminium oxide. It zinc ricjh variety or spinel. Pale-blue to dark blue gemstone.

gallery. A term used for a horizontal passagewasy in a cave, it is also used for a horizontal conduit or channel constructed to intercept water.

galaxite. $MnAl_2O_4$. A black mineral of the spinel group, contains oxides of manganese and aluminum.

It occurs as vein with alleghanyite, calcite, spessartite., rhodonite and tephoite.

galena. PbS A bluish-grey to lead-grey mineral with brilliant metallic lustre, the most important ore of lead, found in hydrothermal veins and as a replacement mineral. It can be easily recognized by its unusual weight and its dull colour, not unlike that of lead itself.It has a right-angled cleavage; which can be easily broken into tiny cubes and is quite soft.

It occurs in ore veins, igneous and sedimentary rocks and disseminated through sediments and is generally associated with quartz, siderite, dolomite, fluorite, calcite and borite as gangue materials. Also known as lead glance.

galenobismutite. 4[$PbBi_2S_4$] Lead bismuth sulphide with anitmony substituting in small amounts for bismuth. It occurs as. Orthorhombic, Lathlike, elongated, flattened and striated crystals.

gangue. The mineral part of an ore rock which is not the objective of the mining operation in question.i.e. the first mineral waste product.

ganister. Sedimentary rock with high siliceous sandstone. and consists of medium to fine quartz grains cemented together with silica. It is found in compact form with fine and even granular texture.

ganomalite. $Ca_2Pb_3Si_3O_{11}$. A mineral composed of calcium and silicate occurring as granular masses or prismatic crystal; possesses resinous to vitreous lustre. It is very brittle and shows uneven fracture and cleavage in two directions. Its hardness on scale is 3.0. It occurs mainly with calcite and jakobsite and/or manganophyllite and tephorite.

ganophyllite. $Mn_7 Al_2Si_2 O_{26}$. $6H_2O$. A brown coloured mineral composed of manganese aluminosilicate.It is monoclinic and consists of short prismatic crystals. It possesses brilliant lustre and shows perfect basal cleavage. Its hardness on Mohs scale.is 4.0-4.5.

garewaite. Igneous rock of ultrabasic clan. A dark-coloured porphyritic hypabyssal rock; contains phenocrysts of corroded diopside in fine-grained hollo crystalline groundmass of olivine. pyroxene, chromite, and magnetite with accessory spinel and labradorite.

garnet. A generic name for a group of mineral silicates that are isometric in crystallization; have a structure consisting of independent SiO_4 tetrahedra and AlO_6 octahedra, which are linked by cations in 8-coordination. The general formula for the garnets is: $A_3B_2(SiO_4)_3$; where A is Fe^{2+},Mn^{2+},Mg^{2+} or Ca^{2+} And B is Al^{3+} Fe^{3+} or Ti^{3+} Garnet has a hardness of 7 on Mhss scale. It can be transparent or opaque with a vitreous lustre and the most common colour is reddish brown. Other colours are yellow, pink, green, violet and sometimes/black. Its occurrence is quite common in gneisses, shists, crystalline limestones and metamorphosed igneous rocks: sometimes primary minerals in igneous roks, especially syenites. The series comprises grossular, pyrope, almandine, spessartite and andradite(gemstones) and uvarovite (non-gemstone).

garnetoid. Group name for minerals which are not primarily silicates, but with structures similar to garnet. Series includes graphite, berzelite (of kuhn) hydrogarnet, plazolite and hibschite.

garnierite. $(Ni,Mg)_3Si_2O_5(OH)_4$. Nickel-bearing serpentine.It occurs as an alteration product of serpentine chromite or talc; and is a member of serpentine family.Its varieties are genthite and noumeite.

garronite. $Na_2Ca_5Al_{12}Si_{20}O_{64}$.$27H_2$ O. A Zeolite mineral belonging to hillipsite group; crystallizes in the tetragonal system.

gas cap. Free gas acculmulated above an oil reservoir. It forms whenever the available gas exceeds the amount that will dissolve in the oil under the existing conditions of temperature and pressure.

gas coal. Bituminous coal containing a high percentage of volatile, gas-yielding constituents, thus making it suitable for the manufacture of flammable gas.

gaspeite. Na_2CO_3. An anhydrous mineral containing sodium carbonate with calcite structure.

gas oil-ratio. The volume quantity of gas proudced with the oil from a particular oil well, expressed in terms of cubic feet of gas per barrel of oil. Abbrev. GOR.

gas spurt. An accumlation of organic matter on certain strata caused by escaping gas.

gas well. A well capable of producing natural gas, or one that produces cheifly natural gas. In some areas, the term is defined on the basis of gas-oil ratio.

gas zone. A rock formation containing gas under a pressure large enough to force the gas out if tapped from the surface.

gaussbergite. Igneous rock of the syenite clan. A porphyritic extrusive rock; contains phenocrysts of leucite. augite and olivine in glassy groundmass potentially sanidine.

gauteite. Igneous rock of the diorite clan. A light grey to greenish- grey with brown weathered crusts porphyritic hypabyssal rock; contains phenocrysts of hornblende, augite, occasional biotite and lime-soda feldspar in groundmass of mainly feldspar strips amounts of magnetite, augite, plus small hornblende, and, biotite and commonly a little analcite.

gay-lussite. $Na_2Ca(CO_3)_2 5H_2O$. A mineral composed of hydrous sodium calcium carbonate; has monoclinic crystals often elongated or flattened and wedgeshaped, usually rough and striated surfaces. It is very brittle and possesses conchoidal fracture. Its hardness on scale.

is 2.5-3.O.

gedanite. A brittle wine-yellow variety of amber, containing small amounts of succinic acid.

gedrite. $4[(Mg,Fe(II),Al)_7 (Si,Al)_8. O_{22} (OH)_2]$. An aluminous variety of the mineral anthophyllite having orthorhombic structure.

gehlenite. $4[Ca_2Al_2SiO_7]$. A mineral composed of calcium aluminium silicate with ferric iron replacing aluminium and magnesium replacing calcium. It is brittle and shows inperfect cleavage in one direction. It hardness on Mohs scale is 5.5-6.5. It is a member of melilite group.

Geiklielite. $2[MgTiO_3]$. A mineral composed of magnesium titanate; shows rhombohedral cleavage. It is a member of ilmenite series with magnesium replacing iron.

gelatinisation. Formation of a gelatin, or colloid, as the result of chemical action.

gem. A general term for a precious or semiprecious stone, particularly after it has been cut and polished. It also applies to a particular variety of a mineral that is worn or used decoratively.

gemology. The scientific study concerned with the identification, evaluation, fashioning, and other aspects of gemstones.

genotype. The type species of a genus.

genus A category in plant and animal classification between family and species.

geo. (1) A narrow coastal inlet bordered by steep cliffs. (2) A prefix denoting earth.

geocentric. Related to the earth as a center, or measured from earth's center.

geochemical anomaly. Above average occurrence of a chemical species in a sample of rock, soil, vegetation, sediment–or stream; indicates the deposits in nearby surroundings.

geochemical cycle. A path followed by an individual element or group of elements in the crustal and subcrustal region of earth. During the cycle, there occur processes of both separation of elements and elemental recombination.

geochemical evolution. A change in any constituent of a rock beyond that amount present in the parent rock.

geochemical prospecting. The use of geochemical and biogeochemical principles and data in the search of economic deposits of minerals, petroleum′ and natural gases.

geochemical facies. The region characterized by particular physicochemical conditions that affect the production and accumulation of sediments.

geochemistry. The study of chemical composition of the various phases of the earth and the physical and chemical processes which have produced the observed distribution of the elements and nuclides in these phases.

geochorn. An interval of geological time that corresponds to a lithostratigraphic unit.

geocronite. [$Pb_{27}(As,Sb)_{12}S_{46}$] Lead arsenide antimonide-sulphide with ratio of arsenic to antimonide from 1: 2to 1: 1

geochronology. Measurement of time intervals of the geological past, accomplished in terms of 'absolute age dating by a suitable radiometric procedure′ *e.g.*, carbon-14 dating, realtive age dating by means of sequences of rock strata or varved deposit analysis.

geochronometry. The study of the absolute age of the rocks of the earth based on the radioactive decay of isotopes, such as uranium-238. uranium-235, thorium-232, potassium-40′ and carbon-14′ present in the rocks and minerals.

geocosmogony. The scientific study of the origin of the earth.

geocosmology. The discipline that deals with the origin and geological evolution of the earth, its planetary features, and the influence of the solar system, galaxy, and the universe on the geological development of the earth.

geode. An almost spheroidal, hollow body, lined inside with inward-projecting small crystals; found frequently in limestone beds but may occur in shale.

geodesic line. A line of shortest distance between any two points on any mathematically defined surface.

geodesy. (1) The science concerned with determination of size and shape of the earth and precise location of points on its surface, (2) The determination of gravitational field of the earth and the study of temporal variations such as earth tides, polar motion and rotation.

geodetic coordinates. Values that specify the horizontal position of a point on a reference ellipsoid with respect to a particular geodetic datum; generally expressed as longitude and latitude.

geodynamics. The branch of science that deals with the forces, and processes of the earth's interior.

geognosy. The scientific study dealing with the solid body the earth as a whole, occurrences of minerals and rocks, and the origin of these and their relations.

geological survey. The geological investigations of an area. It is generally in the form of a geological map and sections, and a report.

geological thermometry. A term applied to a group of techniques designed to determine the temperatures at which various geological processes took place.In order to determine temperatures of subsurface events, e.g., thermal metamorphism, crystalization of granite, deposition of ore veins, indirect techniques must be employed.

geologic time. A term encompassing the time extending from the end of the formative period of the earth (*i,e.*, after it was "established" as a planet) to the onset of human history; it is the part of earth's history that is recorded and described in the succession of rocks. It is subdivided into a number of systems,incorporating all the rocks developed on the earth since its formation approximately 4000 million years ago.

These systems are:

pre-cambrian. From the formation of the earth's crust to approximately 550 million years ago. Areas of metamorphic rocks greatly affected by orogenesis, forming the shield areas of Fennoscandia and North America and many minor areas. Remains of life very sparse.

cambrian. 550 million to 500 million years ago. Mainly sedimentary rocks, often later metamorphosed. Abundant marine invertebrates.

ordovician. 500 million to 435 million years ago. Extensive sedimentary and extrusive igneous rocks. Often later metamorphosed. Characterised' by graptolite fauna.

silurian. 435 million to 395 million years ago. Various rock types developed. Extensive marine invertebrate fauna.

devonian. 3 95 million to 345 million years ago. Extensive continental (Old Red Sandstone) and marine sediments developed; also extrusive igneous activity. Fish fauna characterises the Old Red Sand- stone.

carboniferous. 345 million to 280 million years ago. Extensive marine limestones and deltaic/freshwater deposits developed, the latter being characterised by coal deposits.

permian. 280 million to 225 million years ago. Mainly continental desert deposits associated with igneous activity. Fauna and flora very varied.

triassic. 225 million to 195 million years ago. Mainly continental deposits with later marine sediments. Fauna and flora very sparse.

jurassic. 195 million to 135 million years ago. Charcterised by extensive marine and continental sediments with ammonite fauna and dinosaur remains.

cretaceous. 135 milloin to 64 million years ago. Mainly marine deposits including extensive limestones (chalk) with very varied fauna.

eocene, oligocene. miocene and pliocene. 64 million to one million years ago. Sedimentary and igneous rocks developed. Fauna and flora similar to present day, Shwoing the evolution of the mammals.

pleistocene. one million years ago to the present day, deposits representative of the Ice Age.

holocene- recent post-glacial deposits.

These systems may be simplified by grouping into major eras of geological time. These are:

Azoic- The Pre-Cambrain

Palaozoic- the Cambrian. Ordovician. Silurian. Devonian. Carboniferous and permian.

Mesozoic- the Triassic Jurassic and Cretaceous.

Cainozoic- the Eocene. Oligocene. Miocene. Pliocene.Pleistocene and Holocene.

The Cainozoic era is further subdivided into: Tertiary the Eocene. Oligocene. Mio- cene and Pliocene.

Quaternary- the Pleistocene and Holocene.

geological transportation. Shifting of material by the action of moving air, ice, or water.

geologic column. (1) The vertical sequence of the strata of various ages found in an area or region. (2) Geologic time scale as represented by rocks.

geologic hazard. A geologic phenomenon or condition that presents a risk or is a potential danger to life and property, either naturally occurring or man-made.

geologic high. An oil-field term for a structure on which rocks occur at a higher position than in the surrounding area.

geologic log. A graphic representation of the lithologic or stratigraphic units of both traversed by a borehole.

geologic section. Any succession or rock units found at the surface or below ground in an area.

geology. The study of the earth in terms of its development as a planet since its origin. This includes the history of its life forms, the materials of which it is made, processes that affect these materials and products that are formed of them.

geomagnetic. Pertaining to the magnetic field of the earth.

geomagnetic reversal. A change of the earth's magnetic field between normal polarity and reversed polarity.

geomechanics. The branch of geology dealing with the response of earth's materials to the application of deforming forces and embracing the fundamentals of structural geology.

geomorphic. (1) Pertaining to the earth or of its surface features. (2) Pertaining to geomorphology.

geomorphic cycle. The cycle of the change in the surface configuration of the earth. Also known as cycle of erosion, geographical cycle.

geomorphology. (1) The study of the origin of secondary topographic features which are carved by erosion in the primary elements and built up of the erosional debris. (2) The description and interpretation of land forms.

geopetal. Pertaining to top-to-bottom relations in rocks at the time of formation.

geophysical prospecting. The study of variations in the values of the physical parameters of the crust of the earth with the object of gaining information, for economic purposes, about sub-surface structure or the presence of mineral bodies.

geophysic The branch of physics that deals with the study of the earth and its environment, earth, air, and space. Its branche include the studies: seismology, the behaviour of earthquakes, and the shock waves which they produce; the gravity field and rotation of the earth; the magnetic field of the earth;temperature gradients and rate of heat flow; vulcanology-the behaviour of volcanoes and their products; natural radio-activity; oceanography; meterology and climatology.

geosere. A series of ecological climax communities following each other in geologic time and changing in response to changing climate and physical conditions.

geostrophic motion. Fluid flow that is parallel to the isobars (lines of equal pressure) in a rotating system; e.g. the earth.

geosphere. (1) The solid mass of the earth, as distinct form the atmosphere and hydrosphere. (2) The lithospherse, hydrosphere, and atmosphere combined.

geostatic pressure. The vertical pressure at a point in the earth's crust caused by the weight of the overlyig rock.

geosuture. (1) a boundary zone between contrasting tectonic units of the earth's crust; in many places a fault which probably extends through the entire thickness of the earth. (2) A place where two continents have come together.

geosynclinal. The original now obsolte term for geosyncline.

geosynclinal prism. The load of sediments that accumulates often to great thicknesses in the downward part of a geosyncline, having a slope similar to that of a long plano-convex prism whose convexity is at the floor.

geosyncline. A basin-like or elongate subsidence of the earth's crust. Its length may extend for several thousand kilometers and may contain sediments thousands of meters thick, representing millions of years of deposition. A geosyncline generally forms along continental edges, and is destroyed during periods of crustal deformation, which often produces folded mountain ranges.

geotaxis. The movement of an entire plant in response to gravity

geothermal. Pertaining to the heat of the interior of the earth.

geothermal energy. The energy that can be recovered from the earth's internal heat.

geothermal gradient. That rate of increase of temperature in the earth with depth. It differs from place to place depending on the heat flow in the region and the thermal conductivity of the rocks.

geothermal heat flow. The quantity of heat energy leaving the earth, measured in calories/cm^2/sec. Heat flow measurements in igneous rocks have shown a linear correlation between surface heat-flow and heat produced in the rocks. The heat is due to the decay of uranium, thorium, and

gerhardtite. $Cu_2NO_3(OH)_3$ Basic copper nitrate; orthorhombic, thick tabular crystals. The pyramid zone strongly striated and faces often in oscillatory combination.

germamite.8[$Cu_3Ge,Ga,Fe,Zn)(AS.S)_4$] Copper germanium sulphide with germanium from 6-10 per cent and zinc, gallium. arsenic and ferrous iron in smaller amounts .

gersdorffite. 4[NiAsS]. Nickel sulphide arsenide Cubic, octahedral. cubo octahedral pyritohedral, sometimes striated crystals.

geyser. A hot spring that inermittently ejects sprays and jets of hot water and steam. Eruption is caused by the contact between groundwater and rock that is hot enough to create steam; the intermittent activity is due to the increase in the boiling point of water with pressure.

geyser basin. A valley that contains numerous springs, geysers' and steaming fissures fed by the same groundwater flow.

geyserite. A variety of siliceous sinter; porous, stalactite filamentous or cauliflower-like; also massive compact or scaly. Its hardness on Mohs scale is 5.0.

ghizite. Igneous rock of the diorite clan. A dark grey compact, dense aphinitic extrusive rock; contains large phenocrysts of biotite and smaller ones of augite, olivine, analcite and magnetite in a mesostasis of abundent glass.

giannettite. $cNa_3Ca_3\ Mn(Zr,Fe)TiSi_6\ O_{21}\ Cl$ Calcium sodium manganese chloro-zircono-titanosilicate occurs as triclinic,minute crystals.

gibbsite. $Al(OH)_3$. A white mineral containing aluminum hydroxide; crystallizes in monoclinic system; a main constituent of bauxite.It occurs as secondary product from alteration aluminous minerals found chiefly in bauxite deposits.

gibelite. Igneous rock of the syenite clan. A light grey porphyritic extrusive rock; contains large and abundant phenocrysts of soda-microcline, sometimes, microperthite and a few small prisms of augite in groundmass of soda-microcline and accessory augite and amphibole.

gieseckite. Magnesium potassium aluminosilicate. sometimes with appreciable ferrous oxide. Its formula is uncertain. It is an alteration product of nepheline differing in that it contains water.

gilsonite. A variety of natural asphalt.

gipping A stratigraphic stage name for the base of the British Upper Pleistocene.

gismondine. $8[Ca\ Al_2Si_2O_8.\ 4H_2O]$. A mineral composed of hydrated calcium aluminium silicate with some replacement of calcium by potassium; consists of octahedral crystals with terminal angle face rough and composite. Its colour varies from white to bluish-white, and greyish to brownish red. Its hardness on Mohs scale is 4.5.

giumarrite Igneous rock. A variety to monochiquite; contains much amphibole.

glabella. The bony prominence on the frontal bone joining the supraorbital ridges.

glacial. (1) Pertaining to an interval of geologic time which was marked by an equatorward advance of ice during an ice age; the opposite of interglacial; these intervals are variously called glacial periods, glacial epoces, glacial stages and so on. (2) Pertaining to ice, especially in great masses such as sheets of land ice or glaciers.

glacial abrasion. Alterations of portions of the earth's surface as a result of glacial flow.

glacial accretion. Deposition of material as a result of glacial flow.

glacial advance. Increase in the thickness and area of a glacier.

glacial boulder. A large fragment of rock that has been transported by a glacier.

glacial cycle. (1) A major global climatic oscillation of the order of 100,000 years developed with in an ice age. (2) The ideal case of glaciation continuing so long under fixed climatic conditions that glacial erosion would be complete and replaced by normal erosion.

glacial erosion. Reduction of the earth's surface as result of grinding and scouring by glacier ice armed with rock fragments, together with the erosive action of meltwater streams.

glacial geology. The study of the features and effects resulting from glacial erosion and deposition by glaciers and ice sheets. The term also applies to the features of a region that has been subjected to glaciation. Syn: glaciogeology. Cf: glaciology.

glacial lake. (1) A lake fed by meltwater or forming on glacier ice because of differential melting (2) A lake confined by morainal dam.

glacial lobe. A large tongue-like protrusion from the margin of an ice sheet.

glacial period. Generally, an interval of geologic time which was marked by a major equatorward advance of ice.

glacial plucking. An erosion process whereby fragments of bedrock are dislodged and transported by glaciers. The fragments may be plucked when meltwater in the bedrock joints refreezes and the resulting expansion forces blocks out of their places. The fragments may also be pried loose by the plastic flow of ice around blocks, which become part of the moving glacier.

glacial retreat. A condition in which backward melting at the front of a glacier takes place at at a rate exceeding forward motion.

glacial scour. Erosion taking place as a result of glacial action, whereby the surface material is removed and the rock fragments carried by the rock. Also known as scouring.

glaciation. Alteraticn of any part of the earth's surface by passage of a glacier, chiefly by glacial erosion or deposition.

glacier. A large mass perennial ice showing evidence of present or past flow; it occurs where winter snowfall exceeds summer snow melt. Most of the ice of glaciers originates as fresh, light snowflakes, but it can also derive from rime ice and supercooled water vapour. The first requirement of glacier formation is the accumulation of newly fallen snow in hollows above the snowline.

glacier flood. A sudden release of meltwater from a glacier or glacier-dammed lake formed by the melting of a drainage channel or by subglacial volcanic activity. It may result in a catastrophic flood.

glacier flow. The slow downward or outward movement of the ice in a glacier due to the action of gravity. Deformation within the ice, by intragranular gliding, grain-boundary migration and recrystallization is involved, usually with sliding of the glacier on its bed. It is usually expressed in metres per unit time.

glacierization. Covering of a land surface by glaciers or ice sheets.

glacier milk. A stream of turbid, whitish melt-water containing rock flour in suspension.

glacier table. A block of stone supported by an ice pedestal above the suface of glacier.

glaciofluvial. Pertaining to streams fed by melting glaciers or to the deposits and landforms produced by such streams.

glaciolacustrine. Pertaining to lakes fed by melting glaciers or to the deposits forming there in.

glaciology. The study of glaciers and of ice in all of its natual occurrences, including the formation and distribution of ice, its physical and chemical properties, glacier flow and the interrelation between ice and climate.

gladkaite. Igneous rock of the diorite clan. A fine-grained hypabyssal rock; contains plagioclase(oligoclase to andesine) abundant quartz, hornblende and a little biotite with accessory magnetite and apatite and secondary muscovite and epidote.

glance pitch. A variety of asphaltite

having brilliant conchoidal fractuer and resembling gilsonite but having higher specifice gravity and percentage of fixed carbon.

glaserite.c[$K_3Na(SO_4)_2$] Sodium potassium sulphate. Variety of aphthitalite, the pure end member with 3:1 ratio of potassium to sodium.

glass. Igneous rock constituent, the cryptocryst alline or amorphous material which represents the final phase of magma consolidation occurs as part of the ground-mass; may also form extrusive igneous rocks when rapid chilling occurs.

glassy. Glassy or vitreous, substances are solids with absolutely no crystalline structure. Rare geologically but can be formed by a variety of processes

glassy tuff. Consolidated volcanic dust with porous texture. When unaltered has shreds of glass in interstitial mass of fine glass dust iron oxide and secondary chalcedony or hydrous silica. Usually altered by solution of substance or addition of material with cementation and consolidation glass usually becomes devitrified, rocks can become hard and dense resembling hornfels, minerals can be introduced or secondary minerals can form.

glaucochroite. Ca $MnSiO_4$. A bluish-green mineral composed of calcium manganese silicate; occurs in prismatic crystals.

glucodot. 8(Co,Fe)A_5.S. Cobalt-iron arsenide sulphide with ratio of cobalt to iron from 0.5-0.6.

glauconite. $K_{15}(Fe,Mg,Al)_{4\text{-}6}(Si,Al)_8 O_{20}(OH)_4$.. A type of clay mineral of mica group, it is dioctohedral and occurs in flakes, usually found as rounded grains in sedimentary rocks or coating or pebbles. It is dark green, opaque and dull and may be so abundant that its colour is rock green forming a variety of sandstone known as greensand. The presence of glauconite in a sedimentary rock suggests that it was deposited under the sea.

glauconitic sandstone. A quartz sandstone or an arkosic sandstone that has many glauconite grains.

glaucophane.$Na_2(Mg,Fe(II)_3(Al,Fe(III))_2\ Si_8O_{22}(OH,F)_2$. A mineral of the amphibole group found in metamorphic rocks(*e.g.*, glaucophane schists) and in igneous rocks, particularly soda-rich granites.

glaucophane schists. High grade regional metamorphic rocks consist of glaucophane with lawsonite, jadeite-quartz, aegirine and pumpellyite. Ordinary schistose texture.

glenmuirite. An igneous rock of the diorite clan. A mesocratic, coarse to medium-grained plutonic rock. Contains orthoclase, labradorite, analcite, augite and olivine with accessory apatite, biotite and iron ore.

glessite. Fossil resin similar to amber.

gley. A sticky subsurface layer of clay in some waterlogged soils.

gliding. A deformation-related movement that takes place along lattice planes in crystalline substances. Gliding is of two types: translation gliding and twin gliding. In translation gliding, displacement takes place along preferred lattice planes without reorientation or rupture of the deformed parts. In twin gliding, the lattice of the displaced part of the crystal is symmetrically altered with respect to the lower, undisplaced part.

glimmergabbro. Igneous rock of the gabbro clan. A coarse grained plutonic rock; consists of usual gabbroic composition with addition of biotite and optional accessories quartz, orthoclase, hornblende, magnetite, apatite and pyrite.

glimmerite. An ultrabasic rock consisting almost entirely of the micas phlogopite and/or biotite.

global tectonics. Tectonics on a global scale, such as tectonic processes associated with the very large-scale movement of material.

globulite. A small isotropic, globular of spherules like crystallite of dark colour; found in glassy extrusive rocks.

glomeroporphyritic. A porphyritic texture in which groups of crystals (either all of the same mineral or of different minerals) are scattered thorugh a finer-grained ground mass. Also known as cumulophyric.

gloup. An opening in the roof of a sea cave.

gmelinite.$(Na_2Ca)Al_2Si_4O_{12}.6H_2O$. A mineral of zeolite group that is generally colourless and crystallizes in the hexagonal system. It is brittle ; shows uneven fracture and cleavage easy in one direction. Its hardness on Mhos scale is 4.5.

gneiss. A foliated metamorphic rock commonly derived from granite, shale, diorite, mica schist, rhyolite and other rocks by regional metamorphism associated with moderate to high temperatures and pressures. Gneiss is commonly composed of quartz and feldspar with lesser amounts of mica; however, the term gneiss refers more to the particular texture, characteristic minerals present, and origin of a given rock.

gneissic granite. A variety of granite with a gneissic structure.

gneissic granodiorite. Any of the granodiorite rock with gneissic characteristics.

gneissoid granite. An igneous rock of the granite clan; a granitic gneissoid textured plutonic rock. It constituents are bands or parallel arrangements; it possesses gneissoid structure due to constrained movements of a viscous magma during crystallisation.

gobi. sedimentary deposits in a synclinal basin.

goethite. FeO(OH). A yellow, red or dark-brown mineral crystallizing in the orthorhombic system. It is distinguished from limonite by the fact that it is crystalline but is very variable in form and visible crystals are uncommon. Its hardness on Mhos scale is 5.0-5.5. It occurs in velvety masses of yellow-brown acicular crystals and is often found as botryoidal aggregates and stalactites which when broken show radiating fibres and flat plates. Also known as xanthosiderite.

gondite. High grade regional metamorphic rock. A fine-grained granular rock; contains spessartite, quartz and apatite.

gondwanaland. The ancient continent that is supposed to have fragmented and drifted apart to form eventually the present continents.

It is the name given to the hypo thetical southern hemisphere 'super-continent' consisting of South America, Africa, Madagascar, India, Arabia, Malaya and the East Indies, New Guinea, Australia and Antarctica, prior to its break-up under the forces causing continental drift. It corresponds to Laurasia in the northern hemisphere.

gondwana series. Stratigraphic name for the Indian equivalent of the Karroo Series.

gonnardite.$Na_2CaAl_4Si_6O_{20}.7H_2O$. A

zeolite mineral occurring in fibrous, radiating spheurles.

goniometer. A device used to measure the angles between crystal faces.

gonyerite.$[(Mn,Mg,Fe(III)]_{5-91}[(Si, Fe(III),Al)]_4(O,OH)_{18}$. A mineral composed of hydrous magnesium manganese, aluminium, ferric and ferrous iron silicate. It is rich in manganese oxide (33.83 per cent) and poor in aluminium oxide (0.58 per cent).

gooderite. Igneous rock of the diorite clan. A light grey; mineral, occasionally spotted with pink medium to fine-grained plutonic rock; contains mainly albite with biotite nepheline microperthite and calcite.

gondite. High grade regional metamorphic rock; a fine-grained granular rock; contains spessartite, quartz and apatite.

gonnardite.$c2[Na_2CaAl_6Si_6O_{20}.7H_2O]$. Hydrated sodium calcium aluminium silicate., some replacement of sodium and silicon content with calcium and aluminium.

gorceixite. $39[Al_3,P_2O_7(OH)_7]$. A mineral composed of basic hydrated aluminium phosphate; occurs as grains and pebbles in part microcrystalline Its hardness on Mhos scale is 6.0

gordonite. $MgAl_2(PO_4)_2(OH)_2\ 8H_2O$. A colourless mineral composed of the basic phosphates of magnesium and aluminum.

gordumite. Igneous rock of the ultrabasic clan. A coarse-grained porphyritic plutonic rock, massive with indistinct schistosity contains pseudo-phenocrysts of accessory picotite magnetite and rutile needles.

gorge. A narrow passage between mountains or the wall of canyon, espcially one with steep and rocky walls.

goshenite. Variety of beryl. Colourless or white. Rare. Gemstone use.

goslarite $4[ZnSO_4.7H_2O]$ Hydrated zinc sulphate. Orthorhombic. Crystals rare, normally effilorescent crusts or stalactitic masses.

gossan.A rusty, ferruginous deposit filling the upper regions of mineral veins and overlying a sulphide deposit. It is formed by the oxidation of pyrites. This process occasionally concentrates highly insoluble elements, such as gold in the gossan.

grade. The slope of the bed of a stream or of a surface over which water flows, upon which the current can just transport its load without either eroding or depositing.

graben. A crustal unit or block, generally elongate, that has been depressed relative to the blocks on either side.

graded bedding. A stratification in which each stratum displays a gradation in the size of grains from coarse below to fine above.

graded sediment. A sedimentary unit which displays a sorting effect with the 'coarsest' material at the base and the 'finest' material at the top.

'Coarse' and 'fine' in the sense may be fine sand grading up into clay, or large pebbles grading up into coarse sand. The phenomenon can be used as a method of determining the orientation.

grade scale. A continuous scale of particle sizes divided into a series of size classes.

grahamite. Ablack hydrocarbon mass occurring in veinlike masses; soluble in chloroform and carbon disulphide.

grain. The discrete crystals or particles that make up a rock or sediment.

grain size. Average size of mineral particles composing a rock or sediment. It may be approximated to following scale:

microscopic-less than 0.1 mm	
fine-	0.1mm-1.0mm
medium-	5 mm
coarse-more than	5mm

Scale of grains of particle size in sedimentary rocks:

clay-less than	0.005 mm
slit-	0.005 mm-0.05 mm
sand-	0.05 mm-4 mm
gravel-	2 mm-4mm
pebble-	4 mm-65 mm
cobble-	65 mm-250 mm
boulder-more than	250 mm

granite. A coarse-grained crystalline plutonic rock with granular texture; consisting essentially of quartz(20-40%), alkali feldspar and very commonly a mica, biotite and/or muscovite.

granite aplite. Type of aplite with a granitic composition.

granite clan. Igneous rocks containing at last 66 per cent silica and an average of 30 per cent quartz. Also contain alkali-feldspar and plagioclase with numerous ferromagnesian minerals. Subdivided using feldspar composition add environment of formation. Alkali-feldspar over 60 per cent of total feldspar: granite (plutonic), microgranite (hypabyssal) and rhyolite (extrusive).

Alkali-feldspar and plagioclase in equal amount adamellite (plutonic) microadamellite (hypabyssal) and toscanite (extrusive Plagioclase over 60 per cent of total feldspar: granodiodite (plutonic),microgranodiorite (hypabyssal) and dacite (extrusive)

granite-gneiss. A banded metamorphic rock derived from igneous or sedimentary rocks; mineralogically equivalent to granite.

granite greisen. An Igneous rock of the granite clan; a coarse grained granular plutonic rock. It contains quartz(50 per cent). orthoclase, microcline, oligoclase and muscovite.

granite pegmatite. A variety of pegmatite with a granitic composition.

granite porphyry. A hypabyssal rock characterized by sparse phenocrysts of mica, amphibole, or pyroxene in a medium-to fine-grained groundmass.

granitization. (1) A process wherby various types of rock may be converted to granite or closely related material. (2) The name given to a group of theories which have as a common basis the concept that pre-existing rock can be transformed to granite or granite magma by the action of energetic fluids of appropriate composition arising from the depths.

granitite. Igneous rock of the granite clan. A coarse-grained granular plutonic rock; contains quartz, orthoclase, oligoclase and biotite.

granoblastic. A term for an isotropic fabric consisting of a mosaic of equidimensional anhedral grains. If inequant grains are present, they are randomly arrayed.

granodiorite. Igneous rock of the granite clan. A coarse grained granular plutonic rock. Contains quartz, plagioclase, orthoclase, biotite hornblende and pyroxene.

granogabbro. Type of granodiorite containing plagioclase in the range labrodorite bytowrite.

granophyre. Igneous rock of the granite clan. A fine-grained porphyritic hypabyssal or extrusive rock; contains phenocrysts of feldspar, augite and biotite in a groundmass of intergrown quartz and alkali-feldspar.

granophyric texture. Small-scale graphic texture.

granular texture. A rock texture characterized by the aggregation of mineral grains of approximately equal size. The term may be applied to any type of rock, but is esp. applied to an equigranular, holocrystalline igneous rock whose particles range from 0.05 to 10 mm in diameter.

granule. A rock fragment of diameter 2 to 4mm; larger than a coarse sand grain and smaller than a pebble.

granulite A metamorphic rock composed of equal-sized, interlocking grains. Also, a coarse granular metamorphic rock of the granulite facies.

granulitefacies. The metammorphic facies in which basic rocks are represented by diopside+hypersthene+plagioclase. It is characteristic of deepseated regional dynamothermal metamorphism at temperatures exceeding 920k.

grapestone. A cluster of sand-size trains, such as calcareous pellets, held together by incipient cementation shortly after deposition; the outer surface in lumpy, resembling the bunch of grapes.

graphic texture. In certain pegmatites, granties and microgranites, and more rarely in other rocks, there occurs an intergrowth of quartz an alkali feldspar which present on flat surfaces the appearance of Hebrew, cuneiform, runic or hieroglyphic writing. This results from the simultaneous crystallization of the two minerals, in many cases (but not always) from a eutectic mixture. Smallscale graphic texture is called micrographic or granophric texture.

graphite. A soft black allotorpic form of native carbon, C (hexagonal), found in metamorphic rocks, crystalline limestones, igneous rocks, veins and pegmatites. It is a lead-grey, opaque, dull to submetallic mineral with a black streak. It occurs usually as scales or columner masses and in rare case as hexagonal crystals with perfect basal cleavage. Its hardness on Mhos scale is 1-2, and is sectile and light in weight. It can be recognized by its softness, feel and weight and is distinguished from molybdnite by its black streak.

Graphite is usually found in metamorphosed sediments which originally contained a high proportion of organic material such as plant remains. Hexagonal crystals occur in some marbles; and graphite deposits in some areas are the result of contact metamorphism of coal. Also known as blacklead; plumbago.

graptolite shale. Shale containing a abundance of extinct colonial marine organisms known as graptolites.

graptolithina. The graptolites. Extinct branching organism that lived closely together. They are found in the rocks of Lower Ordovicain to Lower Devonian ages.

gratonite.$Pb_9 As_4 S_{15}$. A grey mineral composed of lead arsenic sulphide; occurs as prismatic crystals as well as massive; brittle fracture and shows no cleavage. Its hardness on Mohs scale in 2.5.

gravel. A loose or uncosolidaed deposit of rounded pebbles, cobbles, or boulders. This term is used in geology with a precise meaning of a grain size, and not implying lossely compacted coarse sediment. In general, the term is applied to grains larger, than coarse sand and finer than pebbles, *i,e.*, 2 mm to 4 mm.

gravitation The mutual attraction between all bodies in the universe.

gravitational collapse. The implosion of star or other astronomical body from an initial size to a size hundreds or thousands times smaller.

gravitational constant. The fundamental constant that apperas in Newton's law of gravitation. It has the value 6.664 10^{-11} $Nm^2\ kg^{-2.}$

gravitational displacement The gravitational field strength times the grvitational constant. Also known as gravitational flux denstity.

gravitational field . The region in space in which a test body would experience a gravitational force; quantitatively the gravitational force per unit mass on the body at a particular point.

gravitational force The force on a body due to its gravitational attraction on other bodies.

gravitational interaction. [Phy] The interaction between all massive particles. It is the weakest of all known interactions, being some 10^{40} th times weaker than the electromagnetic interaction.

gravitational law. [Mech] Every particle in the universe attracts every other particles with a force directly proportional to the product of the masses of particles and the invershy proportional to the particles and square of the distance between them. Thus, the force of attraction between two masses m_1 and m_2 separated by distance of d is given by

$$F=Gm_1m_2/d,^2$$

where G is the gravitational constant.

gravitational mass. [Phy] The mass of a body as it determines the force it experiences in a gravitational field; equal to inertial mass according to the equivalence princple.

gravity. [Phy] The gravitational attraction at the surface of a planet or other celestial body. As gravity is proporitional to the mass of the body and inversely poroportional to the square of the distance from its centre, the gravity on a planet or satellite in teams of the earth's gravity is given by.

$$(dp/de)^2\ Mp,$$

where Mp is the mass of the planet in earth masses, and dp and de are the diameters of the planet and earth respectively.

gravity anomaly. A deviation of the gravity value from the theoretical. value, after a correction has been applied to the point of the earth's surface where the measurement was made. Gravity anomalies are the result of the earth's noniform density.

gravity separation. Separation of immiscible phases by allowing the denser phase to settle down under the influence of gravity.

gravity sliding. Also called gravity gliding and gravitaional gliding, this is a downslope movement of a portion of the earth's crust, usually on the flanks of an uplifted region. This mechanism, activated by gravitational forces, seems to be responsible for many shallow faults. It is also thought to be closely associated with many major mountain systems such as the Alpine belt.

gravity tectonics. Tectonics in which the primary activating mechanism is believed to be downslope sliding due to gravitational effect.

graywacke. An argillaceous sandstone characeterized by an abundance of unstable mineral and rock fragments and a finegrained clay matrix binding the larger, sand-size detrital fragments. Grraywackes often have irgularities on the bedding planes which are known as sole structures are scour marks hollows dug into a bed by the current which deposited the overlying layer. Some graywackes were deposited by turbidity currents on the sea floor and contain occasional shells and other marine fossils. Also spelled as greywacke.

greenalite. A hydrous iron silicate which is an important ore of iron

green mud. Oceanic argillaceous deposite containing detrital material enriched with glauconite and choleite.

greenockite. 2[CdS]. A mineral composed of cadmium sulphide consists of pyramidal crystals, also occurs as earthy masses. It shows good cleavage in one direction and has hardness of 3.0-35 Mohs scale.

greenovite. A red of pinkish variety of sphene containing manganese dioxide.

greensand. (1) Sandstone composed of greensand with little or no cement. (2) A lithostratigraphic unit of the creataceous in Britain and north-west Europe.

greenschist. A schistose metamorphic rock with abundant chlorite epidote or actionlite present, imparting it a green colour.

greenstone. A field terms for any slightly altered basic igneous rock, *e.g.*, diabase epidiorite etc.

greenstone schist. Greenstone with a foliated structure.

greisen. A rock formed granite by the action to fluorine rich gases. The feldspar are altered to muscovite and fluorine rich minerals are formed produing a crumbly white or grey rock consiting essentially of muscovite and quartz with topaz and fluorite.

Grenville. A stratigratphic division of the Canadian Precambrain Lower Ontarian.

greywacke. *See graywacke.*

grike. A cleft in the bare limestone pavements of karst scenery. Also known as gryke.

griquaite. A hypabyssal rock that contains garnet and diopside and sometimes olivine or phlogopite and is found in kimberlite pipes and dikes.

grit. Type of coarse-grained sandstone with angular grains.

gronlandite. Igneous rock of ultrabasic clan. A coarse-grained granular plutonic rock; contains hornblende and hypersthene.

groove. Glaciated marks of large size on rocks.

groove casts. Rounded or sharp, crested, rectilinear ridges that are a few millimeters high and a few centimeters long; found on the undersurfaces of sandstone layers lying on mudstone.

grorudite. Igneous rock of the granite clan. A fine to medium-grained porphyrite hypabyssal rock. Contains phenocrysts of microcline, microcline-microperthite, aegirine and hornblende in a groundmass of quartz, microcline and aegirine.

ground morine. Rock material carried and deposited in the base of a glacier.

groundmass. The fine grained material that holds the phenocrysts in a porphyritic igneous rock. Syn: matrix. Also used for the matrix of sedimentary rock.

groundwater. (1) Subsurface water in the zone of saturation, including water below the water table and water occupying cavities, pores, and openings in underlying rocks. Syn: phreatic water. (2) In an unrestricted sense, all subsurface water as distinct from surface water.

groutite. 4 [MnO,OH] A black coloured mineral composed of basic manganese oxide; shows perfect cleavage.

grunerite. $(Mg,Fe)_7Si_8O_{22}$ $(OH)_2$. A monoclinic mineral of the amphibole group; forms monoclinic crystals.

gruss. A loose accumulation fragmental products formed as a result of the weathering of granite. Also spelled grus.

guano. Phosphatic deposit. Greyish brown. Contains calcium phosphate and various nitrates and carbonates. Compact or friable.

gully erosion. Erosion of soil caused by running water.

gumbo. A special type of soil which yields a characteristic sticky mud when wet.

gumboti. Deoxidised, leached clay that contains siliceous stones.

gummite. Hydrated uranium lead thorium and rare earths oxide. Formula and crystal system ncertain.

gypsum. A variety of gypsum consisting of dirt and sand; found as an efforescent deposit in arid regions, overlying gysum. Also known as gypsum earth.

gypsum. $CaSO_4$ $2H_2O$. A monoclinic mineral consisting of hydrated sulphate of calcium found in clays and limestons; sometimes associated with sulphur. Gypsum is found in three forms: *(i)* selenite is the name given for the crystals, which are generally tabular and often form swallowtail turns. Prismatic crystals also occur, and these are often curved of bent. Crystals of selenite have a platy cleavage which produces thin flexoble sheets. *(ii)* This form is fibrous and found when gypsum occurs in veins and is also known as satinspar or 'beef' *(iii)* This form is found in evaporite deposit, gypsum is granular and albaster is the name for the fine-grained granular form. Its hardness on Mohs scale is 2.0. and can be scrateched with finger nail which distinguishes it form harder calcite and anhydrite. It is a colourless mineral but may be stained a veriety of colours. Gypsum is a common mineral which forms thick beds in evaporite deposits where it is often associated with hallite.

gypsum earth. See gypsite.

gyttja. A fresh water anaerobic mud containing an abunance of organic matter and is capable of supporting aerobic life.

habit. The characteristic shape of crystals that are due to variations in the number, size, and shape of the crystal faces.
habit plane. The crystallographic plane or system of planes along which phenomena like twinning occur.
hachure. A shore line at some angles to contours; used to denote slopes of the ground in geologic maps.
hackly. Irregular surface fracture with sharp angular projections.
hackmanite. A calcium rich variety of sodalite containing a small amount of sulphur.
hacksaw structure. The irregular, saw-toothed termination of a crystal or mineral particle due to intrastratal situation.
hadal. Pertaining to the deepest oceanic environment; in particular to ocean trenches, where depths exceed 6-7km.
hade. The compliment of the dip; the angle that a structural surface makes with the vartical, measured perpendicular to the strike.
hailstone bort. A veariety of diamond occurring as a grey or greyish- black, rounded stone. It is composed of concentric shells of clouded diamond and cement-like material.
hairstone. Quartz embedded in hairlike crystals of rutile, actinolite or any other mineral.
halite. NaCl. The native salt, occurring in massive granular compact or cubic-crystalline forms. Its crystal are grown by slowly evaporting a concentrated salt solution, often possesses concave hopper-shaped faces but rock salt is massive and granular and superficially resembles marble. It is colourless or white but may contain impurities which stain it red, grey or brown, blue or yellow.
Its crystals are transparent and it shows a vitreous lustre, perfect cubic cleavage and a hardnnss of 2.5 on Mohs scale. Most salt deposits have been formed by the evaporation of sea water which is 2.7 per cent sodium chloride or the evaporation of the salty water of inland lakes.Under stress, rock salt becomes plastic and flows salt domes form when layers of rock salt are pushed upwards by the pressure of overlying bedrock. The salt domes are of importance because in various parts of the world's oil-fields they form oiltraps which can be exploited.
half-life. $t_{1/2}$. The time taken for the activity $\int$ of a radioactive isotope to decay to half of its original value, that is for half of the atoms present to disintegrate. Half-lives vary from fractions of a second to millions of years.
halleflinta. A metamorphosed acid volcanic or pyroclastic rock, with characteristic 'flinity' or 'horny' appearance, sometimes banded or porphyroblastic; a kind of hornfels.
halloysite. $Al_2Si_2O_5(OH)_4$ $2H_2O$. A clay mineral related to kaolinite and essentially having the same chemical composition. The term has also been used for nonhydrated variety.
halmyrolysis. The geochemical reaction of sea-water and sediments in an area of little or no sedimentation.
halo. (1) A circular or crescentic distribution pattern about the source or origin of a mineral ore mineral-association or petrographic feature. (2) Discolouration of a mineral viewed in thin

section in the form of a ring.

hammada. A plateau in a desert region with a rocky surface denuded by wind erosion.

hamrongite. An igneous rock of the granite clan; a fine grained porphyritic hypabyssal rock. It contains phenocrysts of biotite in a groundmass of quartz, biotite and andesine.

hancockite. $(Pb,Ca,Sr)_2(Al,Fe)_3 Si_3O_{12}.OH$. A mineral containing basic lead aluminium iron silicate.

hanging. Situated on steeply sloping ground or on top or other ground or having a discordant junction.

hanging valley. A valley whose floor is higher than the level of the shore or other valley to which it leads.

hanging wall. The rock mass above a fault plane, vein, load, ore body or other strucutre. Also known as hanging side.

hanleite. $8[Mg_3Cr_2Si_3O_{12}]$. A mineral containing magnesium chromium silicate.

haplite. A synonym of aplite.

hardness. (1) A property of hard water, chiefly due to the presence of calcium and magnesium ions. (2) The resistance of a mineral to scratching; one of the properties by which minerals may be described. Various scales of hardness exist; the scale suggested is based on the fact that materials with higher values being able to scratch those of lower values:

10. Diamond. 9 Corundum. 8. Topaz. 7. quartz. 6. Orthoclase. 5. Apatite. 4. Fluorite. 3. Calcite. 2. Gypsum. 1. Talc. The surface of minerals with a hardness of less than 6.5 can be marked (scratched) with a knife. Mineral with a hardness of 2.5 or less can be scratched with a finger-nail.

hardpan. A hard subsurface layer resulting from the deposition of silica, clay, oxides, and hydroxides, or carbonate salts in the layer or deposition.

harmonic folding. Folding in the surface of the earth with no sharp changes with depth in the form of the folds.

harmotome. $(K,Ba)(Al,Si)_2(Si_6O_{16}).6H_2O$. A mineral of zeolite group that forms cruciform crystals and shows ion. exchange properties. Also known as cross-stone.

harpolith. A large sickle-shaped igneous intrusion that was injected into previously deformed strata and later deformed with the host rock by horizontal stretching or orogenic forces.

harringtonite. Amorphous chalky white mixture of mesolite and thomosolite from Country Antrim in Northern Ireland.

harrisite. Igneous rich Variety of the gabbro clan; a coarse-grained granular hypabyssal rock; it contains olivine, anorthite, enstatite and chromite.

harzburgite. An ultrabasic rock composed of olivine and orthopyroxene.

hatchettine. $C_{38}H_{78}$. A yellow-white paraffin wax type mineral hydrocarbon; occurs in masses in ironstone nodules or in cavities in limestone. Also known as mineral tallow; mountain tallow; naphthine.

hauerite. $4[MnS_2]$. A mineral containing manganese sulphide.

haughtonite. A variety of biotite rich in iron.

hausmannite. Mn_3O_4. a brownish-black, opaque mineral composed of manganese tetraoxide and belongs to spinel group of minerals.

hauterivian. A stratigraphic stage name for the European mid -Lower Cretaceous.

hauyne. $(Na,Ca)_{4-8}(Al_6Si_6O_{24})(SO_4, S)_{1-2}$. An isometric silicate mineral of the sodalite group occurring as grains embedded in various igneous rocks. Its hardness on Mohs scale is 5.5-6.0. Also known as hauynite.

hauynite. *See* hauyne.

hauynolith. Name given to any igneous rock contining 100 per cent hauyne.

hawaiite. Name given to andesine basalts due to the wide occurrence of these rocks on the Hawaii Islands.

haydenite. Yellow crystalline variety of chabazite.

hazel sandstone. An arkosic, iron containing red-bed sandstone found in western Texax.

headwall. The steep cliff at the back of a cirque.

headward erosion. Erosion caused by water flowing at the head of a valley.

heavy mineral. A mineral with a density more than 2.9, which is the density of bromoform, the liquid used to separate the heavy minerals from light minerals.

heavy spar. $BaSO_4$. Common name for the mineral barytes.

hectorite. Altered hydrated green pyroxene occurring in serpentinite in the Dun Mountains of New Zealand. Also megnesiumrich variety of montmorillonite.

hedenbergite. $CaFeSi_2O_6$. A black mineral consisting of pyroxene along with calcium and iron; occurs at the contacts of limestone with granitic masses.

hedrumite. An igneous rock of the syenite clan, a fine to medium-grained porphyritic hypabyssal rock; contains phenocrysts of microperthite and lepidomelane in a groundmass of microperthite, nepheline, analçine, algirine, lepidomelane and hornblende.

Helderbergian. A stratigraphic stage name for the base of the North American Lower Devonian.

helvine. $(Mn,Fe,Zn)_4Be_3(SiO_4)_3S$. A silicate mineral isomorphous with danalite and genthelvite. Also known as helvite.

helvite. *See* helvine.

hematite. Fe_2O_3. A mineral of iron crystallizing in the rhombohedral system; occurs in black metallic massess or in reddish earthy forms. Also known as bloodstone.

hemimorphic. A crystallographic term used for a crystal having a polar symmetry, so that the two ends have different forms.

hemimorphite. $2[Zn_4Si_2O_7(OH)_2.H_2O]$. A mineral consisting of basic hydrated zinc silicate.

hemipelagic sediment. Deposits containing terrestrial material and the remains of pelagic organisms, found in the ocean depths.

hemiprism. A pinacoid that cuts two crystallographic axis.

hemitropic. Pertaining to a twinned structure in which, if one part is roated by 180° the two parts become parallel.

heptorite. An igneous rock of the gabbro clan. a fine-grained porphyritic hypabyssal rock; it contains phenocrysts of augite, hornblende, hauyne and, labradorite in a groundmass of labrtadorite, olivine, magnetite and glass.

herderite. $Ca,Be(PO_4)(F,OH)$. A mineral consisting of fluoride and phosphate of calcium and beryllium.

hermatypic. A term applied to a reef-building organisms.

heronite. An igneous rock of the syenite clan; a fine-grained hypabyssal rock; contains analcite, orthoclase, labradorite, and aegirine.

hartree. Atomic unit of energy equal to e^2/l_0, where e is the charge of an electron and and l_0 is the atomic unit of length. It is equal to 4.360 X 10^{-18} joule or 27.21 eV.

herzenbergite. 4[SnS] Ein sulphide.

heterogenite. CoO.OH, A mineral consisting of hydrous cobalt oxide.

heteropygous. A term applied to trilobites which have a cephalon larger than the pygidium.

heulandite. $(Na,Ca)_{4-6}Al_6(Al,Si)_4Si_{26}O_{72}$. 24 H_2O. Hydrated sodium calcium aluminosilicate; widespread in basalt and gneiss, silicon-rich variety is clinoptilolite; barium bearing variety is know as beaumontite.

heumite. An igneous rock of the syenite clan; a fine-grained granular hypabyssal rock; contains soda-orthoclase, soda-microcline, barkevikite, biotite, nepheline and sodalite.

hexacoralla. Corals in which the polyps have tentacles arranged in multiples of six. They include the majority of large reef forming cords.

hexagonal close-packed structure. Closed packed crystal structure characterized by regular alternation of two layers; the atoms in each layer lie at the vertices of a series of equilateral triangles and the atoms in one layer lie directly above the centres of the triangles in neighbouring layers.

hexhedrite. An iron meteorite composed of single crystals or aggregates of kamacite; usually contains 4-6% nickel in the metallic phase.

hiddenite. A green variety of spodumene.

high grade metamorphism. Intense level of metamorphism (regional) or contact) resulting in the growth of new minerals and the development of completely new textures. *e.g.*; gneiss or hornfels.

highwoodite. An igneous rock of the syenite clan; a coarse-grained granular plutonic rock; it contains soda-orthoclase, labradorite, diopside, biotite, and apatite.

hilairite. An igneous rock of the syenite clan; a coarse-grained porphyritic plutonic rock; it contains phenocrysts of sodaclase, nepheline and levenite in ground mass of orthoclase, aegirine and sodalite.

hinge fault. A fault whose movement is angular or rotational one on a side of an axis that is normal to the fault plane.

hinge line. A line in a plane of a hinge fault separting the part of a fault along which thrust or reverse movement occurred from that having normal movement.

hirantite. An igneous rock of the diorite clan; a fine grained granular plutonic rock. It contains albite, chlorite and augite.

hinterland. (1) The region behind the coastal district. (2) The terrain on the back of a folded mountain chain.

hitch. A minor dislocation of a vein or stratum not exceeding in extent the thickness of the vein or stratum.

hoegbomite. $Mg(Al,Fe,Ti)_4O_7$. Amineral containing magnesium aluminium iron and titanium oxide.

hogback. Alternate ridges and ravines in certain regions of mountains, caused by the erosive action of mountain torrents.

holmite. An igneous rock of the gabbro clan; a fine-grained porphyritic hypabyssal rock; it contains phenocrysts of augite and olivine in a groundmass of augite, biotite and mellite.

holmquistite. Lithium rich variety of glauocophane.

holoaxial. Having all possible axes of symmetry.

holoblast. A mineral crystal that

originates during metamorphism and is completely formed during that process.
holoclastic. Being or belonging to ordinary clastic rock.
holocrystalline Pertaining to igneous rocks that are entirely crystalline and are without glass.
holohyaline. Pertaining to entirely glassy rocks.
holomictic lake. A lake whose waters undergo complete mixing during periods of circulation overturn.
holyokeite. An igneous rock of the diorite clan; a white, fine-grained granular hypabyssal rock; it contains sodaclase and calcite.
homeoblastic. Said of a metamorphic crystalloblastic texture having constituent minerals of approximately the same size.
homeomorphism. Crystalline substances that have a dissimilar composition but similar habit.
homo. A prefix denoting same.
homocline. Any rock unit in which the strata exhibit the same dip.
homodesmic. Said of a crystal, having atoms bonded in a single manner.
homometric pair. A pair of crystal structures for which X-ray diffraction patterns are identical.
homomorphy. Superficial similarities in the morphology of members of different phyla.
homonym. If two distinct organisms or groups of organisms are given the same name then becomes the name of a homonym. The more recent of the two name is referred to as the junior homonym and is usually invalid.
homopycnal inflow. Flowing water that has the same density as that of the body it enters, resulting in easy mixing of the two system.
hoodoo. A pillar, column, rock pedestal or toadstool rock, often of bizarre shape, produced by rainwash and differential mechanical and chemical weathering.
hopeite. $Zn_3(PO_4)_2.4H_2O$. A grey, orthorhombic mineral composed of hydrated zinc phosphate.
horizontal separation. In faulting, the distance, measured in any specified horizontal direction, between the two sections of a disrupted unit.
horizontal slip. In a fault, the horizontal factor of the net slip.
hornblende. $NaCa_2[(Mg,Fe(II)_4Al,Fe(III)](Si,Al)O_{22}(OH,F)_2$.A mineral of the amphibole group, widespread in metamorphic and igneous rocks
hornblendite. An igneous rock of the ultrabasic clan; a coarse-grained granular hypabyssal or plutonic rock consisting mainly of hornblende.
horseback. A low and sharp ridge of sand, gravel or rock.
horsetail ore. Occuring in fractures that diverge from a larger fracture.
horst. (1) An upthrown area between two parallel faults. (2) One of the older mountain masses limiting the Alps on the west and north. (3) A knobby ledge of limestone beneath a thin soil mantle.
hortonolite. $(Fe,Me,Mn)_2SiO_4$. A dark mineral composed of silicate of iron, magnesium and manganese; a member of the olivine series.
howlite. $Ca_2B_5SiO_9(OH)_5$. A mineral composed of basic calcium borosilicate; shows no cleavage. Its hardness on Mohs scale is 3.5.
hullite. A mineral composed of iron, magnesium, calcium and alkaline aluminosilicate of uncertain crystal system. Its hardness on Mohs scale is 2.0.
humboldtilite. A synonym of melilite.
humic coal. A variety of coal whose attritus is composed mainly of transparent

humic degradation material.

humite. A series of magnesium neosilicate minerals closely related in crystal structure and chemical composition.

hummock. A term variously used to mean a knoll or mound above the general level of a surface or a knob of turf or earth in alpine regions. It is also applied to a mound of broken, floating ice pushed upward by pressure.

humus. The amorphous, colloidal matter present in the soil.

hungarite. An igneous rock of the diorite clan; a fine grained porphyritic extrusive rock. It contains phenocrysts of hornblende and plagioclase in groundmass of plagioclase.

huntite. $2[Mg_3Ca(CO_3)_4.]$. An ore containing magnesium calcium carbonate.

Huronian. A stratigraphic division of the Candian Precambrian pre-Algonkian.

husebyite. An igneous rock of the syenite clan; a coarse-grained granular plutonic rock; contains orthoclase, andesine, nephaline and titanaugite.

huttonite. Monoclinic polymorph of thorite.

hyaline. A term denoting glassy, i.e., non-crystalline.

hyalite. A colourless variety of opal whose appearance varies from glass like transparency to translucent or whitish.

hyalo. A prefix denoting glassy.

hyaloandesite. Volcanic glass with the composition of andesite.

hyalobasalt. Volcanic glass with the composition of basalt.

hyaloclastic. A tufflike deposit formed by the flowing of basalt under water and ice and its consequent fragmentation.

hyalocrysalline. A textural term used for porphyritic rocks in which the volume occupied by phenocrysts is equal or nearly equal to the volume of groundmass.

hyalodiabase. A variety of volcanic glass with the composition of diabase.

hyalophane. $4[(K,Na,Ba)(Al,Si)_4O_8]$ A mineral containing potassium, sodium, barium, aluminosilicate.

hyalopilitic. A textural term for an igneous rock consisting of acicular microlites in a glassy groundmass.

hyalosiderite. Variety of chrysolite with 30-50 per cent molar proportion of ferric silicate.

hybrid. A term applied to rocks formed by the mixing of two magmas or the assimilation of one igneous rock by another.

hydraulic fracturing. A process of injecting sand-charged liquids under such high pressure as to fracture a rock and inject the newly made crevices with coarse sand to deep them open.

hydraulic fill. Earth or waste materials that have been excavated, transported, and flushed into place by flowing water.

hydraulic head. The height of the exposed surface of a body of water above a specified subsurface point. Also, the difference in water level at a point upstream from a given point downstream.

hydraulic limestone. A variety of limestone containing sufficient siliceous aluminous impurities to yield a true cement (*i.e.*, one which will set under water) when calcined.

hydrobiotite. A hydrated variety of biotite.

hydrocastorite. An amorphous mixture of heulandite and stilbite.

hydroclastic rock. Any rock that has been broken by the action of water or deposited by it.

hydrogarnet. A group of minerals closely related to garnets but with

($2H_2O$) replacing silica in the lattice; main types are hibschite and plazolite.

hydrogeology. The study of the geological factors relating to the earth's water.

hydrogoethite. The alteration product intermediate between goethite and limonite.

hydrohalite. NaCl. $2H_2O$. A variety of halite; a product of crystallisation of sea water or saline spring water in an extremely cold climate.

hydrohematite. A massive reddish-black to red alteration product intermediate between hematite and goethite. Also known as turgite; turite.

hydrolith. (1) A rock that is free from organic substances. (2) A chemically precipitated aqueous rock, such as rock salt.

hydrology. The study of all waters in and upon the earth. It includes underground water, surface water and rainfall and embraces the concept of the hydrological cycle.

hydrolyzates. A geochemical division of the sedimentary rocks corresponding approximately to the argillaceous rocks.

hydrometamorphism. Alteration of rocks of material carried in solution by water without the influence of high pressure or temperature.

hydromica. Group name for hydrated minerals with properties and compositions similar to micas; main types are pinite and sericite.

hydrophane. White variety of opal which has a low reractive index and is virtually transparent when placed in water.

hydrosphere. The total body of water of the earth, i.e., the oceans, rivers, lakes, underground and atmospheric water.

hydrotachylyte. Type of techylyte containing upto 13 per cent water.

hydrotalcite.$3[Mg_6Al_2CO_3(OH)_{16}$-$4H_2O]$. A mineral composed of hydrous basic ma gnesium aluminium carbonate; shows perfect basal cleavage; soluble in hydrochloric acid. Its hardness on Mohs scale is 2.0.

hydrothermal alteration. Rock or mineral phase changes that are caused by the interaction of hydrothermal liquids and wall rock.

hydrothermal deposit. A mineral deposit precipitated from a hot, aqueous solution. Hydrothermal deposits are commonly found arranged concentrically around an igneous mass, the high-temperature (hypothermal) minerals being closest to or actually within the igneous rock, and the low temperature (epithermal) ones being formed farthest away.

hydrous. A term denoting containing water.

hydroxyl-annite. A hydrous variety of lepidomelene.

hydroxyl-apatite. $Ca_5(PO_4)_3OH$. A variety of apatite.

hydrozincite. $2[Zn_5(CO_3)_2(OH)_6]$. A white monoclinic mineral composed of zinc carbonate hydroxide; it shows perfect cleavage in one direction and its hardness on Mohs scale is 2.0-2.5.

hygroscopic. Having the property of readily absorbing moisture from the atmosphere.

hyp- A prefix denoting under, nearly; sometimes 'hypo-'.

hypautomorphic. Having some trace of crystal form.

hyper- A prefix denoting very, greatly.

hyperfusible. Any substance capable of lowering the melting ranges in end stage magmatic fluids.

hyperite. An obsolete name for gabbro

containing hypersthene and augite.

hypermelanic. A term applied to igneous rocks which consist almost exclusively of dark minerals with a colour index greater than 90 on Shand's scale.

hyperpycnal inflow. Flowing water that is denser than the body of water it enters resulting in the formation of a turbidity current.

hypersaline. Geological mineral with high salinity.

hypersolvus granites. Granites containing only one feldspar.

hypersthene. (Mg,Fe)SiO_3. A rock-forming mineral of the orthopyroxene group; contains silicates of magnesium and irom.

hypersthenfels. A synonym of norite.

hypersthenite. An igneous rock of the ultrabasic clan; coarse-grained granular plutonic rock composed of hypersthene.

hypidiotopic. Said of the fabric of a crystalline sedimentary rock in which the majority of the constituent crystals are subhedral. Also said of a rock with such a fabric.

hypo-,hyp- A prefix denoting under, nearly.

hypohyaline. A term used for the igneous rocks that have crystalline components in a glassy groundmass, in which the crystal-to-glass ratio is between 3:5 and 1:7.

hypopycnal flow. An inflow of water whose density is less than that of the body of water it enters.

hypolimnion. The lowermost layer of water in a lake, characterized by an essentially uniform temperature that is generally colder than elsewhere in the lake and often by relatively stagnant or/and oxygen-poor water.

hypothermal. A term used to denote high temperature environment (300-500°C) of hypothermal deposits.

hypothermal deposit. Mineral deposit formed at great depths and (300-500°) high temperatures.

hypsometry. The measuring of elevation with reference to sea level.

ianthinite. $2UO_2.7H_2O$. A violet mineral composed of hydrous uranium (IV) dioxide; occurs as orthorhombic crystals.

ice age. A major interval of geologic time during which glacial ice spreads over regions which are not normally ice covered. The emphasis is on a large-scale regional spread of the ice, not merely a local advance of valley glaciers due to minor climatic fluctuations.

iceberg. A large mass of detached land ice floating in the sea or stranded in shallow water.

ice cap. A perennial cover of ice and snow in the shape of a dome or plate on the summit area of a mountain through which the mountain peaks emerge. An ice cap is considerably smaller than an ice sheet.

ice cliff. Any vertical wall of ice, e.g., a glacier or mass or shelt ice.

ice front. The leading stage of a glacier.

ice erosion. Erosion due to freezing of water in rock fractures.

ice island. A large tabular iceberg separated from an ice shelf and found in the Arctic Ocean. The surface commonly has a ribbed appearance when viewed from the top.

iceland spar. A pure and transparent variety of calcite; easily cleaved to form rhombohedral crystals that are doubly refracting. Also known as iceland crystal.

ice pan. A large flat place of first year ice protruding several centimetres above the sea surface.

ice patch. An area of pack ice smaller than an ice field.

ice period. The period of time from freeze up to breakup of ice.

ice pillar. Any tall, narrow mass of ice, such as ice pedestal or mushroom ice.

ice plateaue. An ice-covered highland area whose upper surface is nearly level and whose sides descend steeply to lowland or to the ocean.

ice rafting. Transportation of sediments and rocks fragments of all sizes by floating ice, usually applied to sea ice and icebergs.

ice shelf. A sheet of very thick ice, most of which is afloat but is attached to land along one side. Ice shelves are fed by snow accumulations and the seaward projection of land glaciers.

ice spicule. A small needlelike ice crystal that grows in water.

ice table. A mass of level ice.

ice wedge. Wedge-shaped, foliated ground ice formed in permafrost. It occurs as a vertical or inclined sheet, dike or vein and originates by the growth of hoar frost or by freezing of water in a narrow crevice.

ichor. A fluid thought to be responsible for granitization. The term initially bore the connotation of derivation from magma.

ichthyol. An oil obtained by distilation and sulphonation of bituminous shales, later neutralized with ammonia.

iddingsite. A mineral containing iron calcium magnesium silicate; shows good cleavage in one direction. Its hardness on Mohs scale is 3.5.

idio-. A prefix denoting of itself.

idiogeosyncline. A type of lake cycle geosyncline between stable and mobile areas of the crust the sediments of which are only weakly folded.

idiotopic. Said of the fabric of a crystalline sedimentary rock *e.g.*, an evaporite ore is crystallized limestone in which the majority of the constituent crystals are euhedral. Also said of the rock with such a fabric.

igneous. Pertaining to rock which have congealed from a molten mass.

igneous breccia. A breccia composed of fragments of igneous rock, or a breccia produced by igneous processes; e.g., volcanic breccia.

igneous complex. An assemblage of igneous rocks that are intimately associated and roughly contemporaneous.

igneous facies. A part of an igneous rock differing in structure, texture, or composition from the main mass.

igneous mineral. Any mineral material that forms igneous rock.

igneous petrology. The scientific study of igneous rocks, their occurrence, origin and composition.

igneous rock. One of the main groups of rocks that comprise the earth's crust. They constitute about 15% of the earth's surface, and are formed of molten material (magma) that flows up from the deeper part of the crust. When this material, which is composed largely of silicate minerals, solidifies beneath the surface, the resultant rock formation are referred to as intrusive or plutonic. Material that solidifies at or above the surface is called extrusive or volcanic. their classification is generally based on mineral composition and texture and bound in extrusions and in intrusions.

ignimbrite. A silicic volcanic rock that forms thick, compact, lava-like sheets over a large area of New Zealand. Also known as flood tuff.

ijolite. A plutonic rock of nepheline and 30-60% mafic materials, generally sodic pyroxene, with accessory apatite, sphene, calcite and titaniferous garnet.

ijolite pegmatite. Type of ijolite occurring as a pegmatite.

Illinoisian. Pertaining to the classical third glacial stage of the Pleistocene Epoch in North America.

illite. A group of grey, green or brown mica-like clay mineral found in argillaceous sediments.

illuvial. Pertaining to a material or region characterized by the accumulation of soil by the illuviation of another zone of material.

illuviation. The deposition of colloids, soluble salts and small particles of minerals in an underlying layer of soil.

illuvium. Material leached by chemical or physical processes from one soil horizon and deposited in another.

ilmenite. $FeTiO_3$. An iron-black, opaque rhombohedral mineral found as an accessory in basic igneous rocks, as deposits and in veins.

imbibition. The tendency of granular rock or any porous medium to absorb a fluid usually water under the force of capillary action and in the absence of any pressure.

Imbrian. Pertaining to lunar topographic features and lithologic map units constituting a system of rocks formed during the period of the formation of the Mare Imbrian basin and of deposition of Mare material of the Procellarum Group or during any time between these two events. Imbrian rocks are older than the post-mare craters and associatrd ejecta of the eratosthenian and Copernican systems.

imbricate structure. A sedimentary structure characterized by shingling of pebbles all inclined in the same direction with the upper edge of each leaning downstream or toward the sea. When these consist of fragments having a noticeable elongation, deposited under the influence of a powerful current, the long axes of the pebbles tend to lie more or less parallel with one another 'leaning' in the direction of current flow. It has been suggested that under certain circumstances sand grains might show a micro-imbricate structure.

imbrication. Formation of an imbricate structure. Also known as shingling.

immiscible. Incapable of being mixed to form a homogeneous substance; it is usually applied to liquids; e.g., oil and water are immiscible.

impact. A forceful collision between two bodies which is sufficient to cause an apreciable change in the momentum of the system on which it acts.

impact crater. A depression formed by the impact of an unspecified projectile especially a crater formed on the earth or moon surface where the nature of impacting body is unknown.

impactite. A vesicular glassy to finely crystalline material produced by complete or partial fusion of target rock by the heat generated from the impact of a large meteorite and occurring in and around the resulting crater.

impermeable. Not permitting water or any other fluid to pass through. Also known as impervious.

implosion. The sudden reduction of pressure by chemical or physical reaction which cause an inrushing of the surronding medium..

impoundment. The process of forming a lake or pond by some sort of barrier such as a dam or dike. The body of water formed in this manner is also called an impoundment.

impregnation. Filling of rock pores after formation by liquids such as oil or brine or by mineral material, sometimes with displacement of existing impregnating substances.

impression. A form or shape left on a soft soil surface by plant parts; the soil hardens and usually the imprint is a concave feature.

impsonite. A black, asphaltic pyrobitumen with a high fixed-carbon content derived from metamorphosis of petvroleum.

incandescence. The emission of light caused by high temperatures; white or bright red heat.

incised meander. In a broad sense, an incised meander is meander that has become deepened by renewed downcutting as a result of environmental change. It is usually bordered on one or both banks by vertical walls.

inclination. The angle between the earth's magnetic field and the horizontal at a given point on the earth's surface.

incompetent bed. A sedimentary bed of low rigidity tending to flow when subjected to stress rather than flex or fracture; used in relative sense with regard to contiguous beds.

incompressibility. Property of a substance which maintains its original volume under increased pressure.

incrusiation. Covering with a hard surfaced layer.

indent. To form a depression by forcing inward.

index fossil. The ancient remains and traces of an organism that lived during a particular geologic time period, and that characterises a specific horizon by its abundance, although not restricted to it in time.

index horizon. A structural surface selected as a reference in studying the geological structure of an area.

index mineral. A mineral that occurs

throughout rocks of similar bulk composition and is exposed over a distinct area of the metamorphic terrane.
index plane. A surface used as a reference point in determining the geological structures.
indialite. Granular variety of anorthite occurring in India.
indicatrix. An ellipsoid whose three perpendicular axes are proportional in length to the principal values of the index of refraction of light in an anisotropic medium and point in the direction of the correspoding electric vector.
indicolite. $Mg_2Al_2(CO_3)_4(OH)_2$-$15H_2O$. A blue sodium-rich variety of the mineral tourmaline; occurs in association with gypsum, quartz, gibbosite, and amorphous iron oxide in the oxidation zone of the Sarylakh gold antimony deposit in the basin of the upper part of the Indigirbi river, north-eastern Yakutia. USSR.
induced fission. Fission taking place only as a result of bombardment of nucleus with neutrons, gramma rays or other carriers of energy.
induration. The hardening of a sediment by heat, pressure and/ or lithification and as such grading into true metamorphism.
inertial flow. The flow in which no external forces are exerted on a fluid.
inertinite. A carbon-rich mineeral group, which includes micrinite, sclerotinite, fusinite and semifusinite.
inesite. $Ca_2Mn_7Si_{10}O_{28}(OH)_2.15H_2O$. A pale-red variety of hydrous manganese calcium silicate occurring as triclinic crystals or in massive form.
influent. (1) A stream that flows into a lake or a stream or branch that flows into a larger stream. (2) A stream that flows into a cave.
influent stream. A stream or reach of a stream that contributes water to the zone of saturation and develops bank storage; its channel lies above the water table.
informal unit. A body of rock or other feature that is referred to in casually descriptive but nondefinitive terms; e.g., sandy beds,mucky marsh.
infra. A prefix denoting below.
infrastructure. Structure produced in a plutonic environment at high pressure and temperature. It is characterized by plastic folding and the emplacement of granite and other magmatic rocks.
inground meander. A continually growing or expanding incised meander formed during a single cycle of erosion by the enlargement or accensuation of an initial minor curve while the stream was actively down cutting; a meander that grows in place.
injection, igneous. Nearly synomynous with intrusion but usually restricted to intrusions with narrow feeding channels as dikes, sills, lacoliths etc.
injection gneiss. A composite rock with banding entirely or partially caused by layer-by-layer injection of granitic magma into rock layers.
injection metamorphism. Metamorphism accompaned by intimate injection of sheets and streaks of zones near plutonic rocks.
inland ice. (1) A continental glacier or ice sheet in its entirely. (2) The ice forming the inner part of a continental glacier or large icd, sheet. The term is applied especially to the ice on Greenland.
inlet. (1) A small, narrow groove in a shoreline, through which water penetrates into the land. (2) An inflowing stream, as into a lake.

inlier. A circular of elliptical area older rocks completely surrounded by younger rocks; may be produced by erosion.

innage. Refers either to the volume of liquid or the measured height of the liquid in a tank or container as measured from the bottom of the tank.

inner core. The central part of the earth's core extending from a depth of about 5100 km to the centre (6371 km) of the earth. Its radius is about one third of the whole core. It is equivalent to G layer.

inosilicate. A class of structural type of silicates in which the SiO_4 teahedrons are linked together by the sharing of oxygens to form linear chains of indefinite length.

in place. Refers to a rock, mineral, fossil, exposure or structure not moved from its place or origin.

inselberg. Isolated vegular uplands standing above the general level of the surrounding plains in tropical regions; they may be ridges, domes or hills. Like monadnock., they are remnants of the erosion cycle, but even though there may occasionally be great morphological similarity between the two, they are not the tropical equivalents of monadnocks, which are features of temperate zones.

insequent. Said of a stream or drainage system that is controlled by the associated rock structure or surface features and wanders irregularly across the region of low relief.

insequent stream. A stream developed on the present surface but not controlled or adjusted by the rock structure and surface featuraes; a self-guided stream whose resulting drainage pattern is dendritic as a yong stream wandering irregularly on a nearly level plain under lain by homogeneous or horizontally stratified rocks.

inset. A synonym of phenocryst.

insolation. The effect of the solar heat on rocks at the surface of the earth, especially the effect of changes in temperature on the mechanical weathering of rocks. Heating by the sun during the day causes the rocks to expand and they contract during night on cooling. This causes the rocks to break up.

insular shelf. An area of the ocean floor analogous to the continental shelf but surrounding an island.

insular slope. An area of the ocean floor analogous to a continental slope but surrounding an island.

inter. A prefix denoting between.

interbedded. Having beds laying between other beds with different characteristics; e.g., a lava flow may occur interbedded between two layers of sediments; a limestone may be interbedded between two layers of shale.

intercalation. A layer located between layers of different character.

intercepts. Distances cut off a crystal's reference axis by planes.

interface. (1) The boundary separating the top of the uppermost layer of a sediment from the water in which the sedimentation process is taking place. (2) The contact between fluids in a reservoir.

interference figure. A pattern of light and dark areas observed with a conoscope when a birefringent crystal is placed in a convergent beam of linearly polarized light.

interfingering. Laterally contiguous rock units with extensions of each projecting into the other so as to make a zig-zag boundary between them.

interfluve. The comparatively undissected upland between adjacent

streams flowing in the same direction.
interformational conglomerate. Consolidated pebble to cobble bed within a formation whose constituents have a source outside the formation.
interglacial. Refers to the time period between glaciations (ice ages). The warmer period of time that precedes and follows an ice age.
intergrade. A soil that is transitional between two other soils.
intergranular. Refers to the fabric or texture of an igneous rock characterized by an interlocking network of randomly oriented mineral grains.
intergranular movement. A movement that involves displacements between individual mineral grains. If rocks that are composed of both mineral grains and crystals are subjected to stress, the individual crystals and grains may move independently, and individual grains may maintain their shape and size.
intergranular porosity. The porosity, between the grains or particles of a rock, such as that between the lithoclasts or the bioclasts of a carbonate sedimentary rock.
intergrowth. The interlocking of grains of two different minerals because of their simultaneous crystallization.
interior basin. A depression surrounded by higher land from which no stream flows outward to ocean.
interior valley. A large flat-floored closed depression. Its drainage is subsurface and its floor is covered by alluvium. Interior valleys may become intermittent lakes during periods of heavy rainfall when the sinking stream that drain them cannot manage to runoff.
interior wall. A body wall in bryozoans that partitions pre-existing body-cavity into zooids, parts of zooids, or extazooidal parts.
intermediate moraine. A type of lateral moraine formed at the junction of two adjacent glacial lobes. Also known as interlobat moraine.
intermediate rock. Any of a group of rocks which contain less than 10% of quartz together with either a plagioclase in the andesine-oligoclase range or an alkali feldspar or both the feldspars. Feldspathoids may or may not be present.
intermittent stream. A stream not flowing continuously, flowing only during rainfall and high runoff.
intermontane. Located between or surrounded by mountains.
intermontane glacier. A glacier formed by the confluence of several alpine glacier and occupying a depression between ridges or mountain ranges.
internal drainage. Surface drainage in which the water does not reach the ocean, such as drainage toward the central part of an interior or basin.
internal wave. A submerged wave occurring on a density surface *e.g.*, the thermocline in density stratified water.
interpluvial. Pertaining to an episode or period of geologic time that was drier than the pluvial period occurring before of after it.
intersertal. A term applied to refer the texture of a porphyritric igneous rock in which the groundmass forms a small proportion of rock, filling the interstices between unoriented feldspar laths.
interstadial. Pertaining to a period during a glacial stage, which registered a pause in the advance of a glacier or ice sheet.
intra . A prefix denoting within or inside.

intraclast. A fragment of limestone formed by erosion within a basin of deposition and redeposition there to form a new sediment. The particles may be fragments of partially consolidated calcium carbonate mud or composite particles consisting of aggregates of smaller intraclasts, pellets, ooliths and skeletal material.

intraformational breccia. A rock resulting from cracking and dessication-shrinking of a mud after withdrawal of water followed by almost contemporaneous sedimentation.

intratelluric. (1) Pertaining to a phenocryst that is formed earlier than its matrix. (2) Located, formed, or originating at great depth within the earth.

intrazonal soil. A soil with more or less well developed characteristics that reflect the dominating influence of some local factor of relief, parent material or age over the normal effects of the climate and vegetation.

intrusion. Igneous formation in which the rock is forced through and into existing rock; perhaps along lines of weakness such as faults and bedding planes.

intrusive rock. Igneous rock formed of magma that consolidated beneath the earth's surface. The texture of the intrusive rock depends partly upon the depth at which it has cooled.

inverted plunge. The plunge of folds or sets of folds, whose inclination has been carried past the vertical so that the plunge is now less than 90° in a direction opposite from the original altitude.

inverted relief. A topographic surface that is out of phase with the geologic structure as where valleys are underlain by anticlines and mountains by syncline.

inyoite. $Ca_2B_6O_{11}.13H_2O$. A colourless and transparent monoclinic consisting of hydrous calcium borate.

ionosphere. The region of the earth's upper atmosphere in which free electrons arising from ionization occur, mainly as a result of ultraviolet radiation and x-ray from the sun. The ionosphere is useful in that it enables intercontinental radio transmission round the curved surface of the earth to be achieved, as a result of its property of reflecting electromagnetic waves of radio frequencies; but it is an obstacle to radio astronomy because it reflects a large proportion of the radiation that arrives from extra terrestrial sources.

iridescent. Many coloured, often the result of interference effects in the light reflected by the surface .

irinite. Basic sodium cerium and thorium niobate (columbate) and titanate; of uncertain formula.

iron red. Red varieties of ferric oxide used as pigments.

iron glance. Fe_2O_3. A black or grey variety of hematite with brilliant metallic lustre. Also known as grey hematite; specularite.

iron hat. *See* gossan.

iron pan. Hard-pan in which the cementing material is ferruginous.

iron rose. Type of crystal habit displayed by hematite or ilmenite where large crysals radiate from a centre in a rose-like pattern.

ironstone. Name applied to agrillaceous, arenaceous or calcareous sedimentary rocks which contain a high percentage of iron minerals. The minerals may be a result of primary deposition or later alteration.

irradiation. Exposure to radiation of wavelengths shorter than those of visible light (gamma, X-ray or ultraviolet) either for medical purposes, for destruction of bacteria or for inducing polymerization of monomers.

irreversible reaction. A chemical reaction that proceeds to completion; the resulting products do not react to form the original substance.

isenthalic. Of equal or constant enthapy.

isentropic. Of equal or constant entropy.

isinglass. Muscovite in thin transparent sheets.

isochemical series. A series of rocks with identical chemical compositions.

Isochemical metamorphism. A metamorphism (theoretical) involving no great change in its chemical composition. Also know as treptomorphitionsm.

isochore. (1) A line joining points of equal vertical interval between two datum planes. (2) A graph that shows the variation of one quantity with another. Also known as isochor; isometric. 3. A line that graphically represents the relationship between the pressure and the temperature of a liquid or gas, the volume of the system being kept constant.

island. A body of land smaller than a continent and surrounded by water of an ocean, sea, lake, ore stream. Islands may be formed in many ways, such as by submarine volcanic eruption, or where a mountainous coastal area submerges and peaks remain above water.

island arc. A curved chain of volcanic islands; commonly associated with deep-sea trenches occur on the landward side of the trench and are generally curved, with their convex side facing oceanward.

iso. A prefix meaning equal.

isoanomaly. A line joining points of equal geophysical anomaly; *e.g.*, magnetic anomaly, gravity anomaly.

isochromatic. Pertaining to a variation of certain quantities related to light, in which the colour or wavelength of the light remains same.

isochronon. A clock designed to maintain very accurate time.

isocline. A line connecting points of equal angle of magnetic dip.

isocon. A line connecting points of equal geochemical concentration, e.g., salinity.

isodiapheres. Nuclides in which the difference between the number of neutrons and protons is the same e.g., a nuclide and its decay product after it has emitted an alpha-particle are isodiapheres.

isodiamorphism. The phenomenon of a dimorphous substance being isomorphous (see isomorphism) with another dimorphous substance in both its forms.

isodynamic line. A line passing through points of equal horizontal intensity of the earth's magnetic field.

isoelectric. Pertaining to a constant electric potential.

isofacial. (1) Pertaining to rocks belonging to the same stratigraphic facies. (2) Pertaining to rocks belonging to the same metamorphic facies and having reached equilibrium under the set of physical conditions.

isoflor. A line connecting points of similar flora. A concept used by climatologists in an attempt to determine climatic zones.

isogonic line. A line connecting points of equal magnetic declination.

isogor. A term denoting constant gas-oil ratio.

isogor map. Oil reservoir contour-line map that shows constant gas-oil ratios.

isograd. A line joining points where the rocks have the same facies-especially rocks having attained the same grade of metamorphism.

isohaline. A line on a chart connecting all points of equal salinity.

isohume. A line on a map of chart connecting all points of equal moisture content in a coal bed.

isokom. A line joining points of equal viscosity on a phase diagram.

isoline. A synonym of isopleth.

isolith. A line on a contour-map that denotes the aggregate thickness of a single lithology in a stratigraphic succession composed of one or more lithologies.

isomegethic solution. Solutions formed of solute molecules of almost same size.

isomerism. (1) The existence of two or more chemical compounds with the same molecular formula but having different properties owing to a different arrangement of atoms within the molecule. (2) Nuclei having the same atomic number and the same mass number, but which exist in different energy states, are said to be isomeric. *e.g.*, a nucleus in its ground state and a nucleus in a metastable excited state are isomers.

isometric. (1) Referring to a system of crystalliztion in which the axes are at right angles to each other. (2) A line on a graph (isometric line) showing change of temperature with pressure, when the volume is kept constant.

isomorphism. Similarity or identity of crystalline form, usually indicating similar or analogous chemical composition; *e.g.*, most of the alums are isomorphous.

isoperm. One of the lines of equal *i.e.*, constant, permeability plotted on a reservoir map.

isophysical. A synonym of metamorphic facies.

isopleth. Any line joining points of equal value, abundance, frequency etc., on a map, *e.g.*, contour lines, isobars, isotherms; (isoline is a synonym).

isoseismal. Pertaining to points having equal intensity of earth quake shock or to a line on a map of the earth's surface connecting such points.

isostasy. A theory of the condition of approximate equilibrium in the outer part of the earth such that the gravitation effect of masses extending above the surface of the geoid in continental areas is approximately counterbalanced by a deficiency of density in the material beneath those masses, while the deficiency of density in ocean waters is counterbalaced by the excess in density of the material under the oceans. The most obvious example of this is the deep roots of mountain chains; as erosion of the mountains occurs, reducing their height, compensation in the forms of renewed uplift occurs.

isostatic compensation. The process in which lateral transport at the surface of the earth by erosion or deposition is compensated by lateral-movements in a crustal layer. Also knwon as isotatic adjustment; isostalic correction.

isosterism. The phenomenon in substances having molecules with the same number of atoms and the same total number of electrons; this leads to similarity in some physical properties, *e.g.*, carbon dioxide, CO_2 and nitrous oxide, N_2O.

isotherm line. A line connecting points at an equal temperature in graphs of thermodynamic relation.

isothermal change. A change that takes place at constant temperature, *e.g.*, the isothermal expansion of a gas.

See adiabatic.

isotones. Atoms whose nuclei contain the same number of neutrons but have a different atomic number.

isotonic solutions. Solutions having the same osmotic pressure, being of the same molar concentration, for example human blood and physiological salt solution.

isotope exchange reaction. A chemical reaction in which interchange of atoms of a given element between two or more chemical forms of the element occurs, the atoms in one form being isotopically lebelled so as to distinguish them from atoms in the other form.

isotopes. Atoms of the same element that have the same atomic number, i.e., the same position in the periodic table, but different masses. The difference in mass arised due to the presence of one or more extra neutron in the nucleus.

isotopic number. The difference between the number of neutrons in an isotope and the number of protons.

isotopic tracer. An isotope of an element, either radioactive or stable, a small amount of which is incorporated into a sample material (the carrier) in order to follow the course of that element through a process. Also known as isotopic label; label; tag.

isotropic. (1) Having a tendency of equal growth in all directions. (2) Having identical property in all directions.

isotropic material. A material whose properties are not dependent on the direction along which they are measured.

isotropic radiator. An energy source that radiates energy uniformly in all directions.

isotropy. The quality of a property which does not depend on the direction along which it is measured or of a medium or entity whose properties do not depend on the direction along which they are measured.

isotype. Pertaining to crystalline substance whose chemical formula is analogous to and whose structure is like, that of another specified (reference) compound.

issite. An igneous rock of the gabbro clan. a medium-grained granular hypabyssal rock; contains hornblende, pyroxene, and labradorite.

itabirite. Regional metamorphic rock. A schistose rock containing quartz, hematite, oligoclase and muscovite.

itacolumite. A fine-grained sandstone or thin beds, or a schistose quartzite that contains chlorite, mica and talc, and that exhibits flexibility when split into slabs. Also known as articulate.

italite. Feldspathid igneous rock; a coarse-grained granular plutonic rock. It contains leucite, augite and andradite.

insindrite .An igneous rock of the syenite clan; a fine to coarse grained granular hypabyssl rock; contains microcline, nepheline, biotite and aegirine.

I wave. A longitudinal or *P* wave in the earth's inner core.

ixiolite. $(Ta,Nb,Sn,Mn,Fe)_4O_8$. A mineral previously considered to be a mixture of cassiterite with columbite or tapiolite.

J

jack. *See* sphalerite.

jackpost. Timber used where a coal seam is separated by rock band and one bench is load out before the other.

jacobsite. $MnFe_2O_4$. A black magnetic mineral containing oxides of iron and manganese, a member of magnetite series.

jacupirangite. An Igneous rock of the ultrabasic clan; a fine-grained granular plutonic rock consisting mainly of titan-augite and aegirine-augite with accessory nepheline.

jade. (1) An extremely hard gemstone that may be either the pyroxene mineral jadeite or the amphibole mineral nephrite. Both forms are found in white and various shades of green, brown, yellow and lavender. (2) A term frequently applied to various hard green mineral.

jadeite. $NaAl(SiO_3)_2$. A clinopyroxene mineral occurring as green, fibrous monoclinic crystals; the most valuable variety of jade.

jadestone. *See* jade.

jargoon. Colourless of smoky variety of zircon from Sri Lanka.

jarlite. $NaSr_3Al_3F_{16}$. A colourless to brownish mineral containing aluminofluorides of sodium and strontium.

jarosite. $KFe_3(SO_4)_2(OH)_6$. An ocher-yellow or brown alunite mineral consisting of rhombohedral crystals. Also known as utahite.

jasper. A dense, opaque to slighly transluscent, red, crystalline quartz containing iron oxide as impurity. Also known as jasperite; jaspis.

jasperite. *See* jasper.

jasperoid. A dense, chertlike siliceous rock in which chalcedony or cryptocrystalline quartz has replaced the carbonate materials of limestone or dolomite; a silicified limestone.

jaspilite. A silceous rock resembling jasper and in bands consisting of alternating layers of red jasper and iron oxides.

jaspis. *See* jasper.

jaspoid. A black, green or brown volcanic glass formed from basaltic magma. Also known as basalt glass; sordawalite; tachylite.

jefferisite. A mineral composed of magnesium aluminium silicate of uncertain crystal system and formula; shows perfect basal cleavage. It hardness on Mohs scale is 1.5.

jeffersonite $4[Ca(Mn,Zn,Fe)\ Si_2O_6]$.Acmonoclinic mineral composed of calcium manganese zinc and iron silicate; shows uneven to conchoidal fracture and good cleavage in one direction. Its hardness on Mohs scale is 5.0-6.0.

jetcoal. A special type of compact homogeneous cannel coal or black lignite which can be carved, turned and polished for jewellery and other decorative purposes. Also known as black amber.

jetty. A pier or structure of stones, piles or the like, projecting into the ocean or another body of water to protect a harbour or a pier, or to deflect currents.

joint. A surface fracture (vertical or horizontal) in a rock without displacement. A joint set consists of a group of more or less parallel joints. A joint system consists of two or more joint sets with a characteristic pattern.

joint block. A body of rocks that is bounded by joints; the rock that occurs between adjacent joint.

joint cavity. A solutional hollow

whose position is controlled by a joint on the inner surface of a cave.
joint set. A group of parallel joint in a geologic formation.
joint system. Two or more joint sets.
joint vein. A small vein in a joint.
jordanite. $2[Pb_{14}As_7S_{24}]$. A lead-grey, monoclinic mineral composed of lead arsenic sulphide; occurs as tabular crystals or reinform masses; shows conchoidal fracture and perfect cleavage in one direction. Its hardness on Mohs scale is 3.
joseite. $Bi_3Te(Se,S)$. A mineral composed of telluride of bismuth, along with small amounts of sulphur and selenium.
jotnian. The youngest division of the Baltic Precambrian.
juanite. $Ca_{10}Mg_4Al_2Si_{12}O_{39}$. $4H_2O$. A white orthorhombic mineral composed of hydrated calcium magnesium aluminium silicate. Its hardness on Mohs scale is 5.5.
juddite. Name given to manganese amphibole occurring in India.
jumillite. An Igneous rock of the syenite clan; a brownish-grey, fine-grained porhyritic extrusive rock. It contains phenocrysts of sanidine, olivine, and phlogopite in a groundmass of sanidine, diopside, aegirine-augite and leucite.
junction. (1) The meeting of two or more streams, also the place of such a meeting; a confluence. (2) A point common to two or more survey lines.
junitoite. $(CaZn_2)Si_2O_7$. An orthorhombic mineral consisting of silicate of calcium and zinc.
junoite. $Pb_3Cu_2Bi_8(S,Se)_{16}$. An orthorhombic mineral consisting of sulphide. selenide of lead, copper and bismuth..
Jurrasian relief. A type of relief found in young mountains that consists of many parallel anticlines and synclines; characterized by primary structural forms or by features upon which erosion has had relatively little influence.
Jurassic system. One of the three periods of the Mesozoic Era, ranging from about 195 to 140 million years ago. The Jurassic is divided into the Lower Jurassic (or Lias), the Middle Jurassic (or Dogger), and the Upper Jurassic (or Malm): the boundaries are dated at 172 and 162 million years respectively. The Jurassic has been further sub-divided, on a very minute scale, into a number of local sub-divisions. Over 100 fossil zones have been erected grouped into 11 stages, the lower limit being the pre-Planorbis zone, while the upper limit is marked by the zone of Cypridea punctata.
jurupaite. $(Ca.Mg)_2(Si_2O_5)(OH)_2$. A mineral composed of hydrous calcium magnesium silicate; a variety of xonotlite.
juvenile. (1) Refers to an ore-forming fluid that is derived from magma, via fractional crystallization or other plutonic mechanism. (2) A term applied to water and gases that are supposed to have been derived directly from magma and are thought to have come to the earth's surface for the first time.
juvite. An Igneous rock of the syenite clan; a medium-grained granular plutonic rock; contains orthoclase, nepheline, cancrinite, muscovite, and aegirine.
juxtaposed ice stream. Ice from a tributary glacier that is set into the surface of a glacier and extends to the bed.
jwave. A transverse or *S* wave traveling through the earth's inner core.

K-A age. The radioactive age of a rock determined from the ratio of potassium-40 to argon- 40, present in the rock.
kaersutite. Black titanium rich variety of hornblende.
kainite. $MgSO_4KCl. 3H_2O$; or, $KMgSO_4Cl.3H_2O$. A colourless to yellowish reddish, grey, or blue, monoclinic mineral, occuring in bedded salt deposits of oceanic origin associated with alite, rylvite, polyhalite, carnallite and other salt minerals; used as a fertilizer and as a source of magnesium and potassium compounds.
kaiwekite. An igneous rock of the dirite clan; a dull-green, medium-grained porphyritic extrusive rock; contains phenocrysts of an orthoclase, aegirine-augite, hornblende, nepheline, and serpentaine in a groundmass of augite, oligoclase and magnetite.
kajanite. Feldspathoid igneous rock; a fine-grained extrusive rock; contains phenocrysts of olivine and mica in a groundmass of leucite, diopside and magnetite.
kakirite. Dynamic metamorphic rock; contains augens of residual material surrounded by zones of granulation and recrystallization.
kalinite. $KAl(SO_4)_2.11H_2O$. A colourless or white, transparent birefringent mineral of the alum group ;composed of hydrous sulphate of aluminium and potassium.
kaliophilite. $4[KAlSiO_4]$. A rare hexagonal tectosilicate colourless and transparent mineral, found in ejected blocks of biotite, pyroxenite and augite; melilite-calcite rock. Also known as facellite, phacellite.
kalirhyolite. Plagioclase poor or free type of rhyolite.
kalisilite. High temperature polymorph of kailophilite.
kalisyenite. Anigneous rock of the syenite clan; a medium to coarse-grained granular plutonic rock; contains orthoclase, microperthite, biotite, anorthoclase, hornblende, diopside and minor quartz.
kalistronite. $K_2Sr(SO_4)_2$. A colourless, transparent vitreous mineral composed of sulphates of potassium and strontium; occurs in hexagonal form.
kalitordrillite. An igneous rock of the granite clan; a fine-grained porhyritic extrusive or hypabyssal rock; contains phenocrysts of quartz, orthoclase, anorthoclase, and albite in a groundmass of intergrown quartz and orthoclase.
kalsilite. $KAlSiO_4$. A colourless to white or grey, transparent or translucent mineral occurring in hexagonal form; composed of potassium aluminium silicate; dimorphous with kaliophite.
kame. A mound composed chiefly of stratified sand and gravel, formed at or near the snout of an ice mass or deposited at the margin of a melting glacier.
kame complex. An assemblage of kames, constituting the hilly landscape.
kame moraine. An end moraine that contains numerous kames.
kame terrace. Terrace-like body of stratified drift, left by a glacier against an adjacent valley wall after the ice has melted. *See* kame.
kamperite. An igneous rock of the syenite clan; a fine to medium grained granular hypabyssal rock; contains orthoclase, oligoclase and biotite.
kandite. Group name for kaolinite, dickite, chamosite, greenalite and halloysite.

kansan. The classical second glacial stage of the Pleistocene Eoch in North America, after the Aftonian.

kaolinization. The formation of kaolin by weathering of aluminium silicate minerals or other clay minerals.

kaolin. (1) A group of clay minerals with a two layer crystal structure in which each silicon oxygen sheet is alternately linked with one aluminium hydroxyl sheet and having the approximate composition $Al_2Si_2O_5(OH)_4$. The kaolin minerals are generally derived from alteration of alkali feldspars and micas. (2) A soft white nonplastic clay composed principally of kaolinite.

kaolinite. (1) $Al_2Si_2O_5(OH)_4$. A common clay mineral of the kaolin group. It is the characteristic mineral of most kaolins and is polymorphous with dickite and nacrite. Kaolinite is a high-alumina clay mineral that does not appreciably expand under varying water contact and does not exchange iron or magnesium. The mineral was formely known as kaolin. (2) A name sometimes applied to the kaolin group of clay minerals and formerly applied to indiviual minerals of that group, *e.g.*, dickite and nacrite.

Karelian. A stratigraphic division of the Baltic Precambrian Post Svecco fennian.

Karnian. A stratigraphic stage name for the base East European and Asian Upper Triassic.

karst. A type of topography characterized by sinkholes, caves and cavaerns, dry valleys and underground drainage.

kasoite. Potassium rich variety of calcium from Japan.

kasolite. $PbUO_2SiO_4$. H_2O. A ocher-yellow, greyish green, or greenish black, transparent or translucent, also opaque mineral having monolinic crystals, composed of hydrous lead uranium silicate. Its crystals are associated with metatorbernite, curite soddyite and other secondary uranium minerals.

kassaite. An igneous rock of the diorite clan; a fine-grained porphyritic hypabyssal rock; contains phenocrysts of hauyne, labradorite, barkevikite, and augite in a groundmass of hastingsite, andesine, oligoclase, and orthoclase.

kassite. $CaTiO_2O_4(OH)_2$. A yellowish to pale-yellow mineral consisting of orthorhmbic crystals.

kata. A prefix denoting deepest, greatest or maximum depth.

katamorphism. A process of metamorphism occurring at or near the earth's surface that results in the production of simple minerals from complex minerals as a result of oxidation, hydration, solution and other processes. Also spelled catamorphism.

katatectic layer. A layer of solution residue, usually composed of gypsum and/or anhydrite in salt-dome cap rock.

katungite. An igneous rock of the gabbro clan; a very fine-grained porphyritic extrusive rock; contains phenocrysts of olivine melilite, leucite, zeolites, fluorite and apatite in a cryptocrystalline groundmass.

katzenbuckelite. Feldspathoid igneous rock; a fine-grained porphyritic hypabyssal rock; contains phenocrysts of nepheline, biotite, olivine, mosean, leucite and apatite in a groundmass of nepheline, leucite and aegirine.

Kazanian. A stratigraphic stage name for the base of the Ease European Upper Permian.

kazanskite. An igneous rock of the gabbro clan; a black fine grained granular hypabyssal rock; contains olivine plagioclase, magnetite and spinel.

kedabekite. An igneous rock of the gabbro clan; a medium-grained granular plutonic rock; contains bytownite, andradite and hedenbergite.

keewatin. A division of the Archeozonic rocks of the Candian shield.

Keelvin.K The SI unit of thermodynamic temperature defined as the fraction 1/273.16 of thermodynameic temperature of the triple point of water, i.e., the triple point of water exists exactly at 273.16 kelvins. The units of kelvin and celsius (centigrade) temperature interval are identical. A temperature expressed in degrees celsius is equal to the temperature in kelvins less 273.15^{0}C. This is true both for thermodynamic temperatures and on the International Practical Temperature Scale. The name 'degree kelvin' (Symbol ^{0}K) was discontinued by international agreement in 1967.

kelyphgtic border. See kelyphytic rime.

kelyphytic rime. A peripheral zone of pyroxene or amphibole developed around olivine in some igneous rocks. Also known as kelyphytic border.

Kenoram. A Canadian Precambrian orogeny.

kentallenite. (1) An igneous rock of diorite clan; a medium-grained granular plutonic rock; contains olivine, augite, biotite orthoclase and plagioclase. (2) A special variety of olivine monzonite.

karalite. Type of hornfels containing quartz and biotite as essential minerals.

keratophyre. An altered igneous rock of the syenite clan. A brownish-green, fine-grained porphyritic extrusive rock; contains phenocrysts of orthoclase, sodaclase, biotite, ketophorite, riebekite, barkevikite, diopside, and aegirine in a groundmass of sanidine, quartz and minor ferromagnesium minerals.

kernite. $Na_2B_4O_7$. $4H_2O$. A colourless and transparent variety of hydrated sodium borate occurring as monoclinic crystals; an important source of boron and its compounds. Also known as rasorite.

kerogen. The complex fossilized organic material which yields petroleum-type hydrocarbons on heating and distillation. Also known as petroleum.

kersantite. An igneous rock of the diorite clan; a fine-grained porphyritic hypabyssal rock; contains phenocrysts of oligoclase, biotite and hornblende in a groundmass of plagioclase and diopside.

kerstenite. $PbSeO_4$. The colourless to greenish- yellow or sulphur- yellow mineral composed of lead selenate, having orthorhombic crystals; occurs as oxidation product of primary copper and lead selenides in veins.

Keweenawan. A provincial series of the Precambrian in Michigan and Wisconsin.

khagiarite. An igneous rock of the granite clan; a black, very fine-grained glassy extrusive rock; contains phenocrysts of soda microcline, diopside, aegirine and cossyrite in a glassy ground mass.

kick. A surge or thrust against the normal fluid circulation in an oil well, as a result of a well pressure that exceeds the pressure exerted by the drilling mud Ariva.

kieserite. $MgSO_4$. H_2O. A white mineral that crystallizes is monoclinic system and is composed of hydrous megnesium sulphate; occurs chiefly in sedimentary deposits of oceanic origin associated with halite and other salts and rarely as a product of volcanic activity.

kirunavarite. An igneous rock of the ultrabasic clan; a coarse grained granular plutonic rock; contains magnetite and apatite.

kinetic metamorphism. A cataclastic metamorphism wherein rocks are shattered and deformed by external forces. The result is the formation of a new rock without alteration of the original chemical and mineralogical constituents.

kink band. A deformation band in a single crystal or in foliated rocks in which the orientation is changed due to slipping on several parallel slip planes. Also known as knick zone.

kinzigite. A coarse grained metamorphic rock that is formed principally of garnet and biotite with potassium feldspar, quartz, mica, cordierite and silimanite.

kivite. An igneous rock of the gabbro clan; a fine-grained porphyritic extrusive rock; contains phenocrysts of leucite, bytownite, olivine, biotite, augite and magnetite in a groundmass of labradorite, leucite and augite.

kjelsasite. Calcium-rich and alkali-poor type of larvikite.

klint. An exhumed coral reef or bihern that is more resistant to the processes like erosion than the rocks that enclose it so that the core remains in relief as hills and ridges.

klintite. The dense hard dolomite composing a klint gives to the core a strength and resistance to erosion.

knbelite. [$4(Fe,Mn)_2SiO_4$]. A mineral composed of iron manganese silicate; occurs as orthorhombic crystals of grey-black or brownish black colour, colourless or yellow in thin section.

knick point. 1. A point of sharp change of slope especially in the longitudinal profile of a stream or of its valley.

knoll. A small, rounded hill on a land surface or a submerged elevation of rounded form rising from the ocean floor.

knopite. The name given to Black, cerium rich variety of perovskite.

kodurite. An igneous rock of the syenite clan; a medium-grained granular plutonic rock; contains microcline, spandite and apatite.

komatiite. An igneous suite of Archean rocks that were among the first documented peridotitic magmas.

kongsbergite. A silver-rich variety of a native amalgam composed of about 95% silver.

koppite. A mineral composed of a pyrochlore containing iron potassium and cerium.

korneiite. $Fe_2(SO_4)3.7H2O$. A colourless to brown mineral composed of hydrous ferric sulphate; occurs as monoclinic crystals or fibrous silky masses.

koswite. An igneous rock of the ultrabasic clan; a coarse-grained granular plutonic rock; contains diallage, olivine, hornblende and magnetite.

kozulite. $(Na,K,Ca)_3(Mn,Mg,Fe,Al)_5Si_8O_{22}((OH,F)_2$. A mineral of amphibole group; associated with braunite, shodonite and quartz where sedimentary beds have been metamorphosed by the intrusion of granodiorite in the manganese deposits.

krageroite. An igneous rock of the diorite clan; a fine-grained porphyritic hypabyssal rock; contains phenocrysts of olivine, diopside and hornblene in a groundmass of bytownite, orthoclase, nepheline, diopside and magnetite.

kraton. A major structural unit of the

earth's crust consisting of a large stable mass of rock generally igneous and/or metamorphic sometimes with a thin veneer of sediment. A typical kraton is the Candian (Precambrian) Shield. Blocks which have been termed 'oceanic' or 'submarine' kratons probably do not conform to this defination. Kraton is practically synonymous with shield. Also spelled as craton.

kulaite. An igneous rock of the diorite clan; a fine-grained porphyritic hypabyssal rock; contains phenocrysts of olivine, diopside and hornblende in a groundmass of bytownite, orthoclase, nepheline, diopside and magnetite.

kullaite. An igneous rock of the diorite clan; a medium-grained porphyritic hypabyssal rock; contains phenocrysts of plagioclase and soda-orthoclase in a groundmass of plagioclase, sodamicrocline and chlorite.

kunzite. Lilac coloured, gem-quality variety of spodumene from California, U.S.A.

kvellite .An igneous rock of the ultrabasic clan. A medium grained porpoyritic hypabyssal rock; contains phenocrysts of lepidomelane, olivine, barkevikite, apatite, ilmenite, and magnetite in a groundmass of anorthoclase.

kwave. A longitudinal or *P* wave in the earth's outer core.

kyanite. Al_2SiO_5. A blue or light green coloured neosilicate mineral that crystallizes in triclinic system; occurs chiefly in paragonite schist, muscovite, schist gnesis and in granite and granitic pegmatites. Also known as disthene.

kynophilite. $(K,Na)Al_2Si_2O_7(OH)$. An apple-green aluminosilicate mineral; occurs as an kyanite graphite schists and sillimanite biotite geneiss. Its hardness on Mohs scale is 2.5-4.0

kylite. An igneous rock of the gabbro clan; a fine-grained granular hypabyssal rock; contains augite labradorite, olivine, ilmenite and nepheline.

kymoclastic rock. A hydroclastic rock containing marine and other wave formed fragments.

kyschtymite. A medium to fine grained hypabyssal rock composed of euhedral corrundum crystals and some biotite and green spinel in a groundmass of calcic plagiaclase.

laagte. A braod, almost level drainage courseintheveldofsouthwestern Africa, dry for most of the year, less well defined them a valley.

laanilite. Igneous rock, a form of pegmatite. A very coarse-grained rock; contains quartz, biotite and garnet with accessory iron ore minerals.

labite. $MgSi_3O_6(OH)_2.H_2O$. A mineral composed of hydrous basic silicate of magnesium.

labradite. Igneous rock of the gabbbro clan. An anorthosite, a grey plutonic rock; composed mainly of labradorite and clinopyroxenes.

labradorescence. Form of iridescence observedon cleavage planes (especially when wet) with blues, greens, and reds; opticalinterferencecausedbyminute inclusions in crystal.

labradorite.$xCaAl_2Si_2O_8+yNaAlSi_3O_8$. A blue, grey, green, or brown mineral of plagioclase feldspar group with composition ranging from $Ab_{30}An_{70}$ to Ab_{50} An_{50}, where Ab-$NaAlSi_3O_8$ and $An=CaAl_2Si_2O_8$; occurs widespread as a constituent of anorthosite, andesite, nortite, basalt, gabbro, diorite, amphibolite, and other igneous and metamorphic rocks; especially the more basic types.

laccolith. Abodyofigneousrockintruding into sedimentary rocks so that the overlyingstratahavebeeennotablylifted by the force of intrusion. It may be composite or multiple but the term multiple laccolith (sometimes cedar-tree laccolith) is also commonly used

lacuna (1) A chornostratigraphic unit representing a breach in the record, specifically,. the absent interval at a nonconformity.

lacustrine. (1) Related to, inhabiting, or produced by lakes; *e.g.*, lacustrine environment. (2) A region populated by lakes is said to be lacustrine. for a series of laccoliths 'stacked' one upon the other in a connected complex. Fundamentally a laccolith has the shape of a plano-convex lens lying flat side down. In sills the lateral dimension are very great compared with the thickness; but in laccolithic type of intrusion the latter varies betweenone-thirdandone-seventhofthe diameter.

lacroixite. A pale to yellowishgreen mineral composed of basic phosphate of aluminum calcium mar

lacustrine sediments. Sediments that are deposited in lakes.

lade. (1) The mouth of a river. (2) A water course.

ladinian. European stage; upper middle Triassic.

ladu. The basal avalanche-like part of a pyroclastic flow.

leffittite. $AgHgAsS_3$. A monoclinic mineral.

lag. A thrust fault in which the inverted limbofarecumbentfoldhasbeencutout.

lagg. The depressed marginal areas, often covered body.

lag gravel. A residual accumulation of coarse, usually hard rock fragments remaining on a surface after the finer material has been blown away by winds.

lag mounol. A remmnant of thin unconsolidated surface material left on a limestone pavement when the cover is partlally removed at underlying solution.

lagoon beach. The sandy fringe on the inner protected side of a reef island facing toward the lagoon.

lagoon channel. A pass through a reef, and into and through a lagoon.

lagoon cliff. A very steep to vertical lagoon slope.
lagoon cycle. The sequence of events, andintervaloftime,involvedinthefilling of a lagoon with sediments followed by eventual erosion by wave action and refilling by deposition.
lagoon flat. The nearly horizontal reef flat located lagoonward of the lagoon beach.
lagoonlet. A small lagoon; especially a shallow pool of water on a platform reef.
lagoon marjin. The lagoonward margin of the lagoon shelf or of the reef flat along those parts of the reef that lack islands.
lagoon. (1) A body of shallow water between a barrer island and the hore. (2) A body of water inside the coral reefs of an atoll. (3) A shallow pond or lake generally near but separated from or communicatiing with the open sea.
lahar. A destructive landslide or mudflow of hot volcaniclastic material on the flanks of a volcano, formed when water from any source combines with the hot volcanic debris and slides downward, under its weight. The water source may be water ejected from a craater lake, rapidly melted ice or snow from subglacial eruptions, or very heavy rainfall.
lakarpite. Igneous rock of the syenite clan. A course-grained plutonic rock similar in appearance to dirite; contains orthoclase or microcline with albite, amphiboles and minor aegirine, apatite, and altered nepheline.
lake. (1) A term loosely applied to a sheet of water lying along a coast and connected with the sea. (2) Any inland body of standing water occupying a depression is the earth's surface, generally larger than a part and too deep to permit vegetation. The term includes an expanded part of a river, a reservoir behind a dam, or a lake basin intermittently or formerly covered by water. (3) An English term for a brook or small stream.
lake deposit. A sedimentary deposit laid down conformably on the floor of a lake, usually consisting of coarse material near the shore and sometimes passing rapidly into clay and limestone in deeper water.
lake gage. A gage for measuring the elevation of water surface of a lake.
lakelet. A small lake.
lake loam. A term applied to loesss that may have been formed by deposition in lakes.
lamellar flow. A movement of a liquid in which individual particles move in parallel sheets that slide over one another. Magma beneath the surface of the earth shows lamellar flow.
lamina. A thin, clearly differentiated layer of sedimentary rock or sediment, usually less than 1 centimetre thick.
laminar flow. Adjacent layers of fluid that move smoothly without mixing. Laminar flow occurs in conditions of low velocities, small channel sizes, and high viscosities.
lamination The formation of laminae or the state of being laminated; specifically the finest stratification, such as found in shale lamination. The development of thin, discrete layers of rock. The individual layers may differ from one another or may be homogeneous, the layers being separated by a physical discontinuity, *e.g.*, bedding planes.
lampadite. A variety of wad containing copper; composed of hydrous manganese oxides with about 18% copper oxides and often associated with cobalt oxide.
lamprobolite. A product obtained by heating common hornblende to about 800°C.
lamprophyllite. $Na_2(Sr,Ba)_2Ti_3(SiO_4)_4(OH,F)_2$. A golden brown to dark brown

mineral belonging to orthorhombic and monoclinic systems; occurs chiefly in nepheline syenites, nepheline, aegirine, eudialyte, and other minerals.

lamprophyre. A group of mafic dike rocks. Althoughencompassingarangeof compostions, lamprophyres exhibit in common a porphyritic fabric with mafic minerals as constituents of both phenocrysts and matrix. Feldspars and feldspathoids are confined to the aphanitic matrix, and thus frequently intersect granite plutons in such a manner as to be in some way genetically related to granitic magmatism. Lamprophyresareoftenhighlyaltered, thefeldsparsinparticularbeingheavily decomposed. Fresh olivine is exceedingly rare and it its almost always altered to serpentine and carbonate. Somelamprophyresareveryrichincarbonate content and give effervescence withmineralacids,likelimestone.This widespread alteration is presumably due to late-stage hydrothermal activity. Since lamprophyres are melanocratic varieties of microsyenite and microdiorite, their chemistry is more of less identical with these types, although the high ferromagnesian mineral.

lanarkite. $PbO.PbSO_4$. A white, greenish, or grey monoclinic mineral with prismatics crystals that may be transparent or translucent; occurs as a secondary mineral in the oxidation zone of lead ore deposits; found associated withcaledonite,leadhilliteandcerussite.

land bridge. A strip of land linking two landmasses, often subject to temporary submergence, but permittent migration of organisms.

landerite. A pink to rose-pink variety of grassular garnet.

landerite. $Mn_3(II), Fe(III)(PO_4)_2 (OH)_3.3H_2O$. A brown mineral, occurringasanalterationproductofreddingite.

landocked. Said of a body of water that is nearly or entirely enclosed by land.

landmass. A land area studied as a unit, withoutregardnecessarilytosizeorrelif, onthebasisofthesedimentsderviedfrom it or the paleogeographic evidence indicated by the change in shorlines.

landscape. The distinct association of landforms that can be seen in a single view.

landscape agate. A type of chalcedony thatistranslucentandcontainsinclusions which give it an appearance reminiscent of familiar natural scences. Also knwon as fortification agate.

landscape marble. A limestone which shows patterns resembling a landscape scene when cut at right angles to the bedding; a popular term for a Jurrasic argillaceous limestone of Bristol.

landslide. The movement of a mass of rock or soil, or both down a slope under the influence of gravity. A landslide differs from an earth flow in that the mass of rock remain more or less in one piece. Also known as landslip.

landslide breccia. A breccia that is largely fragmented and whooly assembled by the force gravity, as by rockfall or a rockclide.

landslidelake. A lake resulting from the damming of a stream valley by the material in a landslide.

landslidesapping. The process of causing landslides by a stream undermining a canyon wall.

landslide scar. A bare or relatively bare surfacesornicheontheside of a mountain or other steep slope, left by the removal of earth material from the place where a landslide started.

landslide terrace. A short, rough-surfaced terrace resulting from a landslide.

landtiedisland. A tied island connected

with the mainland by a tombolo.
landslip. See landslide.
langbanite.(Mn(II),Sb(III)4-(Mn(IV),Fe (III),M)$_3$SiO$_{12}$. An ironblack, opaque, hexagonal mineral composed of silicate and oxide of manganese, iron and antimony, occurring in prismatic crystals.
langbeinite.4[$K_2Mg_2(SO_4)_3$]. A cubic, colourless mineral composed of potassium magnesium sulphate; shows no cleavage. Its harness on Mohs scale is 3.5-4.0.
langite. $Cu_4SO_4(OH)_6.2H_2O$. A blue to green mineral composed of hydrous basic copper, sulphate : occurs as a secondary mineral, associated with gypsum.
landsfordite.$MgCO_3.5H_2O$. A mineral composed of hydrate carbonate of magnesium. It changes to nesquehovite when exposed to air. It shows indistinct fracture and cleavage in two directions. Its hardness on Mohs scale is 2.5.
lanthanite. $(La,Ce)_2(CO_3)_3.8H_2O$. A colourless, white, pink or yellow mineral; composed of hydrous carbonate of lanthanum and cerium.
lapies. A limestone pavement.
lapilli. Fragments of pyroclastic rocks measuring between 4 and 32 mm in diameter.
lapilli tuff. A pyroclastic deposit that is indurated; consists of lapilli in a fine tuff matrix.
lapis Lazuli. An azure-blue, violet blue, or greenish-blue, translucent to opaque crystalline rock chiefly composed of lazurite and calcite with some haiiyne, sodalite, and other minerals; used as a semi-precious stone. Also known as lazuli.
lardalite. A nepheline-bearing larvikite.
Larsen variation diagram. Weight per cent of each oxide constituent in a rock analysis is plotted. As the ordinate against the abscissa position; defined as $1/3SiO_2 + K_2O\text{-}FeO\text{-}MgO\text{-}CaO$; a smooth curve is drawn through the points representing a given constituent in a series of analysis.
larnite.4[Ca_2SiO_4]. A monoclinic white mineral composed of calcium silicate.
larsenite. $PbZnSiO_4$. An orthohombic, colourless mineral composed of lead ,zinc silicate ; member of the chrysotile family. **larvikite.** An igneous rock of the syenite clan ; a greyish black plutonic rock. It contains *sodaorthoclase, anorthococlase,* titaniferous *pyroxene, biotite* and *hornblende* with minor nepteline. The coarse-grained *feldspar* rhombs give a fine lustre ; used as an ornamental stone. Also known as laurvigite.
lassallite. $Mg_3Al_4Si_{12}O_{33}.8H_2O$. A fibrous member of the palygorskite family.
lassenite. Unaltered form of volcanic glass.
lateral accretion. Lateral growth of a sedimentary deposit, *e.g.*, the building up of the inner bank of a stream meander by deposition of material brought their by rolling of pushing along the bottom.
lateral secretion. The theory that contents of a vein or load are derived from the adjacent wall rock.
laterite. A sedimentary rock; white to cream to red, amorphous or crystalline residual deposit. It consists of hydrated aluminium iron oxides.
lateritite. A sedimentary rock; redeposited eroded laterite normally mixed with quartz or gneiss fragments. It is formed by mechanical deposition.
lateroid. The metasomatic replacement product. It contains hydrated iron or magnesium oxides.
latite. A propyritic extrusive rock having phenocrysts of plagioclase and potas-

sium feldspar in almost equal amounts, with little or no quartz, and a finely crystalline to glassy groundmass ; the extrusive equivalent of monzonite. It is usually considered synonsmous with trachyandesite and trachytasalt, depending upon the colour.

latitude. Angular distance of a point on the earth's surface north or south of the equator measured along a meridian, the equator being lattitude zero degrees.

latiumite. Rare calcium potassium aluminuum sulphatic silicate.

latosol. A great group of zonal soils characterized by deep weathering and abundant hydrous-oxide material. They are developed under forested humid tropical conditions.

lattice. A regular periodic arrangement of points in three dimensional space; it consists of all those points *p* for which the vector from a given fixed point to *p* has the form $n_1a+n_2b+n_3c$, where n_1, n_2, and n_3 are integers, and *a*, *b*, and c are fixed, linearly independent vectors. Also known as periodic lattice ; space lattice. *See* also unit cell.

lattice constant. A parameter defining the unit cell of a crystal lattice *i.e.*, the length of the edges of the cell or an angle between edges. Also known as lattice parameter.

lattice parameter. *See* lattice constant.

lattice-preferred orientation. The preferred orientation of crystallographic axes or planes in metamorphic rocks, it results from crystal gliding and /or dynamic crystallization and is dependent on the mineral structure and on the pressure, temperature and stress during deformation. In igneous rock it is mainly related to the original shape of the crystals during settling or flow. Also known as shape-preferred orientation.

lattorfian. A stratiigraphic name for the European Lower Oligocene.

Laurasia. The northern protocontinent of the Mesozoic Era, corresponding to *Gondwana* in the Southern Hemisphere north of the Himalayas, prior to its break up under the forces causing Continental drift. It corresponds to Gondwanaland in the southern hemisphere.

laurdalite. An alkalic syenite containing more than 10 percent modal feldspathoids and characterized by porphyritic texture. Igneous rock of the syenite clan. A very coarse-grained pseudoporphyritic plutonic rock; contains nepheline, anorthoclase, soda-orthoclase, diallage and diopside.

laurentian A stratigraphic division of the Canadian precambriam, the pre Huronian granite intrusion into the Ontarian in Canada.

laurvigite. *See* larvikite

lava. Molten rock that issues from openings at the earth's surface or on the ocean floor; the term also applies to the rock formed from the solidified material. Such openings may be located in craters or along flanks of volcanoes, or in fissures not associated with volcanic cones. Its chemical compostion and gas content vary with the particular volcano form which it is ejected; moreover, lava extruded at an early eruptive stage may differ from later extrusions. Lava generally contains varying proportion of oxides of Si, Al, Ca, Na, Mg, K, Fe, and water, together with smaller quantities of other elements, large areas, while basic lavas have a much lower viscosity and flow readily. Lavas which are extruded under water commonly take on the form of distorted globular masses, the characteristic so-called pillow shape. They apparently form as a result of the rapid chilling of the outer skin of the lava which forms at the liquid/liquid interface, resulting in a more or less spherical 'balloon' of lava. As this grows in size, it flattens under its own weight. The pillows have highly vesicular cores.

lava blister. A small hollow steepsided swelling raised on the surface of a lava flow by gas bubbles puffing up the viscous crust of the flow.

lava cascade. A cascade of fluid incandescent lava formed when a lava river passes over or over a precipitious part of course.

lava cave. *See* lava tube.

lavacone. Avolcaniccone formedoflava flows.

lave dome. A done-shaped mountain of solidified lava in the form of many individual flows formed by the extrusion of highly fluid lava.

lava flow. A lateral surface out pouring of molten lava from a vent or fissure; also the solidified body of rock that is formed. Lava flows interbedded with sedimentary rocks may be confused with concordant sills. The following criteria may be used for distinguishing purposes:

Any evidence that the rock layer transgresses the bedding planes indicates that it is a sill.

lava fountain. A Jetlike eruption of lava that issues vertically from a vent or fissure; shot into the air magma reaches the surface by the hydrostatic pressure on the liquid and the expansion of gas bubbles forming in it. Fountains usuallly rangs from 100 metres in heights and sometimes reaching 300 metres,. Also known as fire fountain.

lava lake A lake of lava that is molten and fluid; usually contained within summit volcanic crater in a pit crater on the flanks of a shield volcano.

lava plain. A broad stretch of level or nearly lvel land usually many hundreds of square kilometers in extent underlain by a relatively thin succession of lava flows most of which are basaltic and the product of fissure eruptions.

lava plateau. An elevated tableland or flat-topped highland that is several hundreds to several thousands of square kilometre in area; underlain by a thick succession of lava flows.

lava shield. *See* shield volcano.

lava tube. A hollow space beneath the surface of a solidified lava flow formed by the withdrawal or molten lava after the surficial crust. Also known as lava cave; lava tunnel.

lava tunnel. *See* lave tube.

lavialite. Low grade regional metamorphic rock; contains relict phenocrysts of labradorite in a recrystallized groundmass of green hornblende; veined with alteration passages of quartz. microcline, biotite and hornblende.

law. A regularity which applies to all members of a broad class of phenomena.

law of equal declivities. Where homogenous rocks are maturely dissected by consequent streams all hillside slopes of valleys cut by the streams tend to develop at the same slope angel thereby producing symmertical profiles of ridges spars and valleys.

law of faunal assemblages. A general law of geology which states that similar asemblages of fossil organisms indicate similar geologic ages for the rocks that contain them. **law of faunal succession.** A general law of geology which states that fossil organisms succeed one another in a definite and recognizable order each geologic formation having a different total aspect of life from that in the formation above it and below it; or the relative ages of the rocks can be determined from their fossil content.

law of homonymy. A principle in toxonomy stating that any name that is a junior homonym of another name must be rejected and replaced.

lawofincludedfragments. Thelawstating that if recognizable fragments of one rock are found enclosed in another rock, then the enclosing rock must have been formed later than the rock material of the included fragments.

law of nature. A generalization of science representing on intrinisc orderliness of natural phenomena or their necessary conformity to reason.

lawoforiginalcontinuity. Agenerallaw of geology which states that a water-laid stratum at the time it was formed must continue laterally in all directions untill it thinsout as a result of nondeposition or until it abouts against the edge of the original basin of deposition.

law of original horizontality. Water laid sediments are deposited in strata that are horizontal of nearly horizontal and parallel or nearly parallel to the earth's surface.

lawofpriority. Aprinciple in toxonomy stating that nomenclature of a toxonomic group is based on priority of publication. Thus the valid name of a genus or species can be only that name under which it was first designated.

law of rational indices. The statement incrystallographythatcrystalfacesmake simple rational intercepts on suitablecrystal axes, *i.e.*, the axes of reference or the three axes forming the edges of the unit cell of each crystal lattice.

lawofstream gradients. A general law expressingtheinversegeometricrelation between stream order and the mean stream gradientofagivenorderinagiven drainage basin.

lawofsuperposition. A law upon which all geologicchronology is based. It states thatinanysequenceofsedimentary strata thathasnotbeenoverturned the youngest stratum is at the top and the oldest at the base; each bed is younger than the bed beneath, but older than the bed above it.

laxfordian. A startigraphic division of the Scottish Precambrian the younger Lweisian.

layer. A tabular body of rock ice, sediment, or soil lying parallel to the supporting surface and distinetly limited above and below it.

layer depth. The depth to the top of the

Sill	Lava flow
Uniform and regular cindery.	Top generally irregular and often
No fragments in adjoining sediments.	Fragments may occur in overlying rocks, as pebbles or xenoliths.
Rarely vesicular or amygdaloidal.	Commonly vesicular or amygdaloidal.
Does not show weathering of upper surfaces.	May show weathering of upper surfaces.
Medium to fine-grained rarely glassy owing to slower cooling. rapid cooling	Fine-grained or glassy owing to
Metamorphoses the upper and lower contact rocks. only.	Can metamorphose the lower contact
Stringers and veinlets of rock may penetrate both overlying and under lying rocks.	Small veins in underlying rocks only Care must be taken when flows have irregular tops.

thermocline, *i.e.*,to the bottom of the mixed layer.

layer-lattice minerals. The sheet silicates or phyllosilicates. The layer-lattic group of minerals.

layeredintrusion. An intrusive body in which there are layeres of varying mineralogical composition.

layered pernafrost. Ground consisting of perennially frozen layers alternating with unfrozen layers or taliks.

layered series. A body of igneous rocks showing banding that simulates the stratification of a sedimentary sequence andhavingaparallalismofplatyminerals thatresembles the beddingof sediments.

loyering. A tabular succession of the different components of igneous, sedimentary or matamorphic rocks, or the formation of layers of material, one upon the other, in a particular rock.

layer stripping. A procedure that removes the effects of an upper crystal layer. It modifies seismic and gravity data to make than appear as though the seismic source and receivers, or the gravity stations, were at the base of the stripped layers and unaffected it.

layer structure. A type of crystal structure built up by the distinct layer writs; as in mica, clay and graphites.

layer tinting. A method of depicting relief on a map by the distinctive shading or coluring of the areas between contour lines in a manner suggestive of progressivechange,sothatthepatternofdistribution of high and low areas is revealed or emphesized at a glance.

lazuli. A synonym of lapis lazuli. Also spelled lazule.

lazulite.(Mg,Fe(III) $Al_2(OH)_2(PO_4)_2$. A violet-blue or azureblue mineral with vitreous lustre; composed of basic aluminium phosphate and occurring in masses or monoclnic crystals. Also known as false lapis.

lazurite. $(Na,Ca)_{7-8}(Al,Si)O_{24}S,SO_4,C_{12},(OH)_{22}]$. A blue or violet-blue felspathoid mineral; the principal constituent of lapis lazuli.

leachate. A solution obtained by leaching.

leaching. (1) The dissolution of soluble constituentsofamixturebyaliquiedsolvent. (2) The extraction of soluble metalsorsaltsfromanorebymeansofslowly percolating solutions.

lead glance. PbS. A synonym of galena.

leadhillite. $Pb_4(SO_4)(CO_3)_2(OH)_2$. A yellowish, greenish or greyish-white, monoclinic mineral consisting of basic carbonate and sulphate of lead.

leading edge. The frontal edge of a thrust sheet with respect to the given stratigraphic unit.

lead-lead age. An age in years calculated from the ratio of lead-207 to lead-206, a byproduct of the uranium thorium-lead age method.

lead ocher. A yellow or red, sealy or earthy lead mononide.

leadratio. The ratio of one lead isotope to another, *e.g.*, Pb-206/Pb-204; Pb-207/Pb-209, Pb-208/Pb-204; and Pb-207/Pb-206.

leaked fossil. A fossil deposited as a result of a stratigraphic leak.

leaking mode. A seismic wave that is imperfectly trapped, so that its energy dissipatesorescapesacrossalayersboundary, causing some attenuation.

leakyequifer. A confined aquifer whose confining beds will conduct significant quantities of water into or out of aquifer.

leaky confining bed. A confining bed through which water can move in or out of the adjacent acquifer.

lean cannel coal. Cannel coal that is low in hydrogen and transitional to bituminous coal in rank.

lean clay. A clay of low plasticity owing to a relatively high content of slit or sand.

leaping. The sudden and radical shifting of a divide from one possible to another, as where the valley system drained by a captured stream is transferred and added in a moment to that of the master stream.

lead-uranium ratio. The ratio of lead-206 to uranium-238 and/or lead-207 to uranium-235, formed by the decay of radioactive uranium within a mineral. The ratios are frequentaly used as a part of the uranium-thorium lead age maethod of estimating the geologic age of a rock of mineral.

lechatelierite. Naturally occurring fused silica glass found in quartzose enclosures in volcanic rocks.

lectotype. A syntype, chosen if needed after the original description, to take the place of the holotype.

ledian. Cription, to take the place of the holotype.

ledian. A stratigraphic stage name of the base of the European Upper Eocene.

ledmorite. Igneous rock of the syenite clan. A melanocratic holocrystalline-granular plutonic rock; contains orthoclase, and radite, biotite, pyroxene, and apatite.

lee. The side of hill, dune or other object that is sheltered or turned away form the wind.

left-lateral fault. Displacement along a fault such that, in plan view, the side opposite the observer appears displaced to the left.

leidleite. Igneous rock of the dioriteclan. A porphyritie hypabyssal pitchstone; contains phenocrysts of plagioclase in a groundmass of feldspars, augite and glass.

Leintwardinian. A stratigraphic stage name for thebase of the British Upper Ludlovian (Silurian System).

legend. Explanation of the symbols and patterns shown on a diagram of map.

lenticular. Having the shape of a biconvex lens.

Leonardian. A statigraphic stage name for the North American Lower permian.

lepidomelane. Iron rich variety of biotite. Its Varieties are alurgite, annite (of Dana) and hydrodyal-annite..

leonite. $K_2Mg(SO_4)_2.4H_2O$. A colourless, white, or yellowish mineral composed of hydrous magnesium potassium sulphate occurs as monoclinic crystals.

leopardite. Igneous rock of the granite clan. A streaky fine-grained porphyritic hypabyssal rock; contains phenocrysts of quartz in a groundmass of albite, quartz, orthoclase, and mica; often stained by hydroxides of iron and aluminium.

lepidoblastic. A textural term for a schistose or foliated rock caused by parallel alingnment during the recrystallization of platy minerals such as micas, chlorite, talc and graphite.

ledipocrosite. FeO(OH). A ruby or blood-red mineral crystallizing in orthorhombic system; associated with limonite in iron ores and is a component of meteroties.

lepidolite. $K(Li,Al)_3(Si,Al)_4O_{10}(F,OH)_2$. A rose or pink coloured mineral of mica group crystallizing in monoclinic system. Also known as lithium mica..

leptite. A quartz feldspathic metamorphic rock that is finegrained with little or no foliation; formed as a result of regional metamorphism of the highest grade.

leptynite. Medium grade regional metamorphic rock. A light coloured coares grained granular rock; contains quartz, feldspar, mica and chlorite.

leuc-, leuco. A prefix denoting white, colourless.

leucite. $KAlSi_2O_6$. A white or grey mineral of feldspathoid group. It is an important rock forming mineral in alkalic rocks, and usually occurs in trapezohedral crystals with a glassy fracture.

leucitite. A fine-grained or porphyritic extrusive rock or hypabyssal igneous rock, mostly composed of pyroxene and leucite.

leucitophyres. (1) Ignerous rock of the gabbro clan. A porphyritic extrustive rock; contains phnocrysts of leucite in a groundmass of leucite, nepheline and aegirine. (2) Undersaturated trachytes containing dominant leucite.

leucocratic. A term applied to light-coloured rocks, especially igenous ones that contain between 0 and 30% of dark minerals; *i.e.*, rocks of color index between 0 and 30.

leucophyre. An igenous rock of the *gabbro* clan; a porphyritic plutonic rock. It contains *feldspar* phenocrysts (often partly altered to saussurite and kaolinite) in a groundmass of pyroxene, ilmenite and quartz.

leucotephrite. A geological term used to denote leucite rich, nepheline free, tephrites.

leucophanite. $(Na,Ca)_2BeSi_7(O,F,OH)_7$. A greenish mineral composed of beryllium, calcium, sodium silicate containing fluorine and occuring in glassy, tabular crystals.

Leucoxene. A general term for fine-graned opaque, whitish, alteration products of ilmenite, commonly consisting mostly of rutile and partly of anatase of sphene and occurring in some igneous rocks. Sometimes, the term is also used to designate avariety of sphere.

Levantain. A stratigrahic stage name for the top of the European Pliocene (Tertiary System).

levee. A raised bank formed by deposition, which allows the level of the river to rise above its flood plain.

Lewisian. A stratigrphic division of the Scottish Precambrian, the oldest in Britain, covered, uncoformably by Torridonian.

lag wave. A short-period, higher mode surface wave, with a group velocity of about 3.5 Km/sec, that travels over longpaths in continental crusts only.

lherzite. Igneous rock of the ultrabasic clan. A coarse-grained hypabyssal rock; contains biotite, hornblende, ilmenite, and garnet.

lherzolite. An ultrabasic rock consisting of olivine ad orthoand clino-pyroxenes.

liebenerite porphyry. Altered nepheline porphyry; contians nepheline phenocrysts replaced by a sealy scricite aggregate.

liesegang rings. Secondary, rested rings or bands caused by rhythmic precipitation within a fluid-saturated rock.

ligerian. A stratigraphic stage for the European Lower Turonian (Cretaceous System).

light mineral. A rock-forming mineral of a detrital sedimentary rock, having a specifc gravity lower than 2.85; *e.g.*, dolomite, feldspar, feldspathoids, muscovite, quartz.

lignite. A brown black coal that is intermediate in coalification between peat and sub-bituminous coal; consolidated coal with a calorific value less than 8300 BTU/Ib, on a moist, mineral-matter-free basis.

limb. That area of fold between adjacent fold hinges; it may be planar or slightly curved.

limaite. Tin rich variety of gahrite penocrysts with little or no feldspar. Analysis suggests that it is commonly undersaturated.

lime feldspar. A misnomer for calcium feldpar, *i.e.* anorthite.

limestone. Any sedimentary rock consisting essentially of carbonates; common impurites include chert and clay. It is the most common, important, and widely distributed of the carbonate rocks and is the consolidated equivalent of limy mud, calcareous sand, and/or shell fragments. The two most important constituents are calcite and dolomite, but small amounts of iron-bearing carbonates may also occur. Limestones can be classified into three major groups organic; chemical; and detrital or clastic. Organic and chemical limestones are known as autochthonous limestones, whereas the clastic limestones are known as allochthonous. Limestones may be fresh water or marine, and usually indicate deposition in a warm, clear water environment.

lime. (1) CaO. A general name for calcium oxide. (2) A term commonly missued for calcuim in deplorable expressions such as line feldspar.

limeclast. A litholast derived by erosion from one older limestone, also an intraclast disrupted from partly consolidated calcareous mud on the bottom of a sea or lake.

lime concretion. A conceretion is soil, having a variable shape and size and consisting of an aggregate of precipitated calcium carbonate or of other material cemented by it.

lime mud. The unconsolidated micritic component of a limestone.

limepan. (1) A playa with a smooth, hard surface of calcium carbonate, commonly tafa. (2) A type of hardpan cemented chiefly with calcium carbonate resulting in uncertainties as to its mode of origin. This is particularly true of the ancient limestones.

organic limestones. Many plants and animals secrete $CaCO_3$, and, after the death of the organism, their skeletons, either complete or broken up, may in the absence of any other sedimentary material accumulate to form a limestone. Some special types of organic limestone are :

Reef limestones	biohermal limeetones
Shelly limestones	
Coral limestones	collectively known as Algal
lsmestones	biostromal limestones
Crinoidal limestones	(Dchinodermata)
Foraminiferal(Protozoa)	some biostromal, somelimestones pelagic varieties

Chalk (more or less synonymous with pelagic limestone) is the characteristic Upper Cretaceous white, fine-grained limestone, but 'chalks' of other ages have been recorded, e.g., Tertiary. (When used in the stratigraphical sense 'C') The type 'Chalk' from south-east England consists of a small percentage of foraminiferal material, some finely comminuted shell material, and a high proportion of calcareous skeletal material from unicellular planktonic algae.

Chemically prrecipitated limestones. Mainly these are of three types : (a) limestones belonging to an evaporite sequence (often dolomite); (b) oolitic and pisolitic limestones (c) cal tufa. *Clastic (detrital) limestones.* These mechanically deposited carbonate rocks are made up essentially of fragments of organic carbonate or a preexisting limestone rock. (Oolitic limestone is included in this group by some authors.) The classification of clastic limestones is based upon the same grain-size scale as used for the detrital sedimentary rocks :

calci-rudite	2 mm and above
Calcarenite	1/16 mm-mm
Calci-lutite	Less than 1/16 mm

If the debris is dominantly organic in

origin the prefix 'bio' may be used ; bioclasticlimestonesarenotuncommonly found interbedded with biostromal limestones. The term coquina is applied tocemented coarse shell debris. Some of the shelly 'grits' of the English Mid-Jurassic are coquinas. Note that the 'calcite is synonymous with calci-lutite, and doesnot imply an argillaceous limestone or calcareous clay or shale.

The three-fold division of limestones given, though theoretically ideal, is increasingly being replaced by a more practical rock nomenclature. This terminology is based upon the nature of the limestone matrix and of the particle types present. Two matrix types are distinguished. Their names form the suffix of the resultant rock names :

Micrite; micro-crystalline calcite (grain size less than 0.01 mm)Sparite ; coarse calcite tshan 0.01 mm) forming a mineral cement.

The prefixes of the limestone names are determined by the particle types present. These may include (prefixes are in parenthese) :

Intraclasts (Intra-), Ooliths (Oo-), Pellets (Pel-), Organic hard parts, whole or fragmented (Bio-).

Thus an oolitic rock with a sparite cement is termed as oosparite, whereas one composed of intraclasts set in micrite is an intramicrite. A pure calcilutite may be termed a micrite, and a limestone composed entirely of reef-building organisms, a biolithite. Rocks consisting of only micrite and sparite are called dismicrites. This system of naming limestones has many practical advantages, since it does not depend upon the determination of the origin of the sediment.

limnetic. (1) Relating to the pelagic or open part of a body of fresh water. (2) Said of lake-dwelling organisms and communities that are free from direct dependence on the bottom or shore.

limnic. (1) A term used for a body of fresh water. (2) Said of coal deposits formed inland in peat bogs, as distinguished from paralic deposits.

limnobios. The life of the freshwater environment.

limnology. The scientific study of fresh waters, especially of ponds and lakes.

limonite. A field term for a group of brown amorphous hydrous ferric oxides. Originally it was thought to be a mineral with a definite composition-$2Fe_2O_3$, $3H_2O$, or similar, but is now known to consist of various amorphous and cryptocrystalline constituents. The following are the most important : goethite ($\alpha FeO.OH$), which on heating yields haematite (Fe_2O_3); lepidocrosite ($_yFe_2O_3$) ; turrgite-haematite with absorbed water.

Limonite also includes various varieties of amorphous iron hydroxides. It occurs widely as coatings, earthy masses, and in a number of other forms as the weathering product of all iron-containing minerals, and is frequently intermixed with many other substances including colloidal silica, clay minerals, aluminium hydroxides, *etc*. The most characteristic property of the mineral is its yellow or brownish mineral is its yellow or brownish streak. Its colour ranges from yellow, various shades orange, brown, and red, to black, and this is a function of particle size, the proportions of the various constituents, and the extent of hydration. Amongst well-known types of limonite are deposits of gossan red and yellow ocheres, bog iron ore, and laterite. Limonite may be considered as a minor ore of iron. Also known as brown hematite.

limurite. High grade contact metamorphic rock. A coarse granular rock ; contains axinite, diopside, actinolite, zoisite, albite, and quartz ; produced at granite, calcareous rock contacts.

limy. (1) Containing significant amount of lime or limestone. (2) Containing calcite, *e.g.*, limy dolomite.

lindoite. An Igneous rock of the granite clan; a medium to coarse-grained hypabyssal rock. It contains microcline-microperthite, arfvedsonite, quartz and zircon.

lineament. A linear topographic feature of regional extent that is believed to reflect crystal structure and which is in itself an expression of the underlying structural features. Such features may include: valleys controlled by faulting or jointing; fronts of mountain ranges; narrow, straight mountain or hill ranges and ridges; lines of isolated hills; straight coastlines, modern and ancient. Structural features associated with lineaments include fault zones, fracture (joint) zones, fold axes, linear igneous intrusions, lines of volcanoes or volcanic fissure eruptions. Lineaments are often located by means of a detailed study of maps (both geological and topographic) and serial photographs (photogeology). The correlation of these major structural trends is important for the elucidation of regional tectonic history (Fig. 58).

lineation. A general term for any rock feature showing a linear structure ; *e.g.*, parallel crystal arrangement, slickensides, boudinage, and fold axes.

lingulid. Any lingulacean brachiopon belonging to the family Lingulidae, characterized mainly by an elongate oval to spatulate outline and a biconvex shell.

link. An unbroken section f a stream channel between two notes in a drainage network.

linkage. The jointing at a sharp angle of two branches of a mountain range.

linked veins. An ore deposit pattern in which adjacent, more or less parallel veins are connected by diagonal veins or veinlets.

links. (1) A winding course of a river. (2) A term used for a narrow area of flat or indulateral land built up along a coast by drifting sand, and covered with treaf or coarse grass.

linn. (1) A pool of water, especially a deep on below a fall. (2) A torrent running over rocks. (3) A precise or a steep ravine.

liparite. A synonym of rhyolite.

lipogenesis. In evolution, accelerated development as a result of the omission of certain ancestral stage.

lipotextite. Nonligufied basic material with in anatectic magma.

liptite. A microlithotype of coal that contains is combination of eximite marcerds totalling at least 95%.

liplobiolith. A combustible organic rock formed by the accumulation of liptobiolites; *e.g.*, sport coal.

liplodatrinite. A maceral of coal within the exinite group, having no recognizable structure and low reflectance and florescence.

liquation. The separation of a solid mixture by heating till one of the constituents melts and can be drained away.

liquefaction. A change in phase of a substance to the liquid state, especially of a substance which is a gas at normal pressure and temperature ; *e.g.*, liquefaction of air.

liquefaction of gases. A gas possessing a critical temperature above room temperature may be liquefied merely by increasing the pressure on it. Otherwise, the gas must first be cooled to below its critical temperature and then compressed ; or if desired, cooled directly to its boiling point under normal pressure.

liquefied gas. A gaseous substance converted to liquid phase by cooling or compression ; *e.g.*, liquid air; liquefied petroleum gas (LPG).

liquid. A state of matter intermediate between a solid and a gas, in which the molecules are relatively free to move with respect to each other but are restricted by cohesive forces to the extent that the liquid maintains fixed volume. Liquids assume the shape of the vessel containing them, and are only slightly compressible.

liquid-crystal display. A digital display in an electronic device like calculator, etc., based on liquid crystal cells that change their reflectivity in an applied electric field.

liquid crystals. Compounds in transition state between solid and liquid forms. They are viscous, jelly-like materials that are not isotropic and exhibit interference patterns in polarized light. The various esters of cholesterol are the examples of liquid crystals. They have the ability to indicate small temperature differences by changing colour. They are of three types: nematic, smectic, cholesteric. Liquid crystals are used in colour TV and electronic display tubes ; temperature sensitive tape ; and electronic drives for clocks and calculators. They can align with dichroic dye molecules in a thin-layer cell to produces colour changes in transmitted light.

lithic arkose. (1) A term for an arkose containing appreciable rock fragments. (2) A term used for a sandstone containing less than 75% quartz and metamorphic quartzite and having or F/2 ratio between 1:1 and 3:1, where F denotes feldspars and fragments of gneiss and granite and R denotes all other fine-grained fragments.

lithic crystal tuff. A tuff that is intermediate between crystal tuff and lithic tuff but is predominantly the latter.

lithic subarkose. A term used for a sandstone composed of subequal amounts of feldspar and rock fragments; specifically a sandstone containing 10-25% feldspar, 10-25% rock fragments and 50-80% quartz, quartzite and chart.

liroconite. $4[Cu_2Al(AsO_4)(OH)_4.4H_2O]$. A sky-blue to green monoclinic mineral composed of basic hydrated copper arsenate; shows uneven fracture and indistinct cleavage. Its hadness on Mohs scale is 2.0-2.5.

listric. A term applied to fracture planes which curve, either to near horizontal ones which steepen, or to near-vertical ones which become less steep. Those which are concave upwards conform to the definition most strictly, and the term 'negatively listric' has been applied to those which are concave downwards. The term 'shoulder-thurst' has also been used for the latter type.

listwanite. Low grade regional metamorphic rock. A yellowish green schistose rock; contains dolomite, magnesite, quartz, talc and limonite.

lith- A prefix meaning rock, rock-like.

lithic. Pertaining to stone. A descriptive term applied to rock fragments occurring in a later formed rock, e.g., lithic tuff.

lithic tuff. Type of vocanic tuff. Dominant constituent of later formed rock fragments.

lithifaction. (1) The conversion of a newly deposited sediment into a solid rock, involving such processes as cementation, and crystallization. (2) The lateral termination of coal bedowing to an increase in impurities.

lithification. Completion of process of diagenesis in which the original loose sediment becomes a compact rock.

lithify. To change to stone, or to petrify, especially to consolidate from a loose sediment to a solid rock.

lithiophilite. $4[LiMn(PO_4)]$. An orthorhombic mineral composed of lithium manganese phosphate; shows uneven fracture and perfect cleavagg in one direction. Its hardness on Mohs scale is 4.5.

lithiophorite. Lithium rich variety of wad. It has metallic lustre and shows blackish-grey streak. Its hardness on Mohs scale is 3.0.

litho. A prefix meaning rock or stone.

lithofacies. A facies particularly characterized by its rock type.

lithofraction. The breaking of rock fragments during transportation in streams or by wave action on beaches.

lithographic limestone. A dense, homogeneous, very fine-textured limestone; a micrite.

lithographic texture. A texture of certain calcareous rocks characterized by uniform, clay-sized particles and a very smooth appearance resembling that of a lithographic stone.

lithohorizon. A surface evidencing lithostratigraphic change, or having a particular lithostratigraphic character; highly useful for correlation. It is usually the boundary of a lithostratigraphic unit, but may be a thin marker layer within a lithostratigraphic unit.

lithology. The description and study of rocks, as seen in hand-specimens and outcrops, on the basis of colour, grain size, and composition.

lithophile. Elements having a strong affinity for oxygen, concentrated in the silicate minerals.

lithophysae. Hollow bubble like structures composed of concentric shells of finely crystalline alkali feldspar, quartz and other minerals; occurring in certain silicic volcanic rocks.

lithosol. An azonal group of shallow soils lacking well-defined horizons and composed of imperfectly weathered fragments or rock. It usually develops on steep slopes.

lithosome. A body of sediment deposited under uniform physiochemical conditions; the lithostratigraphic equivalent of a biosome.

lithosphere. (1) The rigid outer crust of rock on the earth about 80 kilometres thick, above the asthenosphere. (2) The solid portion of the earth, as compared with the atmosphere and the hydrosphere.

lithostratigraphy. The element of stratigraphy that deals with the lithology of strata, their organization into units based on lithologic character, and their correlation.

lithotype. A visible band in humic coals, recognized by physical characteristics rather than botanical origin. The four lithotypes of banded bituminous coal are vitrain, clarain. durain, and fusain.

lithozone. A term to indicate a body of strata that is unified in a general manner by lithologic features but for which there is insufficient need or information to justify its desigantion as a formal unit.

littoral. Pertaining to the benethic environment or depth zone between high water and low water, or to the organisms of that environment.

littoral drift. See longshore drift.

lizardite. $Mg(Si_2O_5)(OH)_4$. A greenish-white, monoclinic mineral composed of hydrous magnesium silicate; a variety of serpentine.

Llandeilian. A stratigraphic stage name for the European Lower Lower Ordovician.

liandoverian. A stratigraphic stage name for the European Lower Silurian.

llano. A term for extensive plane, with or without vegetation.

Llanvirnian. A stratigraphic stage name for the European Lower Ordovician.

load. The material that is carried or moved by a natural transporting agent such as a glacier, stream or wind.

loades stream. A stream that has all the sediment it can carry.

loam. A soil containing sand, silt, humus and clay in roughly equal proportions.

lode. A mineral deposit consisting of a zone of veins, veintets, or disseminations also a mineral deposit in solid rock as opposed to a placer deposit.

lodestone. Name given to any magnetic variety of magnetite.

lode tin. Tin ore occurring in veins, as distinguished from stream tin.

lodranite. A strong iron meteorite composed of bronzite and olivine, within a fine network of nickel-iron.

loellingite. 2[$FeAs_2$]. A silver-white to steel-grey mineral composed of iron arsenide with small amounts of cobalt, nickel, antimony and sulphur. Also known as leucopyrite.

loess. Argillaceous sedimentary rock. A soft, fine-grained, loosely coherent blanket deposit composed of angular particles of quartz, feldspar, hornblende, mica and bits of clay. Because it contains vertical tubules (remains of rotted-out grassroots), loess stands in nearly vertical walls, despite its weak cohesion.

loess doll. A nodule or concretion of calcium carbonate found in loess; it bears a resemblance to a doll, a child's head, or a potato developed vertical jointing. Loess reworked by river action is called brickearth.

log. A record of the rocks passed through by a borehole or a well.

lomonosovite. $Na_5Ti_2Si_2PO_{13}$. A dark brown mineral of uncertain crystal system; composed of sodium titanium phosphorosilicate; shows uneven fracture and poor cleavage. Its hardness on Mohs scale is 3.0-4.0.

longitude. Angular distance between the meridian of a given place and the prime meridian of Greenwich, England; measured east or west to a maximum value of 180 degrees.

longitudinal fault. A fault whose strike is parallel to that of the geheral structural trend of the region.

longitudinal joint. A steeply dipping joint plane in a pluton that is oriented parallel to the lines of flow.

longitudinal stream. A stream that follows the strike of the underlying rocks.

longitudinal valley. The valley of a subsequent stream, developed parallel to the general strike of the underlying strata.

Longmyndian. A stratigrphic division of the English Precambrian, the Shropshire Post-Uriconian sediments.

longitudinal crevasse. A crevasse roughly parallel to the direction of ice movement.

longitudinal drift. A long, tap ored, sharp crested sand drift formed on the lee side of a narrow gap in a ridge or sharp oriented transversely to the prevailing wind, especially in a desert or steppe region where ridges interrupt flat plains or plateaus.

longitudinal dune. A long, narrow sand dune, usually symmetrical in cross profile oriented parallel with the direction of the prevailing wind; it is wider and deeper on the windward side but tapers to a point on the leeward side and generally form behind an obstacle in an area where sand is abundant and the wind is strong and constant longitudinal dunes may be a few metres high and several kilometres long.

longshore bar. A low sand embankment formed primarily by wave action. It lies generally parallel with the shoreline at some distance from it, and is submerged during high tide.

loparite. Alkali bearing variety of perovskite from the Kola Peninsular, USSR.

lopolith. A concordant, typically layered igneous intrusion that is planoconvex or lenticular in shape. The central part is sunken because of the sagging of the underlying country rock. The term lopolith is used for the both small tabular intrusions related to sills and for very large intrusions whose dimensions are massive rather than tabular shaped masses of igneous rock lopolith' is here suggested for the larger kind of lopolith. They commonly display layered and banded structures, and clearly their origin must be different from that of the smaller lopoliths mentioned above, which are most nearly related to sills.

lost circulation. The condition during rotary drilling when the drilling mud escapes into porous, fractured, or cavernous rocks penetrated by the bore hole and does not return to the surface.

lostriver. (1) A dried up steam in an arid region. (2) A stream that disappears underground in a darst region.

lovewave. A type of surface wave having horizontal motion that is shear or transverse to the direction of propaga tion. Its velocity depends only on density and rigidity modulus and not on bulk modulus.

low. General term for such features as a structural basin.

lower core. A term that includes the earth's inner core and the transitional zone of the outer core, *i.e.*, equivalent to the F layer and the *G* layer.

lower mantle. The part of the mantle which lies below a depth of about 1000 km and has a density of about 4.7 g/cm^3, in which the seismic velocity increases slowly with depth. It is equivalent to the *D* layer.

Lowestoftian. A stratigraphic stage name for the British mid Middle Pleistocene.

low grade metamorphism. Level of metamorphism (regional or contact) where some new minerals are developed and existing texture are reinforced or partly altered, *e.g.*, schist or slate.

low velocity zone. The zone in the upper mantle, generrally defined as from 60 to 250 km in depth in which velocities are about 6% lower than in the outermost mantle. It is probably due to near melting poin-temperature of the material.

loxoclase. Sodium rich variety of orthoclase.

l-tectonite. A tectonite whose fabric is dominated by the presence of lineatons, such as deformed conglomerate in which the pebbles are strongly elongated.

lublinte. $CaCO_3$ A soft and spongy variety of *calçite* with encrustations on *chalk marl* name for the British Lower Pleistocene.

ludian. A stratigraphic stage name for the base of European Lower Oligoene.

ludlovian. A stratigraphic stage.

lunar crater. A roughly circular depression in the lunar surface, relatively shallow and ranging in diameter up to hundreds of km; lunar crater may have resulted from meteorite impact, uolcanism, or subsidence.

lugarite. Igneous rock intermediate in composition between syenite and gabbro clans; contains barkevikite, titanaugite, labradorite, and nepheline.

lujavrite. Igneous rock of the syenite clan; contains microline-microperthite, oligoclase, nepheline, aegirine and arfvedsinite.

lumachelle. A consolidated shell bank or shell conglomerate, usually of molluscan remains.

lunar. Pertaining to or occurring on the moon.

lunarite. A general term for lighttoned, highly reflecting surface rocks of the lunar highlands or errace.

lunar playa. A relatively small level area on the surface of the moon, occupying a low place in ejacta blankets surrounding lunar craters. It is believed to be either a fallback or small lava flow.

lunar regolith. A thin, grey layer on the repeated meteoritic and secondary impact over a long period of time. It consists of partly cemented or loosely compacted fragmental material.

lunule. A perforation passing through the test of some advanced echinoids.

lusclalite. Igneous rock intermediate in composition between syenite and gabbro clans. A coarse-grained plutonic rock ; contains plagioclase, rthoclase, olivine, biotite, and nepheline.

lusitanite. Igneous rock of the syenite clan. A medium-grained plutonic rock ; contains riebeckite, aegirite, orthoclase, microcline-microperthite and sodaclase.

lustre. The apearance of a surface dependent on reflected light. This is a useful property of minerals, generally described qualitatively and in comparative terms.

lydite. A sedimentary siliceous deposit; very fine-grained quartz and chalcedony. It is coloured greyish-black by carbonaceous matter or greenish brown by chlorite and iron hydroxides.

lustre mottling. The potted shimmering appearance of certain rocks caused by reflection of light from cleavage faces of crystals that contain small in clusions of other minerals.

lutalite. A glassy leucite nephelinite or olivine leucitite, with more than 50% mafic mainerals, and a higher sodium to potassium ratio than most olivine leucitites.

lutecite. Fibrous chodcedony characterized by inclined extinction and by fibres that are seemingly about 30° to the c-axis.

luteous. Having an essential proportion of muddy sediment in a limestone, recognised by the presence of many particles of clastic quartz of slit size.

lutetian. European stage; Eocene.

lutaceous. Said of a sedimentary rock formed from mud pertaining to lutite.

lutite. A general name for rocks composed of material that was once mud, *e.g.*, shale, mudstone, calcilutite.

luxullianite. Igneous rock of the granite clan. A porphyritic plutonic rock ; contains phenocrysts of orthoclase in a groundmass of quartz, tourmaline, orthoclase, mica and cassiterite.

luzonite. Reddish-grey variety of famatinite.

l wave. *See* surface wave.

lyell. (1) A block of rock transported and released by an iceberg; an ice-rafled block. (2) Imperfeetly stratified deposit resembling till, but transitional between till and clay.

lynchert. A bank of earth that accumulates on the downhill side of an ancient plongeo hed field as the disturbed soil moves downslope under gravity.

lyndochite. A variety of aeschynite, having relatively high content of calcium and thorium.

lysocline. The level or occean depth at which the rate of solution of calcium carbonate just exceeds its combined rate deposition and precipitation.

maar. A shallow, flat-floored crater not associated with volcanic cone. Mass appear to be result of multiple shallow explosive eruptions. Craters that interect the water table may form lakes.

macedonite . $PbTiO_3$. (1) A mineral composed of oxides of lead and titaium. (2) A basaltic rock that contains orthoclase, biotite, olivine, sodic plagioclase, and rare pyriboles.

marceral. The macroscopic organic constituents that comprise the coal mass; all petrologic units seen in polished or thin section of coal. Macerals are to coal as minerals are to inorganic rock.

maceration. The process of disintegrating sedimentary rock such as coal and shale in order to extract and concentrate acid insoluble microfossils. It includes many chemical treatment by oxidising agents, and alkalies that can remove extraneous minerals and organic constituents. This process is widely used in palynology.

macgovernite. $Mn_5(AsO_3)SiO_3(OH)_2$. A mineral containing basic manganese arsenite and silicate.

mackayite. $FeTe_2O_5(OH)$. A green mineral composed of basic iron tellurite.

mackensite. $(Fe_6Si_3AlO_{15}(OH)_3$. A greensih-black, monoclnic mineral composed of basic iron aluminosilicate

mackinawite. $(Fe,Ni)S$. A tetragoanal mineral occuring as a corrosion product. Also known as kansite.

macro A prefix meaning large.

macro-axis. The larger lateral axis of an orthorhombic or triclinic crystal; it is usually the b-axis.

macrocrystalline. Pertaining to the texture of holocrystalline rock in which the constituents are visible to the naked eye and having grains or crystals greater than 0.75 mm in diameter in recrystallized sediment.

macrofacies. A system of genetically interconnected sedimentary facies of the same age. It includs the area of erosion from which the sediments of these facies are derived so that an erosional interval represents part of a facies tract. Also known as facies tract.

macrofossil. A fossil large enough to be studied without the aid of a microscope.

macropinacoid. The [100] pinacoid in an orthorhombic, monoclinic or triclinic crystal. Also known as front pinacoid; orthopinacoid.

macropore. A pore in the soil of a large enough size so that water is not held in it by capillary action.

macrospic. Large enough to be visible to the naked eye, *i.e.*, without the aid of microscope.

madeirite. An igneous rock of the ultrabasic clan; a green and black porphyrite plutonic rock; contains phenocrysts of augite and olivine in a very fine groundmass of plaargioclase, calcite and magntite.

madupite. Feldspathoid ignoeous rock; a fine grained porphyritic extrusive rock; contains phenocrysts of diopside and phlogite in a groundmass of leucite compostion of glass.

maenaite. An igneous rock of the diorite clan; a fine grained porphyritic hypabyassal rock; contains phenocrysts of plagioclase in groundmass of oligoclase, orthoclase, pyroxene and hornblende.

mafic index. A parameter used for igneous rocks. equal to to 100× $[FeO+ Fe_2O_3] / [FeO+ Fe_2O_3 + MgO]$

mafite An alternate name for aphanite.

mafratie. An igneous rock intermediate in composition between syenite and gabbro clans; a medium to coarse-grained hypabyssal rock; contains labradorite, sanidine ,hornblende, and pyroxene.

maghemite. y- Fe_2O_3. A member of the megnetite series of minerals; occurs in the form of iron oxide that is strongly magnetic.

magma. A naturally occurring molten rock material formed within the the crust or upper mantle of the earth which may consolidate to form an igenous rock. Magma comprises a complex system of molten silicates with water and other gaseous material in solution. It is unlikely that any igneous rock truly represents the magma from which it was formed as the volatile constituents are almost inevitably lost during or after consolidation. Within the molten mass various processes operate which tend to produce fractions of varying composition.

magma basalt. A term applied to porphyritic, glassy basaltic rock that resembles the ordinary basalt.

magma blister. A pocket of magma whose formation has raised the overlying land surface.

magma chamber. A reservoir of magma in the shallow part of the lithosphere from which the volcanic materials are derived ;the magma has ascended in to the crust from an unknown source.

magmatic. Of pertaining to or derived from magma.

magmatic differentiation. (1) The process by which the different types of igneous rocks are derived from a single parent magma. (2) The process by which ores are found by solidification of magma. The magmatic differentiation can take place by:*(i)* fractional crystallization,*(ii)* liquid immiscibility,*(iii)* thermal diffusion, and *(iv)* solution transport of the material.

magmatic emanation. A mixture of gases and liguid vapours ejected from a magma.

magmatic evolution. The continuing change in composition of magma as a result of magmatic differentation, assimilation, or mixing of magmas.

magmatic segregation. Concentration of crystals of a particular mineral in certain parts of a magma during its cooling and crystallization.

magmatic stoping. A process of igneous intrustion in which magma gradually works its way upward by breaking off and engulfing blocks of the country rock.

magmatic water. Water issueing from an underground magmatic source.

magmatism. The formation of igneous rock form magma.

magnafacies.A major continuous belt of deposits that is homogeneous in lithologic and paleontologic characteristics and that extends obliquely across time planes or through several time stratigraphic units.

magnesian calcite. (Ca, Mg)CO_3. A veriety of clacite consiting of randomly substituted magnesium in place of calcium. It may consist of 4-19% of $MgCO_3$.

magnesian dolomite. A dolomite mineral rock, with an excess of magnesium with Ca/Mg ratio ranging form 1.0 to 1.5; or a dolomite rock containing 50 to 75% dolomite and 20-25% magnesite.

magnesian limestone.A dolomite rock with an appreiable amount of magnesium, a limestone having atleast 90% calcite, no more than 10% dolomite,an appoximate MgO equivalment of 1.1-2.1% and an approximate magnesium carbonate equivalent of 2.3.- 4.4% ; or limestone whose Ca/Mg ratio ranges from 60 to 105.

magnesioarfvedosnite. A magnesium rich variety of arfvedsonite.

magnesioferrite. $MgFeO_3$ Megnesium iron oxide, formed at the mouths of fumeroles.

magnesiohastingsite. A magesium rich variety of hastingsite.

magnesiokatophorite. A magnesium rich nd member of the catophorite series.

magnesiomagnetite. A magnesium rich variety of magnetite.

magnesioribeckite. A magnesium rich variety of riebeckite.

magnesite. $MgCO_3$. A white mineral of the carbonate group found in irregular veins in serpentine and formed by replacement of dolomite and limestone; shows conchoidal fracture and perfect cleavage in one direction. It is used in making refractory magnesia. Its hardness on Mohs scale is 4.0.

magnetic constant. μ_o. Permeability of free space. The fundamental constant that has the value 4.010^{-7} henry per metre. It arises as the constant of proportionality in Ampere's law, its value denending on the choice of units.

magnetic dipole. An object such as a permanent magnet, current loop, or a particle with angular momentum, which experiences a torque in a magnetic field, and itself gives rise to a magnetic field as if it consisted of two magnetic poles of opsite sign separated by small distance.

magnetic domains The fact that ferromagnetic substances are not necessarily always magnetized led to the theory that they consist of separate domains, each of which is spontaneously magnetized, but the magnetic moments of which may not be aligned. If an external magnetic field is applied to the the substance to magnetic moment of the domains (not the domains themselves) are rotated so that they lie parellel to the field; the substace then acts as permanent magnet.

magnetic elements. The three quantities, magnetic declination angle of dip and horizontal component, which define completely the earth's magnetic field at any point.

magnetic equator A line of zero magnetic dip lying fairly near the geographical equator. Also known as aclinic line; geomagnetic equator.

magnetic field. A field of force that is said to exist at any point if a small coil of wire carrying an electric current experiences a couple when placed at the point. A magnetic field may exist at a point as result of the presence of either a permanent magnet or a circuit carrying an electric current, in the neighbourhood of the point.

magnetic field strength The force that would be exerted on a unit north pole at a given point in the magnetic field. The strength of a magnetic field measured in amperes per metre (SI units), or oeresteds (c.g.s. units). It is given by $H=B/\mu_o - M$, Where B is the magnetic flux density, M is the magnetization and μ_o is the magnetic constant.

magnetic The strength of magnetic field through an area, based on the idea of the number of lines of force passing through the area. It is given by the product of a given area and the component of the magnetic field strength at right angles to the area. The c.g.s. unit of magnetic flux is the maxwell. The derived SI unit of magnetic flux is the weber.

magnetic moment (1) The torque experience by a magnetic dipole in a field of unit magnetic field strength. It is measured in weber metres. This is also called magnetic dipole moment. (2) The product IA, where I is the current flowing through a small loop of wire of area A. It is measured in ampere metres square (Am^2). This is also called the electromagnetic moment.

magnetic permeability. μ. The ratio of the magnetic flux density in a medium to the external magnetic field strength that induces it. The 'relative permeability, μr, is the ratio of the permeability of sub-stance to the permeability of free space (*see* magnetic constant). For most of sub-stance μr has a constant small value. When μr is less than 1, the material is said to be diamagnetic; if μr is greater thanI, it is paramagnetic. A few substances, notably iron, have very large values of μr which tend to fall as the field strength in-crease so that the magnetic flux density tends to a limiting value called the satura-tion value. Such substances are said to be ferromagnetic.

magnetic pole. One of the two regions located at the ends of a magnet that generate and respond to magnetic field in much of the same way as electric charge generate and respond to electric field.

magnetic potential. The work which must be done against a magnetic field to bring a magnetic pole of unit strength from a reference point to the point under consideration. Also known as magnetic scalar potential.

magnetometer. An instrument for com-paring strengths of magnetic fields, mag-netic moments and sometimes also the direction of a magnetic field, such as the earth's magnetic field.

magnetosphere. The region where the earth's (or any celestial body) magnetic field is constrained by the streaming solar wind. It has a streaming solar wind. It has a shape similar to that of a comet with a blunt nose towards the sun and a long tail always away from the sun. This magnetic cavity effectively shields the earth and its atmosphere from the direct influx of the solar wind.

magnetite. Fe_3O_4. An opaque, iron-black and streak-black isomeric mineral and member of the spinal structure type. It often contain magnesium oxide and/or titanium oxide and constitutes an impor-tant ore of iron. It is very common and widely distributed accessory mineral in rock of all kind and also occurs as a heavy mineral in sands. Its hardness on Mohs scale is 6.0.

magnetoplumbite. $(Pb,Mn)_2 Fe_6 O_{11}$. A black mineral consisting of ferric oxide along with Pb(II) and Mn(II) oxides and occurs in acute metallic hexagonal crys-tals.

magnophorite. $NaKCaMg_5Si_8O_{23}OH$ A monoclinic mineral composed of a basic silicate of sodium potassium cal-cium and magnesium. It is a member of the amphibole group.

malachite. $Cu_2CO_3 (OH)_2$. A bright green, monoclinic mineral consisting of basic carbonate of copper(II). It is a minor ore of copper and a common secondary mineral associated with azurite in the oxidised zone of copper sulphide deposits. Its hardness on Mohs scale is 3.5.-4.0.

malachite. A fine-grained lamprophyre with small, rare phenocrysts or hornblende, labradorite and sometimes biotite embedded in a matrix of hornblende, andesine, and some quartz.

malignite. An igneous rock of the syente clan; a mottled, light grey, medium to coarse-grained plutonic rock; contains orthoclase, nepheline, apatite, biotite, and aregirine-augite.

malladrite. Na_2SiF_6. A hexagonal mineral composed of sodium fluorosilicate, occurring as small crystals in volcanic holes in Vesuvius.

mallardite. $MnSO_4.7H_2O$. A light pale-rose coloured monoclinic mineral composed of hydrous managaese sulphate.

malmstone. A sedimentary siliceous deposit; it contains sponge spicules in a matrix of globular opal and chalcedony.

malpais. A term used in the southwestern U.S. for a region of rough and barren lava flows. The connotation of the term varies according to the locality.

manandonite. $Li_4Al_{14}B_4Si_6O_{29}$-$(OH)_{24}$. A white or very light coloured mineral composed of basic borosilicate of lithium and aluminium.

manasseite. $sMgAl_2(OH)_{16}(CO_3)$.-$4H_2O$. A hexagonal mineral composed of basic hydrous magnesium and aluminium carbonate; normally forms scales; shows waxy lustre and white streak. Its hardness on Mohs scale is 2.0.

managanese nodule. A small, irregular, brown to black mass of manganese-rich materal mainly manganese oxide mineral, that occurs on the ocean floor as a result of pelagic sedimentation or precipitation.

manganite. $MnO(OH)$. A grey to black orthorhombic mineral; an ore of manganese.

manganosiderite. An iron rich variety of rhodochrosite; intermediate between rhodochrosite and siderite.

mangerite. An igneous rock of the syenite clan; a grey, medium-grained granular plutonic rock; contains microperthite, pyroxene, apatite, biotite and plagoiclase.

manjak. A variety of bitumen possessing briliant lustre.

mansjoite Fluorine rich variety of dioside.

mantle. (1) A general term for an outer covering of a material of one kind or another. (2) The zone of the earth below the crust and above the care, which is divided into the upper mantle and the lower mantle, with a transition zone in between. The upper boundary is marked by the *Mohorovicic* discontinuity, and the boundary between mantle and core by the Gutenberg discontinuity. The upper mantle extends to a depth of 400 km and is characterized by regional dissimilarities in sesmic velocity profile.

mantle-crust mix Rock whose properties are between those of the crust and those of the mantle.

mantled gneiss dome. A dome in metamorphic terrains that has a remobilized core of gneisss surrounded by a concordant sheath of the basal part of the overlying metamorphic sequence.

manto. A flat-lying bedded deposit, either a sedimentary bed or a replacement strata-bound ore body.

marble. Limestone recrystallised by metamorphism to produce a granular mosaic of calcite or dolomite crystals. Colourful streak in marble are due to inclusions of minerals such as mica, pyrite, epidote, quartz, felspar, or iron oxide. Purer varieties, without inclusions, are valued by sculptors; veined varieties are used as ornamental stone.

marcasite. FeS_2. A common pale-yellow or nearly white mineral that crystallizes in orthorhombic system. It is very common as nodules and concertions in sedimentary rock. Its hardness is 6-6.5 on Mohs scale. The 'marcasite' of jewellers is haematite.

marekanite. Alternation product of rhyolite perlite glass; glassy balls with concave identations.

mare material. Dark and relatively smooth, heavily cratered igneous rock, chiefly of mafic, ultramafic composition, underlying the lunar maria.

mareugite. An igneous rock of the gabbro clan; a medium-grained grangular plutonic rock; contains bytownite, hauyne, hornblende and augite.
margarite. (1) A string of bedlike globulities, commonly found in glassy igneous rocks. (2) $CaAl_2(Al_2Si_2)O_{10}(OH)_2$. A yellow, pink or reddish, brittle mica mineral.
margarosanite. $PbCa_2(SiO_3)_3$. A colourless or snow-white triclinic mineral composed of silicate of lead and calcium; occurs in lamellar masses.
marginalescarpment. A seaward slope of a marginal plateau with a gradient of 1 : 10 or more.
marginalfisseau. A fracture, bordering an igneous intrusion, that has become filled with magma.
marginal plateue. A relatively flat sheft adjacent to continent and similar topographically to, but deeper than, a continent sheft.
marginal sea. A semi-closed sea adjacent to a continent, floored by submerged continental mass.
marginalthrust. One of a series of faults bordering an igneous intrusion and crossing both the intrusion and the wall rock.
marialite. $3NaAlSi_3O_5$. A scapolite mineral that is isomorphous with meronite.
marine. Of, or belonging to, or caused by the sea.
marine abrasion. (1) Erosion of a bedrock surface by the to-and-fro movements of an overlying layer of sand under the influence of waves. (2) Erosion of submarine canyons by downslope movement of sediments under the influence of gravity.
marine-built terrace. A wave-built terrce produced by marine processes.
marine-cut terrace. A terrace or platform cut by wave erosion of marine origin. Also known as wave-cut terrace.
marine erosion coast. A Coast whose configration results mainly from marine erosion, as in the straightoning of sea cliffs by wars.
marineerooatie. A sedimentary particle of anomalous size or lithology, transported and deposited in marine sediments.
marine invasion. The spreading of the sea over a land area.
marine limit. The present or the former limit of the sea; a shoreline.
marine peneplain. A abrasion plateform of large areal extent, uplifted above the reach of the waves before wave erosion had succeeded in perfecting a smooth plane.
merine plain. (1) A coastal plain of marine sediments. (2) A hypothetical wave-cut surface produced during the ultimate stage of marine erosion.
marineswamp. An area of low, salty, or blackish water found along the shore and characterized by abundant grasses, trees and similar vegeation. Also known as paralic swamp.
marine terrace. (1) A narrow coastal strip, formed or deposited material, sloping gently seaward. (2) A seacoast terrace formed by the merging of a wave-built terrace and a wave-cut platform. Also known as sea terrace; shore terrace.
mariupolite. An igneous rock of the diorite clan; a coasre-grained porphyrtic plutonic rock; contains phenocrysts of nepheline in a groundmass of sodaclase, nepheline, aegirite and lapidomelane.
marker. (1) A rock unit or stratigraphic feature that is distinictive and easily recognised over long distances, especially in the subsurface ; *e.g.*, a bed or beds

readily identfied on a electric log. (2) A layer that accounts for a characteristic segmentofasesimic-refractiontime-distance curve and can be followed over reasonably extensive areas.

marker bed. A geologic formation serving as a marker.

marker horizon. A marker represented by a rock surface stratified level, such as a boundary based on electric or other mechanically recorded logs, that may serve to delineate lithostratigraphic units.

markfieldite. An igneous rock of the granite clan; a fine-grained porphyritic hypabyssalrock;containsphenocrystsof plagioclaseinagroundmassofquartz,orthoclase, and muscovite.

marl. An old term lossely applied to a variety of materials, mostly unconsolidated earty deposits, composed of clay with magnesuim and calcium carbonate, usually including shell fragment and sometimes glauconite; used as a fertilizer for acid and lime-deficient soils.

marlstone. An indurated rock of about thesamecompositionasmarl;i.e.,anearthy limestone; it is less fissile than shale.

marloesite. An igneous rock of the diorite clan; a fine-grained porphyritic extrusive rock; contains phenocrysts of albite and olivine in a groundmass of augite, plagioclase, and magnetite.

marlite. *See* marlstone.\

marlstone. (1) An indurated rock that has about the same composition as marl; may be considered as an earthy or impure argillaceous limestone. Also known as marlite. (2) A hard ferruginous rock of the middle. Lias in

marly. Pertaining to, containing, or resembling marl.

marmatite. A dark brown to black mineral composed of iron-bearing spahlerite. Also known as christophite.

marmolite. A pale-green serpentine mineral, occurring in thin lamination; a variety of chryotile.

marmorize. The process of recrystallization of limestone by heat (with or withoutpressureeffects) to form marble. The term is practically obsolete, but is occasionally used for the slight thermal effects produced by a dyke or still intruded into limestone.

marosite. An igneous rock of the diorite clan, a medium to coarse-grained plutonic rock; contains biotite, augite, apatite, sandine, plagioclase, nepheline and sodalite.

marrite. $PbAgAsS_3$. A monoclinic mineralcomposedofsulphideoflead,silver, and arsenic; occurring as small crystals in Valais, Switzerland.

martite. Hematite occurring in iron-black octahedral crystals; psedomorphous after magnetite.

masanite. An igneous rock of the granite clan;afine-grainedporphyritichypabyssal rock; contains phenocrysts of plagioclase and quartz in a groundmass of quartz, orthocalse and muscovite.

maseagnite. $(NH_4)_2SO_4$. A yellowish-grey mineral found in guano, near burning coal-beds or as lava incrustation; its hardness on Mohs scale is 2.0-2.5.

mascon. A large-scale high density lunar mass concentration below a ringed mare on lunar surface.

massif. A massive blow of rock within an orogenic belt, generally more rigid than the surrounding rocks, and commonlycomposedofcrystallinebasement or younger plutons.

massive. Said of a mineral deposit, havingalargeconcentrationoforeinone place.

mass movement. Unit downslope movementofaportionofthelandsurface, as in creep, landslide or slip.

mass wasting. A synonym of gravity transport.
matidite. 8[$AgBiS_2$]. An orthorhombic mineral composed of silver bismuth sulphide; shows uneven fracture and no cleavage. Its hardness on Mohs scale is 2.5.
matrix. The natural material in which a rock of mineral is embedded.
maucherite. 16[$Ni_{11}As_8$. A tetragonal mineral composed of nickel arsenide; consists of tabular crystals; shows blackish-grey streak, uneven fracture, and no cleavage. Its hardness on Mohs scale is 5.0.
Maysvillian. A stratigraphic stage name for the North American mid Upper Ordovician.
maze cave. A consisting of repeatedly rejoined passages.
mazzite. $H_2CaMg_2(Al,Si)_{36}O_{72}.28H_2O$. A heagonal mineral of zeolite family.
mboziite. $Na_2CaFe(II)Fe_2(II)$ $Th_2Si_6O_{22}$ $(OH)_2$. A mineral amphibole group.
meander. One of a series of regular freely developing sinuous curves, loops, turns in the course of a stream. Meandering stream have fairly light bed loads and banks that are much steeper than those of braided steams. Most are close to base leve, only by becoming ore sinuous can they lower their gradients. Other not near base level become sinuous because of difficulty in carrying fine sediments when there is no bed load. As water rounds any bend, its inertia carries it towards the concave bank, causing a slight uptiliting of the water surface; this tilt tends to make the stream rush towards the opposite bank as it flows downstream, and the curvature of the bank tend to reverse, thus the outside bank of each curve is subject to erosion at the same time as the deficiency of water on the inner bank results in deposition. The combination of these two actions compels meander bends to migrate..
meander belt. The zone along a valley floor across which a meandering stream periodically shifts its channel.
meander core. A hill enclosed by a stream meander. Also known as rock island.
meander cutoff. The shortened channel resulting when a stream cuts through a meander neck.
meandering stream. A mature stream having a pattern of successive meanders. Also known as snaking stream.
meander line. A surveyed line, generally irregular, but not a boundary line, especially, a traverese of the margin or bank of a permanent natural body of water.
meander lobe. The more or less elevated, tongue-shaped area of land enclosed within a stream meander.
meander neck. The narrow strip of land between adjacent loops of a meandering stream.
meander niche. A conical or creascentic opening in the wall of a cave formed by downward and lateral stream erosion.
meander plain. A plain built by meandering process, or a plain of lateral accretion.
meander scar. A crescentic cut mark in a bluff or valley wall, formed by a meandering stream and indicating its formed route.
meander scroll. One of a series of acuate ridges and troughs formed along the inner bank of a stream meander as the channel migreated laterally down-valley and towards the outer bank.
meander spur. An undercut projection of high land that extend into the concave part of and is enclosed by a meander.
measures. A mining term formerly applied to sequences of sedimentary rocks in bore holes or mining shafts.

mcasureout. Measurement of drill pipe or tubing as it is pulled out of the hold usually to determine the depth of the well or the depth to which the pipe or tubing was run.

mechialsediment. Sediment consisting of the broken particles of preexisting rock transported mechanically by water, wind, or ice; synonymous with clastic sediment.

mechanical weathering. Breakdown or weathering of the earth's surface by mechenical means as frost wedging and root wedging.

medial moraine. (1) An elongate moraine carried in or upon the middle of a glacier and parallel to its sides.(2) A moraine formed by glacial abrasion of a rocky proturbance near the middle of a glacier.

Medinan. Obsolete synonym of Alexandrian.

mediterranean. Said of a deep sea that is in or between continents and connects with the ocean by a narrow opening.

Mediterranean suite. A major group of igneous rock having high potassium content.

medium grained. (1) Said of igneous rock, and of its textrue, in which the individual crystals have an average diameter of 1-5 mm or the individual particles have an average diameter of 1/16 2 mm. The termisnalogy used in relative sense to describe rocks that are neither coarsegrained nor fine-granied.

megabreccia. A rock consisting of randomly oriented clests with sharp edges and of very large size.

megagroup. A rock stratigraphic unit made up of stratigraphic groups and representing a major event in geologic history.

megaripple. A ripple of great size developing in fluvial, tidal or marine environments.

megascopic. Said of an object or phenomenon or of its characteristics, that can be observed with naked eye.

megashear. A strike slip fault with a horizontal displacement that exceeds significantly the thickness of the crust.

meionite. $3[CaAl_2Si_2O_8. CaCO_3]$. A mineral of seapolite group; isomorphous with marialite.

meizoseismal. Pertaining to maximum destructive force of an earthquake.

mela. A prefix denoting dark.

malano. A prefix denoting black or dark.

mealanocratic. A term used to described igneous rocks which consists mainly of dark minerals, essentially the ferromaganesian minerals. The corresponding colour index ranges from 60 to 90 on Shand's scale.

mealnophlogite. Pseudomorph of quartz after cristoblite; possesses spherical aggregates with an opaline crust and a fibrous interior.

melaphyre. An altered amygdalidal basalts.

melitite. $(Na,Ca)_2(Mg,Al)$ $(Si,Al)_2O_7$. A white or pale yellow orthosilicate mineral; shows conchoidal fracture and good cleavage in one direction.

melilitite. Feldspathoid igneous rock; a fine-grained extrusive rock; contains augite, melilite and magnetite.

melnikovite. Finely divided black-variety of pyrite.

meltegite. Feldspathoid igneous rock,. a medium-grained granular plutonic rock; contains biotite, nepheline, pyroxen and granet.

member. A lithostratigraphic unit of subordinate rank, comprising some specially developed part of a formation. It may be formally defined and named, informally named or unnamed.

Mercalli scale. An arbitrary sccale of earthquake intensity, ranging form I to XII.

mero-. A prefix denoting part, fraction of.

mecrocrystalline. A term applied to rock which contains both crystals and glass.

merohedral. Said of crystal classes in a system, the general form of which has only one half, one fourth, or one eigth number of equivalent faces the corresponding form in the holohedral class of the same system.

mervinite. $Ca_3MgSi_2O_8$. A white monoclinic mineral composed of calcium magnesium silicate; shows perfect cleavage in one direction. Its hardness on Mohs scale is 6.0.

mesa. An isolated flat-topped hill bounded on at least one side by a steep cliff and having an extensive summit area. Mesas are erosional remnants that persist because of protective cover of more resistant sedimentary rocks, especially sandstones or of lava flows or gravels.

meso. A prefix meaning middle.

mesocratic. A term used for rocks epcially igneous ones that contain between 30 and 60% of dark mineral; *i.e.*, whose colour index lies between leucocratic and melanocratic.

mesocrystalline. Said of the texture of a rock consisting of crystals intermediate in size between those of a microcrystalline and a macrorystalline rock; also said to of a rock with such texture.

mesogene. Said of a mineral deposit or enrichment of mingled hypogene and supergene solutions; also said of such solution and environment.

mesolite. $8[Na_2Ca_2(Al_2Si_3O_{10}).\text{-}8H_2O]$. A white or grey, monoclinic mineral composed of hydrous sodium calcium aluminium silicate; shows perfect prismatic cleavage. Its hardness on Mohs scale is 2.2-2.4.

mesonorm. Theoretical calculatetion of minerals in metamorphic rock of the measozone as indicatied by chemical analysis.

mesopelagic. Pertaining to pelagic environment of the ocean between 100 and 500 fathoms.

mesosiderite. A stony-iron meteorite in which the silicates are mainly pyroxene and calcic plagioclase.

meseilexite. A silaxite in which the dark-coloured components constitute more than 5% of the rock.

mesophere. The lower mantle; probably not involved in the tectonic processes of the earth.

mesostasis. Partly amorphous, partly crystalline interstital material which may occur in igneous rocks, the end product of magma consolidation; forms a glass or a thin coating on earlier minerals.

mesothermal. Deposits formed hydrothermally at tempertures roughly between 475.575K.

mesotil. A semiplastic or semifriable derivative of chemically weathered hill, developed beneath a partially drained area.

mesotourmalite. A variety of tourmalite containing 5-50% of tournaline.

mesotype. A group of zeolite minerals, including natrolite, mesolite, and scolecite.

mesozoic. An era of geologic time, from the end of Paleozoic to the begining of Cenozoic, *i.e.*, about 225 to 65 million years ago. It included in Triassic, Jurassic and Createceous periods. Manne fossils from the world standard for the correlation of Mesozoic strata. Mesozoic rocks are widesparead on every continent, and marine deposits from this era increase in area with successive periods.

mesozone. The intermediate depth zone of metamorphism in metamamorphic rock, characterized by moderate temperatures 575-775K, hydrostatic pressure, and shearing stress.

meta. A prefix denoting metamorphism of the rock so qualified.

meta-anthracite. Coal having at least 98% fixed carbon; the highest rank of anthracite.

metabassite. Any metamorphosed basic igneous rock.

metabolite. Altered volcanic glass.

metrabrucite. Variety of periclase pseudomorphed after brucite (of Beudant).

metabentonite. Altered, metamorphosed, or somewhat indurated bentonite, characterized by clay minerals that no longer have the property of absorbing or absorbing large quantites of water.

metabrucite. Variety of periclase pseudomorphed after brucite.

metacryst. A large crystal developed in a metamorphic rock by recrystallization, such as garnet or staurolite in mica schist.

metalimnion. The horizontal layer of a thermal stratified lake in which the temperature decreases rapidly with depth. The metalimnion lies between epilimnion and the hypolimnion and includes the thermocline.

metalliferous. Metal-bearing; pertaining to mineral deposit from which metals can be extract by metallurgical operations.

metallogenic A term used with the connatation of mineralizaion, especially of ore deposition. Metallogenic epochs are units of geologic time during which particular mineral deposits were formed.

metamict. A term applied to a mineral whose original crystal structure has been disrupted by radiation from radioactive elements. (e.g., uranium orthorium) contained in that mineral, while the original external crystal form is retained.

metamorphic complex. The metamorphic rocks constituting a whole group closely related on a regional and/ or stratigraphic basis.

metamorphic convergence. A term used to indicate two metamorphic processes converging from opposite directions but resulting in the same matamorphic product.

metamorphic correlation. The determination of equivalence of metamorphic features, either between the metamorphic grades of rocks of different original composition, or between a metamorphic unit and its unmetamorphosed representative elsewhere.

metamorphic differentiation. A collective term for the various processes by which minerals or mineral assemblages are locally segregated from an initally uniform parent rock during metamorphism, e.g., garnet porphyroblasts in fine-grained mica schist.

metamorphic facies. A set of metamorphic rocks characterized particular mineral associations indicating origin under restricted temperature and pressure conditions.

metamorphic rock. A rock formed from pre-existing solid rocks by mineralogical, structural, and chemical changes in response to extreme changes in pressure, temperature and shearing stress constituent minerals have been recrystallized and reoriented and new minerals grown without the whole rock melting; may be a result of contact, dynamic, or regional metamorphism.

metamorphism. Range of processes involving combinations of high temperature, high pressure and chemical action by which rocks in the crust change in nature. Some overlap with diagenesis, but weathering is specifically excluded.

metamorphism, auto. Metamorphic process involving changes which occur during the cooling of an igneous mass as a result of the activity of residual fluids within the mass. The term polymetamorphism is applied to cases where two or more successive metamorphic episodes have left their traces in the rocks.

metamorphism, dynamic. Metamorphic process involving intense localized stresses which tend to break up the rocks, sometimes reducing them to fine powder.

metamorphism, regional. Metamorphic process involving large-scale action of both heat and 'pressure', producing a wide range of new minerals and a widespread development of tectonites always associated with orogenies and igneous intrusions.

metamorphism, retrograde. Any reversal of metamorphim which produces rocks of a lower metamorphic grade from rocks of higher grade (excluding weathering processes).

metamorphism, thermal. Metamorphic process involving heat alone, without significant pressure effects. It is normally associated with igneous, when the term contact metamorphism may be used.

metaquartzite. A quartzite formed by metamorphic recrystallization, as distinguished from an orthoquartzite, whose crystalline nature is of diagenetic origin.

metasediment. A sediment or sedimentary rock that shows evidence of having been subjected to metamorphism.

metasomatic. Pertaining to the process of metasomatism and its results. The term is especially used in conection with the origin of ore deposits.

metasomatism. A metamorphic process whereby existing minerals are transformed totally or partially into new minerals by the replacement of their chemical constituents. This occurs by the introduction of capillary solutions, originating either outside or within the rock, that are reactive with the existing minerals.

metastasy. Lateral adjustments of the earth's crust, as opposed to vertical movements.

metavolcanics. An informal term for volcanic rocks that show evidence of having been subjected to metamorphism.

meteor. Any relatively small fragment of solid material associated with a meteor and made luminous as a result of friction during passage through the earth's atmosphere.

meteoric water. A term applied to water which enters the rocks from the atmosphere i.e., rain, dew, hail, snow, and also the water of rivers and streams.

meteorite dust. Small angular or flat particles representing the product of fragmentation or crushing of meteorites.

meteor shower. A large concentration of falling meteors.

meteor swarm. A group of meteorioids that have closely similar orbits arround the sum.

metharmosis. Changes that occur in a buried sediment after uplift or consolidation but before the onset of weathering. Also spelled metaharmosis.

meyerhofferite. $Ca_2B_6O_{11}.7H_2O$. A colourless hydrated borate mineral; crystallizes in triclinic system.

miargyrite. $AgSbS_2$. A grey to black mineral composed of sulphides of silver and antimony; crystallizes in monoclinic system.

miarolithic. A chorismite type of igneous rock having miarolitic cavities or vestiges thereof.

miarolitic. A term applied to small irregular cavities in igneous rocks especially granites into which crystals of the rockforming minerals protrude; pertaining to such cavities or a rock containing them.

miaskite. Igneous rock of the syenite clan; a mottled, medium to coarse-grained plutonic rock ; contains microperthite, nepheline, mica, and oligoclase.

mica. Any of a group of monoclinic minerals of the general formula $(K,Na,Ca)(Mg,Fe,Li,Al)_{2\text{-}3}(Al,Si)_4O_{10}(OH,F)_2$. It consists of complex phyllo-silicates with perfect basal cleavage, which split into thin elastic laminae and range from colourless to black. An essential feature of mica group is the replacement of one silicon by one aluminium in the silicon-oxygen lattice and a perfect basal cleavage.

mica book. A crystal of mica usually large and irregular, whose cleavage plates resemble the leaves of a book.

micaceous. Consisting of; containing, or pertaining to mica.

mica schist. A schist which is composed essentially of mica and quartz and whose characteristic foliation is mainly due to the parallel orientation of the mica flakes.

micrinite. An opaque granular variety of incrinite of medium hardness showing on plant-cell structure.

micrite. (1) A descriptive term for the semi-opaque crystalline matrix of limestones, consisting of chemically precipitated carbonate mud crystals less than 4 microns in diameter, and interpreted as lithified ooze. The term is now used in a descriptive sense without genetic implication (2) A limestone with less then 1% allochems and consisting dominantly of micrite matrix.

micro- A prefix meaning "small"; when joined with a rock name, it usually signifies fine-grained.

microcrystalline. A term applied to a rock texture comprising crystals that are visible only with the aid of a microscope.

microfacies. The composition features or appearance of a rock or mineral in thin section under the microscope.

microfossil. A fossil too small to be studied without the aid of a microscope.

microgranodiorite. An igneous rock of the diorite clan; a medium grained hypabyssal rock ; it contains biotite, quartz, plagioclase, orthoclase, hornblende, and proxene.

microgranular. (1) Said of the texture of a microcrystalline xenomorphic igneous rock, also said of a rock with such texture. (2) Said of the texture of a carbonate sedimentary rock wherein the particles are mostly 10 to 60 microns in diameter and are well sorted while the finer clay-sized matrix is absent, also said of a rock with such texture. (3) Said of a foraminiferal wall made of minute calcite crystals.

microhill. A very rough, small sand column raised by the formation of pipprakes, ranging from a few millimetres to several centimetres high.

micrelinear. Any lineation that is invisable to the naked eye.

mierolite. $(Na,Ca)_2(Ta,Nb)_2O_6(O,OH,F)$. (1) An isometric mineral of the pyrochlore group that occurs in pegmatites and constitutes an ore of tantalum. (2) A microscopic crystal that polarizes light and has some determinable optical properties.

microlithology. Microscopic study of the characteristics of rocks.

microlitic. Said of the texture of a porphyritic igneous rock, having groundmass composed of an aggregate ofmicrolitesinagenerallyglassyorcryptocrystalline interstitial material.

microlog. A drill-hole resistivity log recorded with electrodes mounted at shortdistancesfromeachotherintheface of a rubber-padded microresistivity sonde.

micropegmatite. Very fine-grained intergowth of quartz and orthoclase.

microperthite. A fine-grained intergrowth of potassic and sodic feldspar in whichthelamellae(5-100micronswide) are visible only with the aid of microscope.

micropyric. Said of the texture of an igneous rock in which the phenocrysts are of microscopic size (longest dimension 0.2 mm); also said of a rock having such texture. Also known as microporhyritic.

micropoikilitic. Said of the poikilitic texture of an igneous rock that can be distinguished only with the aid of a microscope ; also said of a rock having such texture.

microporphyritic. *See* micropyric.

microseisms. Ground motion that is not due to earthquakes or explosions; storms at sea are the main source. There is not yet a generally accepted theory of microseisms.

microsomite: Chlorine rich variety of cancrinite.

microspherulitic. Said of the texture of an igneous rock having spherulitic character that can be made visible only with the aid of a microscope.

microstylolite. A stylolite in which the relief along the surface is less than a millimet such as one indicating differential solution betweenmineral grains.

mid-Atlantic ridge. The portion of the mid-oceanic ridge that lies within the limits of the Atlantic Ocean.

mid-oceanic ridge. A long submarine mountain chain running through the Atlantic, Indian and South Pacific oceans and associated with divergent plate bounderies.

miemite. A yellowish-brown variety of dolomite found a Miemo in Tuscany.

migma. A prefix denoting mixed.

migmatite. A rock composed of igneous or igneous-appearing and/or metamorphic materials, which are generally distinguishable megascopically. A term more or less synonymous with injection gneiss. Modern usage preserves the idea thatamigmatiteisamixedrock,i.e.,made up form two different source. Generally the two materials are a pre-existing host rock, which is inevitably a metamorphic rockofsomesortandaninvadinggranitic material which may take the form of (i) amagma;(ii)ahydrothermalsolutionor; (iii) a rather indefinite diffusion of materials ('emanations') through the host rock.

miharaite . An igneous rock of the gabbro clan; a fine-grained porphyrite extrusive rock; contains phenocrysts of bytwownite and hypershthene in a groundmass of a augite, plagioclase, and megnetite.

milarite. $K_2Ca_4Be_4Al_2Si_{24}O_{62}.H_2O$. A colourlesstogreenish,glassy,hexagonal mineral composed of hydrous silicate of potassium, calcium, aluminium and beryllium.

milky quartz. An opaque milky white variety of crystalline quartz, often with a greasy lustre; milkiness is due to the presence of air-filled cavities.

Miller indices. Three integrers identifying a type of crystal plane the intercepts of a plane on the three crystal parameters;

the reciprocals of these fractions reduced to integral proportions are the Miller indices. Also known as crystal indices.

millerite. 3[NiS]. A hexagonal mineral composed of nickel sulphide; shows greenish-black streak, uneven fracture, and perfect cleavage in two directions. It hardness on Mohs scale is 3.0-3.5.

milli. A prefix denoting thousandth part; *i.e.,* 1/1000.

millidarcy. A unit used to measure the reservoir or aquifer

milligal. A submultiple unit used to measure geophystell gravity.

mill rock. A variety of coarse, braccia found in or close to the volcanic units.

mill stone. A coarse grained sandstone or fine grained quartz conglomerates.

mimetie. (1) Pertaining to a twinned or malformed crystal that appears to have a higher grade of symmetry than it actually possesses. (2) Said of a trotonite whose deformation fabric, formed by recrystallization or neominaralization reflects, and is influenced by anisotropic stracture; also said of fabric itself.

mimetic crystallization. Recrystallization and/or neomineralization in metamorphism that reproduces any preexistant anisotropy bedding, schistosity, or other structures.

mimetite. $Pb_5(AsO_4,PO_4)_3Cl$. A yellow to brown mineral of the apatite group, comonly containing calcium arsenate and/or phosphate; a minor ore of lead. Also known as mimetene mimetesite.

mimosite. An igneous rock of the gabbro clan; a very dark, finegrained extrusive rock; contains augite, ilmentite, magnetite, olivine, and plagioclase.

mindel glaciation. The second glacial stage of the Pleistocene in the Alps.\

mineral. A naturally occurring substance with a characteristic chemical composition expressed by chemical formule; may occur as individual crystals or may be disseminated in some other mineral or rock. According to many mineralogists, a minerals is a strucurally homogeneous solid of definite chemical composition formed by the inorganic processes of nature. This definition includes ice as a mineral but excludes coal, natural oil, and gas.

mineral facies. Rocks of any origin whose constituents have been formed within the limits of a pressure-temprea-ture range characterized by the stability of certain index minerals.

mineral filler. A finely pulverized inert mineral or rock that is included in a manufactured product, *e.g.*, rubber, paper and plastics, to impart certain useful properties, such as hardness, smoothness, or strength. Some of the common mineral fillers are asbestos, kaolin, and talc.

mineralization (1) The process by which valuable minerals are introduced into a rock, resulting in an ore deposit, either actual or potential. The term includes various types of this process; e.g; impregnation, fissure filling, replacement. (2) Process of fossilization in which inorganic materials replace organic features.

mineralogy. The study of minerals; including their formation. occurrence, properties, composition and classification.

mineraloid. A naturally occuring, usually inorganic substance that is not considered to be a mineral because it is amorphous and thus lacks characteristic crystal form e.g., opal.

minette. A syentic variety of lamprophyre composed principally of biotite phenocrysts in a matrix of orthoclase and biotite.

mining geology. The geological aspects of mineral deposits, with special regard to problems related to mining.
minium. Pb_3O_4. A scarlet or orange red mineral consisting of an oxide of lead. Also known as red lead.
minnesotaite. A rare iron analogue of talc.
Miocene. The epoch of the Tertiary Period between the Oligocene and Pliocene Epochs.
miogeosyncline. The nonvolcanic portion of an orthogeosyncline, located adjacent to the crator.
mirabilite. $Na_2SO_4.10H_2O$. A yellow or white monoclinic mineral consisting of hydrous sodium sulphate, occurring as a deposit from saline lakes, palys, and springs and as an efflorescence; the pure crystals are known as Glauber's salt.
misenite. $K_8H_6(SO_4)_7$. A white mineral composed of native potassium hydrogen sulphate.
misfitstream. A stream whose meanders care not proportionate in size to the members of the valley or to the member scars preserved in the valley wall; i.e., a stream that is either too large or too small to have croded the valley in which it flows.
mispickel. FeAsS. Old name for the mineral arsenopyrite.
mississippion. A term for the Lower Carboniferons of the Mississippi valley.
Mississippian. A term for the lower carbonferawns of the Mississppi valley although the lower boundaries of the Lower Carboniferous and Mississippian are approximately the same, the upper boundaries vary.
Missourian. Lower part of the Upper Pennsylvanian of North America.
missourite. Feldspathoid igneous rock; a mottled grey, coarse grained granular plutonic rock; contains augite, biotite, leucite, and olivine.
mixed-base crude. A crude oil in which both naphthenic and parafinic hydrocarbons are present in about equal propotions.
mixed crystal. A crystal whose lattice sites are occupied at random by different ions or molecules of two different compounds.
mixed ore. An ore consisting of both oxidised as well as unoxdised minerals.
mixite. $Cu_{11}Bi(AsO_4)_5(OH)_{10}. 6H_2O$. A white to green mineral composed of a hydrous basic arsenate of bisumth and copper
mobilebelt. A belt of rock with consistent tectonic, metamorphic, and chronologic features, typical of Proterozoic terranes.
mobilization. Any process by which solid rock become sufficiently soft and plastic to permit it to flow or to permit geochemical migration of mobile components.
mock lead. *See* sphalerite.
mock ore. *See* sphalerite.
mode. (1) The minerl compostion of a rock, usually expressed as percentages of total weight by volume. (2) The value, or group of value that occurs with the greatest frequency in a set of data; the most typical observation.
modified mercalli scale. An earthquake intensity scale. Having twelve divisions ranging from I to XII.
modulus. A constant factors or multiplier for the conversion of units from one system to another.
modulus of elasticity. The ratio of the increment of some specified form of stress to the increment to some specified from of strain, such as Young's modulus, the bulk modulus, or the shear modulus. Also known as coefficient of elasticity.
moffette. A small opening in an area of late-stage volcanic activity that permits

carbon dioxide to move out. It is a type of fumarole.

mogote. A large residual hill of limestone associated with a karst-erosion landescape.

Moho. An abbreviation for the Mohorovicic discontinity.

mohole. A proposed deep borehole to penetrate into the earth's mantle below the Mohorovicic discontinuity.

Mohorovic discountinuity. The boundary surface that separates the earth's crust from the subjacent mantle. It marks the level at which *P*-wave velocities change abruptly from 6.7-7.2 km/s (in the lower crust to 7.68-6 km/s or average 8.1km/s at the top of above mantle; its depth ranges from about 5-10 km beneath the ocean floor to about 35 km below the continents, although it may reach 60 km or more under some mountain ranges. It is variously estimated to be between 0.2. and 3 km thick.

Moh's scale. A standard of ten minerals by which hardness of a mineral may be measured. The scale includes, from softest (designated 1) to hardest (designated 10): Talc, gypsum, calcite, fluroite, apatite, orthoclase, quartz, topaz, corundum and diamond.

Moine. A stratigraphic division of the Scottish Precambrian, the metamorphic equivalent of the Torridonian in Scotaland which passes up (conformbaly, into the Lower Dalradian.

moissanite. SiC. A silicon carbide mineral found in meteorites; indentical with artifical carborundum.

molasse. A paralic sedimentary consisting mainly of shale, subgrewacke, sandstone, and congolomerate. It is more clastic and less rhythmic than the preceding flysch and is generally postorogenic. It is formed by the erosion of mountain ranges after the final phase of an orogeny (post-tectonic).

molybdenite. MoS_2. A metallic lead-grey mineral that crystallized in hexagonal system and is commonly found in scales or foliated masses; its hardness on Mohs scale is 1.5; important ore of molybdenum.

molybdine. *See* molybdite.

molybdite. M_oO_3. A mineral, most of which is actually ferrimolybdite. Also known as molybdine.

molysite $FeCl_3$. A brownish-red or yellow mineral composed of native ferric chloride.

monadnock. A hill or mountain rising conspicuously above the general level of a peneplain in a temperate climate, representing an isolated remanant in a region that has been largely beveled to its base level. Also known as troso mountain.

monalbite. A modification of albite with monoclinic symmetry that is stable under equilibrium conditions at temperatures (about 1275K) near the melting point.

monazite. (Ce,La,Nd,Th) (SiO_4,PO_4). A yellow to brown, monoclinic rare-earth mineral, found as an accessory mineral in acid igneous rocks, in pegmatite dykes, and in detrital sand. It is a principal ore of rare-earths and of thorium. Also known as cryptolite.

monochiquite. A lamprophyre composed of olivine, pyroxene, and usually mica or amphibole phenocrysts embedded in a glass or analcine groundmass.

mondhaldeite. An igneous rock of the dirite clan; a coarse-grained granular plutonic rock; contains phenocysts of augite, bytownite, hornblende, and leucite in a glassy groundmass.

monimolimnion. The deep, usuall salty layer of a meromictic lake, of hig density.

monoumuthite. An Igneous rock of the syenite clan; a coarse-grained granular plutonic rok; contains apatite, calcte, hasting slate, nepheline and sodaclase.

mono. A prefix denoting single, one.

monoclinal scarp. A scarp resulting from a steep downward flexure between an upland block and a tectonic basin.

monocline. A local steepening in an otherwise uniform gentle dip.

monoclinicsystem. Oneofthesixcrystal systems, characterized b either a single two-foldaxisofsymmetry,asingleplane ofsymmetry,oracombinationofthetwo. Crystals belonging to this system are referred to three unequal, elements of themonoclinicsystemwillincludethere-fore the three parameters and the angle between the vertical and inclined axes.

monogenetic. Resulting from one processofformationordervied from one source, or originating or developing at oneplaceandtime; *e.g.*, said of a volcano built up by a single eruption.

monogeosyncline. A primary geosyncline that is long, narrow, and deeply subsided; it is composed of shallow water and situated along the inner margin or borderlands.

monolith. (1) A piece of unfractured bedrock, generally, a few metres in length. (2) A large upstanding mass of rock, such is volcanic spine. (3) A vertical soil section, chosen to illustrate soil profile.

monomneralic. A term used for a rock composed almost wholly of a single mineral.

monomorphic. A term used for a bryozoan colony in which all zooids in zones of astogenetic repetition are feeding zooids of similar morphology.

monomicitc. (1) Said of a lake with only one yearly overturn. Tropical lakes overturn in the winter and polar lakes in the summer. (2) Said of a clastic sedimentary rock composed of a single mineral species.

nacreous. Having an iridescent lustre resembling that of mother-of-peral. Also known as pearly.

nacrite. $Al_2Si_2O_5(OH)_4$. A crystallized clay mineral of kaolinite group ; most closely stacked in *c*-axis direction.

nadorite. $PbSO_2PbSbO_2Cl$. A yellowish brown mineral containing antimony.

naegite. A variety of zircon containing traces of thorium and uranium.

nagatelite. A black, monoclinic mineral composed of phosphosilicate of aluminium, rare-earth elements, calcium, and iron. It occurs in tabular masses. Its hardness on Mohs scale is 5.5.

nagyagite. $8[Pb_5Au(Te,Sb)_4S_{5-8}]$. A monoclinic mineral composed of lead, gold tellurium and antimony sulphide; occurs as tabular crystals; shows blackish grey streak and good perfect cleavage in one direction. Is hardness on Mohs scale is 1.0-1.5.

nahcolite. $NaHCO_3$. A white monclinic mineral consisting of naturally occurring sodium hydrogen carbonate.

naif. A term used for a gemstone having a true or natural lustre when uncut such as natural, upolished few of a diamond crystal.

nail-head spar. A variety of calcite.

nakauritc. $Cu_8(SO_4)_4(CO_3)(OH)_6.48H_2O$. An orthorhombic mineral of copper.

naked karst. Karst that is developed in a region without soil cover, so that its topographic fcatures arc well cxposed.

nakhlite. An achondritic stony meteorite composed of an aggregate of diopside and olivine.

nambulite. $NaLiMn_8Si_{40}O_{28}$. A mineral of pyrexeoid group.

nannofossil. A collective term for fossil discoasters and coccoliths, both primarily calcareous microfossils, near the limit of resolution of the light microscope and hence best studies with electron microscope. This term is also used in more general sense for other marine fossils, smaller than microfossils.

nannoze. Unconsolidated pelagic deposit consisting of more than 69% calcium carbonate, in which nannoplankton make up more than 30% of the recognizable skeletal remains.

nantokite. CuCl. A colourless, white or greyish isometric mineral containing cuprous chloride.

napoleonite. A synonym of orthoclase and corsite.

nappe. A sheetlike, allochthonous rock unit that has moved in a predominatly horizontal surface. The mechanism may be thrust faulting, recumbent folding, or gravity sliding.

narsarsukite. $Na_2(Ti,Fe)Si_4(O,F)$. A mineral composed of fluoride and silicate of sodium, iron, and titanium.

nasonite. $Ca_4Pb_6Si_6O_{21}Cl_2$. A white mineral composed of silicate and chloride of calcium and lead; occurs in granular masses.

native element. An element which occurs in the free state e.g.. copper, silver, gold, and carbon (graphite and diamond).

natroalunite. $NaAl_3(SO_4)_2(OH)_6$. A mineral composed of basic sulphate of sodium and alumiinium; shows conchoidal fracture and good cleavage in one direction. Its hardness on Mohs scale is 3.5.-4.0.

natrocalcite. $NaCu_2(SO_4)(OH)\text{-}H_2O$. An emrald-green mineral composed of hydrous basic sulphate of copper and sodium.

natrojarosite. $NaFe_3(SO_4)_2(OH)_6$. A yellow to brown hexagonal mineral composed of basic sodium iron sulphate; shows conchoidal fracture and perfect cleavage in one direction. Its hardness on Mohs scale is 3.0.

neap tide A tidal pattern in which the difference in level between high and low tide is a minimum. Neap tides occur when the earth and sun are at right angles with respect to the moon; *i.e.* during the first and third quarter moon.

nearshore. A widely used term denoting a zone begining at the low tide line and continuing offshore for an indefinite but generally short distance. The breaker zone, longshore drift and waves long shore currents are found in the nearshore region.

nebulite. A migmatite in which traces and patches of the original fabric remian.

neck (1) A volcanic plug. (2) A narrow strip of land connecting two larger areas. (3) The narrow bank or rip of water forming the part of a rip current where feeder currents converge and flow swiftly through the incoming breakers or surf and out to the head.

negative element. A large structural feature or part of the earth's crust, characterized through a long geologic time period by frequent and conspicious downward movement or by extensive erosion, or by an uplift that is considerably less rapid or lessr frequent than that of adjacent positive elements. Also known as negative area.

negative elongation. Elongation of an anisotropic crystal that is parallel to the faster of the two plane-polarized light rays.

negative movement. Subsidence of a part of the earth's crust, actual in relation to sea level or relative in relation to adjacent parts of the crust.

neoautochthon. A stable basement or antochthon where a nappe has stopped movement.

neoblast. A grain, especially in a metamorphic rock, that is of more recent formation than the other grains of the, same or other mineral species in the rock.

neceonian. European stage : Lower cretaceous.

necryst. An individual crystal of a secondary mineral in an evapirite.

neocrystallization. Crystallization or recrystallization that involves the development of new minerals among the fabric elements.

neocrystic texture. A secondary non-laminated teature of an evaperite.

neoglaciation. The removal of glacier ice growth in certain mountain areas during the 'Little Ice Age, following its shrinkage or disappearance during the Altithermial interval.

neomagma. A magma formed by partial or complete fusion of pre-existing rock under conditions of plutonic metamorphism.

neomineralization. Chemical intercharge within a rock resulting in alteration of its mineral components with the production of new minerals.

neontology. The scientific study of existing organisms, as opposed to paleontology.

neosilicate. A structural type of silicate mineral characterized by linkage of isolated SiO_4 tetrahedra by ionic bonding only, *e.g.*, olivine.

neotype. A specimen selected as type subsequent to the original description when the primary types are known to be destroyed, a nomenclatural type.

nepheline $(Na,K)AlSiO_4$. A mineral of feldspathoid group crystallizing in the hexagonal system and occurring as glassy or coarse crystals of grains, or green tobrownmassesinalkalicigneousrocks. Its hardness on Mohs scale is 5.5-6.0. Also known as nephelite.

nepheline basalt. Silica deficient variety of basalt, contains feld spathoids in place of feldspars.

nepheline phonolite. The fine graned equivalent of nepheline syenite.

nepheline monzonite. A nephaline syenite in which sodic plagioclase exceeds the quantity of alkali feldspar.

nepheline syenite. A phaneritic plutonicrockwithgranulartexture,composedmainlyofalkalifelspar,nepheline, and dark-coloured materials.

nephelinite. A dark-coloured, aphanitic rock of volcanic origin, composed essentiallyofnephelineandpyroxene; texture is usually phosphyritic with large crystalsofaugiteandnephelineinafine-grained matrix.

nephelite. *See* nepheline.

nephrite. Anexceptionallytough, compact, fine-grained, bluish or greenish premalite or actionalite, constituting the less ra re kind of jade and formerly worn asaremedyofkidneydiseases.crustwere deposited from or crystallized out of water.

neptunite. $(Na,K)_2(Fe,Mn)TiSi_4\text{-}O_{12}$. A black mineral composed of silicate of sodium, potassium, iron, manganese, and titanium.

neritic. Of pertaining to the ocean environmentordepthzonebetweenlowtide level and 100 fathoms, or approximately theedgeofthecontinentalshelf; also pertainingtotheorganismslivinginthatenvironment.

nesosilicates. a mineral such as olivine, composed of indepedent silicon-oxygen tetrahedra bonded by ionic bonds, without sharing of oxygen atoms.

nevadite. An igneous rock of the granite clan; afine-grainedporphyriticextrusive rock; it contains phenocrysts of biotite, hornblende,quartz,andsanidineinavery fine groundmass.

neve. A term used for a mass of hardened snow at the source or head of glacier; it refers to overall snow cover that exists duringthemeltingperiodand sometimes from one year to another.

new ice. A general term for secently formedicelessthan5cmthick,composed of ice crystals that may be weakly frozen togetherandthathaveadefiniteformonly while they flost.

newlandite. An igneous rock of the ultrabasic clan; A mediumgrained granular hypabyssl rock; contains diopside, enstatite, and garnet; it occurs in diamond pipes of kimberlite.

new red sandstone. The red sandstone facies of the permian and Triassic systems.

ngavite. A enondritic stony meteorite composedofbronziteandolivineinafriable, breccia-lake mass of chondrules.

niccolite. NiAs. pale or copered, hexagonal mineral with metallic lustre; itshardnessonMohsscaleis5.5-5.6. Also known as nickline.

niche. A shallow cave re-entrant producedbyweatheringanderosionnear the base of a rock face or cliff, or beneath a waterfall.

niche glacier. A connon type of small mountain glacier, occupying a funnel-shaped hollow or irregular recess in a mountain slope.

nick. (1) A place of abrupt inflection in a stream profit. (2) A sharp angle cut by waves, currents, or ice at the base of cliff.

nickline. See niccolite.

nick point. American spelling of knick point.

Niggliclassification. Aclassificationof rocks on the basis of their chemical composition, similar in some respects to the CIPW classification.

nip. (1) A small, low break in the slope on a beach, produced at the high water marks wavelets. (2) A horizontal cavity formed in soluble rock at the edge of a waterbody. (3) Thinning of a coal seam, particularly if caused by tectonic movements. Also known as want.

node (1) The point on a fault at which the apparent displacement changes. (2) That point on a standing wave at which there is no vibration.

nodular Composed of or having the form of nodules; *e.g.*, a nodular ore.

nodule A term that conveys the idea of a rounded mass or lump; there is no connotation of size. It can be applied to a lump of mineral aggregate that is normally without internal structure and contrasts with its embedding matrix.. The term is also used to describe the concretionary lumps of metals found on the ocean floors; *e.g.*, manganese nodules..

non-conformity. (1) An obsolete synonym of unconformity. (2) An unconformity between stratified rocks above and unstratified igneous or metamorphic rocks below.

nonesite. An igneous rock of the gabbro clan; a fine-grained porphyritic extrusive rock; contains augite, enstatite, and labradorite phenocrysts in a groundmass of augite and plagioclase.

nonflowing artesian well An artesian well in which the hydrostatic pressure is not great enough to elevate the water above the land surface; the water does however, rise above the local water table.

non-plunging fold. A fold whose hinge line is horizontal.

nonrotational strain. Strain at a point in which the orientation of the principal axis of strain remains unchanged.

nonsorted. Said of a nongenetic grow which is dominently circular and has a nonsorted appearance due to the absence of a border of stores developed singly or in groups.

nonsorted crack. A rare form of patterned ground representing the boulder free varient of a sorted crack

nontronite. $Na(Al,Fe,Si)O_{10}(OH)_2$. An iron-rich clay mineral of the montmorillonite group that represents the end member in which the replacement of aluminium by ferric ion is essentially complete. Also known an pintguite.

nordmarkite. Igneous rock of the syenite clan. A reddsh, medium-grained granular plutonic rock; contains biotite, hornblende, microperthite, oligoclase, and quartz.

norite. A coarse-grained plutonic rock containing labradorite as the chief constituent and differing from gabbro by presence of hypersthene as the dominant mafic mineral.

norm. The theoretical mineral composition of a rock expressed in terns of normative mineral molecules that have been determined by specific chemical analysis for the purpose of classification and coomparison.

normal consolidation. Consolidation of sedimentary material in equilibrium with overburden pressure.

normal dispersion. The dispersion of sesmic surface waves in which the recorded wave period increases with time.

normal moveout. The increase in arrival fine of a seismic reflection event resulting from an increase in distance from source to detector, or from dip of the reflector.

nomal sandstone. A term used for a variety of sandstone composed almost exclusively of quartz, with subordinate amounts of other minerals.

normal shift. In a fault, the horizontal componentoftheshift,measuredperpendicular to the strike of the fault.

normalsoil. A soil whose profile is more or less in equ'librium with the environment, and which shows the effects of the environmentonitsdevelopmentfromthe parent material.

normal zoning. Zoning in a crystal of plagioclase in which the zones become progressively more sodic outward.

normal erosion. Wearing away of topographic featuresbyprevailingagencies in the temperate climate. The term is open to criticism becaause erosion as found in temperate areas may in fact be abnormalorbecauseonemodeoferosion is just as normal as another.

northfieldite. An igneous rock of the granite clan, a white, coarsegrained granular plutonic rock; contains muscovite and quarz.

northfieldite. Anigneous rock of the granite clan, a white, coarsegrained granular plutonic rock; contains muscovite and quartz.

northupite. $Na_3MgCl(CO_3)_2$. A colourless, white, grey or yellow isometric mineral consisting of octahedral crystals of sodium magnesium carbonate.

nosean. *See* noselite.

noselite. $Na_4Al_3Si_3O_{12}SO_4$. A blue, brown, or grey coloured mineral of sodalite group; hardness on Mohs scale is 5.5. Also known as nosean.

noumeite. A dark green variety of garnierite.

novacekite. A magnesium enriched form of autunite.

novaculite. A hard, dense, even textured, light-coloured, cryptocrystalline, silicceous sedimentary rock, similar to chertbutforthedominanceofmicrocrystalline quartz over chalcedony. Also known as razor stone.

novaculiticchert. Agreychertthatfragmentintoslightlyroughsplinterypieces.

nugget. (1) A small mass of metal found free in nature. (2) A water worn lump of placer gold or other metal.

nummulitic limestone. A sedimentary rock composed chiefly of nummulite shells.

nunatak. An isolated hill, knob, ridge, peakofbedrock projecting,oranyrock masswhichprojectsthroughanicesheet, generally found at the margins of a sheet where the ice is extremaly thiw.

oasis. A fertile, vegetated area in the midst of a desert, where the water table has come close enough to the surface for wells and springs or seepages to exist, thus making it suitable for human habitation.

ob . A prefix indicating inverted.

oblique fault. A fault that strikes oblique to, rather than parallel or perpendicular to the strike of the constituent rocks or dominant structure.

obliqueslap. In a fault, movement or slip that is intermediate in oriatation between the dip slip and the strike slip.

obsequent. Said of a stream, valley, or drainage system, being in a direction opposite to that of the original consequent drainage.

obsidian. A jet-black volcanic glass, usually of rhyolite composition, characterized by conchoidal fracture; formed by the rapid cooling of viscous lava; gernerally forms the upper part of lava flows. It is used for making jewelry and art object.

obsidianite. An obsolete term for a small rounded glassy object, now known as tektite.

obstruction cave. A glacier cave formed as a result of interrupted ice flow due to a bedrock or other subglacial protuberance.

obstruction. A morine formed where the movement of the ice is obstructed, as by a ridge of bedrock.

ocellar. Said of the texture. of an igneous rock in which aggregates of smaller crystals (*e.g.*, biotite) are arranged radially or tangentially around larger, euhedral crystals (*e.g.*,leucite) or form rounded eyelike branching forms; also said of a rock having such as texture.

ocher. An earthy, powdery, red, brown or yellow oxide of iron which is used as pigment. Also spelled ochre.

ochre. *See* ocher.

occult mineral. A mineral component of rock which cannot be seen through microscope, but whose presense can be detected by chemical analysis.

ocean. A major primary subdivision of intercommunicating body of salt water occupying the depressions of the earth's suface.

ocean basin. The great depression occupied by the ocean on the surface of the lithosphere.

ocean floor. The near-horizontal surface of the ocean.basin.

oceanic basalt. Basalt ocurring in the ocean basin. The lava originates in either the asthenosphere or upper matle. Ocernie basalt are found associated with mantle hot spots or with divergent plate boundaries.

oceanic crust. Mass of basaltic material about 5 km thick, lying under the occon bottom and covered with a thin sedimentary veneer.

oceanic tholeiite. The principal igneous rock encrusting the deeper parts of the ocean basis; believed to be parental to alkaline basalt.

oceanite. A picritic basalt with more than 50% olivine in the mode; usage not recommended.

oceanization. The conversion of continental curst into oceagic crust.

octahedral cleavage. Crystal cleavage in planes parallel to the face of octahedron as in fluorite.

octahedral plane. The plane in a cubic lattice having three numerically equal Miller indices.

octahedrite. The most common iron meteorite, having 6-18% nickel in the metal phase and having intimate intergrowths lying parallel to the octahedral plane.

odinite. An igneous rock of the gabbro clan. A medium-grained porphyritic hypabyssal rock, contains phenocrysts of augite, hornblande, and labradorite in a groundmass of feldspar laths and hornblande needles.

odonotolite. Fossil bone or tooth coloured by vivianite and partly replaced by apatite or calcite. Its hardness on Mohs scale is 5.0.

offlap. The successive lateral contraction extent of strata due to their depostion in a shrinking sea or on the margin of a rising landmass. Also known as regressive overlap.

off-reef. Pertaining to the sea floor away from the margin of a reef.

offretite. $(K_2Ca)(Al_{10}Si_{26})O_{72}$. 30–$32H_2O$. A zeolite mineral.

offset. (1) The horizontal component of displacement a fault, major perpendicular to the disrupted horizon. (2) In seismic prospecting, the horizontal distance from energy source nearest detector, or form a shothole to the line of profile; also, the horizontal displacement, measured form the detector, of a point for which a calculated depth applies. (3) A spur of minor branch from a hill-range or mountain-range . (4) A new corallite formed in a corallum by budding.

offset deposit. A mineral deposit, especially of sulphides, formed partly by magnatic segregation and partly by hydrothermal solution, near the surface rock.

offset ridge. A ridge consisting of resistant sedimentary rock which has been made discontinous as a result of faulting.

offset well. An oil well drilled near the boundary of a property and opposite to producing or completed well on an adjoining property, for the purpose of preventing the drainage of oil or gas by the earlier well. An obligataion to drill such offset wells contained in oil and gas leases.

offshore. The comparatively flat zone of variable width extending from the outer margin of the shoreface to the edge of the contienental shelf.

offshore beach. A single, long, narrow ridge of sand which rises slightly above the level of high tide and lies parallel to the shore form which it is separated by a lagoon. Also known as barrier beach.

offshore drilling. The drilling of oil of gas wells into water covered locations, usually on submerged continental shelves.

offshore gas. Natural gas produced from reservoirs under the offshore continentel shelves.

offshore oil. Oil produced form reservoir under the offshore continantal shelves.

ogive. One of a priodically repeated series of dark, curved structures occurring down a glacier that resemsble a pointed arch.

O horizon. An organic layer formed above a mineral soil incorporating one or both of two subdivisions: the O_1 horizon, in which the original forms of that remains are recognizable, and the O_2, in which these forms cannot be recognized.

oid. A suffix denoting 'like', having the 'form of'.

oil-base mud. A drilling mud with clay particles suspended in oil rather than in water.

oil-cut mud. Drilling mud unitestionally admixed with crude oil; may result from oil entering the mud while drilling or from a drill-stem test of an oil reservoir.

oil field. (1) The surface boundaries of an area from which petroleum is obtained. (2) Two or more oil pools on a single geologic feature or otherwise closely related.

oil pool. A subsurface accumulation of petroleumthatiscapapleofproducingoil ineconomicquantities.Alsoknownasoil reservoir.

oil sand. A term lossely applied to any prous stratum containing petroleum or impregnated with hydrocarbons. The term is also applied to productive limestone and dolomite.

oil shale. A kerogen bearing finely laminated brown or black sedimentary rock that will yield liquid or gaseous hydrocarbons on distillation. It does not containfreeoil.Itapperastoforminaspecial kind of anaerobic environment and is not commonly found in association with coal-bearing strata.

oil trap. Anybarriertotheupwardmovementofoilorgas, allowing either or both toaccumulate. Atrapincludesareservoir rock and an impermeable roof rock; the contact between these is concave downward.

oil water contact. The boundary layer or surface between an oil accumulation and the underlying bottom water.

old age. A term aplied to the final stage inthecycleoferosioninwhichthesurface has been brought to almost base level. It may also refer to that stage of stream development in which sediment builds up and water flow decreases.

old lake. (1) A lake in an advanced stage of filling by sediments or vegetation. (2) A lake whose shoreline exhibits an advanced stage of development.

old mountain. A mountain that has formed prior to the tertiary period, especially a fold mountain formed before the Alpine crogeny.

oldredsandstone. A thick, sequence of nonmarine, predominaantly red sedimentary rocks, mainly sandstones, conglomerates, adn shales, representing the Devonian System in Great Britain and north west Europe.

old snow. Fallen snow that has lost its crystalline appearance, through metamorphlism.

old stream. A stream developed during the stage of old age.

oleic. Pertaining to derived from; or contained in an oil.

ologe-. A prefix meaning small or little.

Oligocene. An epoch of the early Tertiary period, after the Eocens and before the Micene.

oligoclase. A variety of plagioclase feldspar with composition ranging from $Ab_{90}An_{10}$ to Ab_{70} An_{30}; shows perfect cleavage in one direction. It occurs in igneous rocks of intermediate to high silica content.

oligoclasite. An igneous rock of the gabbro clan. A diorite contianing 5% or less ferromagnesian minerals, the remainder being oligoclase.

oligomictic. (1) Saidofaclasticsedimentary rock composed of a single rock type, such as orthopuartzite conglomerate, also said of clasts of such rocks. (2) Said of a lake that circulates only at rare intervels when abnormally cold speels occur.

oligonite. A variety of siderite containing upto 40% manganese carbonate.

olyopelic. Said of a lake-bottom deposit that contains very little clay.

oligophyra. A light coloured diarite containing oligoclase phenocrysts in a groundmass of the same mineral.

oligosiderite. A nebeorite containing and a small amount of metallic iron.

oligotrophic peat. Peat containing a small amount of plant nutrients.

olistolith. An exotic block or other mass transported by submarine gravity sliding or slumping and included the binder of an oliststrone.

olistostrome. A sedimentary deposit consisting of a characterictic mass of intimately mixed heterogensous materials that accumulated as a semifluid body of submarine gravity sliding or slumping of unconsolidated sediments.

olivenite $Cu_2(AsO_4)(OH)$. An olive green, dull brown, yellow,or grey mineral composed of basic arsenate or copper; crystalizes in orthorhombic system. Also known as leucohalcite; wood copper.

olivine. $(Mg,Fe_2)SiO_4$. A neosilicate group of olive-green or brown mineral composed of magnesium iron silicate. It consists of the isomorphous solidsolution series forsterite fayalite. Olivine is a common rock forming mineral of basic, ultra basic, and low-silica, igneous rock (gabbro, basalt, periodotite, dunite). It crystallizes early from a magma, weathers readily at eath's surface, and metamorphoses to serpentine. It has vitreous luster and its hardness on Mohs scale is3.27.3.37. The essential structure consists of a series of isolated SiO_4 tetrahedra which are linked by means of metal cations. The olivines are othorhombic, but well-formed crystals are rare. They show no cleavage but disply conchoidal fracture Common olivine is a member of a continuous series solutions between Mg_3SiO_4 (forsterite, abbreviated Fo) and Fe_2SiO_4 (fayalite, abbreviated Fa). Many names have been given to the intermediate compounds between these pure end members:

Forsterite	0 10% Fa
Charysolite	10-30% Fa
Hyalosiderite	30-50% Fa
Hortonolite	50-70% Fa
Ferrohortonolite	70-90% Fa
Fayalite	90-100% Fa

olivine basalt. Any of a group of olivine bearing basalts.

olivine dibase. An igneous rock composed mainly of olivine and formed from the leritic magmas by differentiations in thick sills.

olivine nephelinite. An extusive igneous rock differing in comosition from nephelinite only by the presence of olivine. Also known as nepheline basalt.

olivinite. An olivine-rich ore-beraing igneous rock that also contains other pyrosenes and/or amphiboles.

olivinoid An olivine- like substance occurring in meteorites.

ollenite. High grade regional metamorphic rock having schistose texture; contains epidote, garnet, horblende, rutile, and sphene.

olmsteadite $K_2Fe_4(Nb,Ta)_2(PO_4)_4$ $4H_2O$. An orthorhombic rare mineral.

olsacherite. $Pb_2(SeO_4)(SO_4)$. A rock mineral.

ombrogenous. Said of peat deposits whose moisture content depends upon rainfall.

ombrotiphic. Pertaining to a short lived pond which gets its water mainly from rain.

omphacite. A dark green to pale green granular or foliated, high temperture aluminous clinopyroxene mineral with a vitrerous lustre; commonly occurs in the rock eclogite; variety of augite.

Onesquethawan. North American stage: Lower and Middle Devonian.

ongonite. Topaz bearing quratz kreratophyre, enriched in lithium and fluorine.

onkilonite. An igneous rock of the gabbro clan. A steel-grey, fine- grained porphritic extrusive rock; contains phenocrysts of olivine in a groundmass of augite, leucite,and nepheline.

onlap. A type of overlap characterized by regular and progressive pinching out of thestratatowardsthemarginsofadeposi-tional basin, each unit transgresses and extendsbeyondthepointofreferecnceof theunderlyingunit. Alsoknownastrans-gressive overlap.

onyx . A banded chalcedonic silica in which the bands are straight and parallel, natural colours are usually red or brown withwhitealthoughblackisoccasionally enountered.

onyax marble. A compact, usually banded, gerally translucent varierty of calcite resembling true onyax in appearance. It is usually deposited from coldwater solutions. Often in the form of stalagmites and stalactites in caves.

onyx opal. Common opal with straight parallel markings.

oovid. (1) An indiviual spherite of an oblitic rock. (2) A general, nongeneric term for a particle that resembles oolith in appearance and shape.

oolite. A sedimentary rock, usually a limestone, made up chiefly of ooliths cemented together.

oolith. Rounded grains up to 2mm in diameter,usually composed of calcium carbonate or silica, hematite, chamosite or limonite. Normally show real or concentric structures around a nucleus of quartz or shell fragment.

oolitic. Pertaining to, including or comprising spherical or nearly spherical paricles (up to few millimeters in diameter) with a resulting fish roe appearance. The 'oolithsareusuallyformed by growth (often of chalk or limestone) round an existing nucleus.

ooliticchert. Chertcomposedchieflyof oolith.

oolitic ironstone. Marine sedimentary rock, ironstone formed of ooliths composed of chamosite, limonite or hematite in fine-grained matrix of calcite, chamosite, chalybite, limonite or hematite.

oolitic limestone. An even-textured limestonemadeupalmostentirelyofcalcareous ooliths with essentially no matrix.

oolitic texure. The texture of sedimentary rock consisting largly or ooliths showing tangential contacts with one another.

oolitization. The act or process of forming ooids or an oolitic rock.

oolitoid A sedimentary particle smaller in size and shape to an ooliths but lacking its internal structure.

oomold. A spheroidal opening in a sediementary rock, or insoluble residue, produced by the dissolution of an oolith.

ooze (1) A pelagic sediment consisting of atleast 30% skeletal remains of calcareous or siliceous pelagic organisms. The rest being clay minerals. (2) A soft and muddy piece of ground such as a bog, usually resulting from the flow of a spring or brook. (3) Soft mud or slime typically ,covering the bottom of a lake or river.

opal. A natural hydrated form of silica which is amorphous and usually occurs in botryodial or stalacitic masses; its hardness on Mohs scale is 5-6.

opal agate. Variety of opal displaying an agate-like banding of different colours.

opal jaspar. Yellow variety of opal with inclusions of iron oxide.

open fold. A fold in which the limbs diverge at a large scale.

open form. A crystal whose faces do not enclose space, *e.g.*, trigonal prism.

ophi-. A prefix for serpentine, *i.e.*, snake-like.

ophiolite. Masses of mafic and ultramafic igneous rocks, such as basalt, gabbro and olivine-rich rocks, whose structure and composition identify them as segments of ocean crust pushed into the continents by plate collisions

Ophiolites found embedded within certain deformed mountain belts-represent thelastvestigesofoceanthatdisappeared ascontinentsapproachedeachother. The term is also applied to their metamorphic equivalents, which are usually rich in albite and amphiboles.

ophite. A general term for diabases that have retained their ophitic structure although the pyroxene is altered to uralite.

ophitic A texture of an igneous rock in which randomly oriented plagioclase crystals are surrounded by a single large crystal of pyroxene.

opoka. A porous, flinty, calcarous sedimentary rock, with conchoidal or irregular fracture, consisting of finegrainedopalinesilicaandhardenedbythe presence of silica of organic origin.

opposite tide. The high tide at a corresponding place on the opposite side of the earth accompanying a direct tide.

optical activity. The property of a substance *e.g.*, quartz, to retate the plane of polarization of light. Such a substance is said to optically active.

optical axis. (1) A line passing through a radially symmetrical optical system such that rotation of the system about this linedoesnotalteritinanydetectableway. (2) The axis in a doubly refracting medium in which the ordinary and extraordinary waves propagate with the same velocity, and double refraction vanishes, Also known as optic axis; principal axis.

optical calcite. Crystalline calcite so clear that it finds use in making Nicol prism. It is usually iceland spar.

opticalcrystal. Any natural or synthetic crystal that is used in infrared and ultravioletopticsandforitspiezoelectric effects.

opticaldensity. The degree of opacity of a translucent medium expresed as long (I_o/I) Where I_o is the intesity of incident ray, while I is the intensity of transmitted ray.

optical element. A part of an optical instrument which acts upon the light passing through the instrument, such as lens, mirror, or prism.

opticaltwinning. Simultaneous growth oftwocrystalswhicharethesameexcept that the structure of one is the mirror image of structure of the other.

optic axis. *See* optical axis.

opticnormal. The axis of a crystal which is perpendicular to the optic axis.

orangite. A bright orange yellow variety of throite.

orarite. A lamellar intergrowth of a potassium feldspar and a plagioclase naear anorthite.

O ray. In uniaxial crystals, the ray which vibrates perpendicular to the optic axis.

orbicule. Large spherical, onionlike masses with distinct concentric shells; the size range is between 2 and 15 cm, but some are 3 m or more in diameter. They are variously distributed through phaneritic rocks of silicic to basic composition. The formation of orbicules usually occurs at early stage of rock consolidation.

orbite. A porphyritic gabbro containing large crysts of hornblende, or plagiclase and hornblendes.

ordonezite. $ZnSb_2O_6$. A brown Ftriaginal mineral.

ordosite. A dark-coloured syenite containing about 60% sodic clinopyroxene.

Ordovician. The second earliest period of Palegoic era extending for 65 million years, from 500 to 435 million years ago. In the older literature, the Ordovician is sometimes known as the Lower Silurian.

ore. Rock including ore minerals, those suitablefortheextractionofrequiredsubstances. Also includes country rock and genague minerals.

ore body. A continuous, well-defined mass of material of sufficient ore content to make ore extraction economically feasible.

ore control. Any tectonic, geochemical or lithologic feature thought to have influenced the formation and placement of ore.

ore cluster. A group of interconnected ore bodies.

ore deposit. Rocks containing minerals of economic value in such amount that can be extracted profitably.

ore district. A combination of several ore deposits into one common whole or system.

ore lead age. Measurement of the age of the earth by comparing the relative progress of the two decay series U-235-Pb207 and U-238-Pb-206.

ore magma. A magma that may crystallize into an ore; the sulphide, oxide, or other metallic facies of a solidified magma.

orendite. An igeous rock of the syenite clan. A reddish-grey, fine-grained prophyritic vesicular extrusive rock; contains phencrysts of phlogopite in a groundmass of leucite and sanidine.

ore shoot. A chimneylike, pipelike, or ribbonlike, mass of ore within a deposit (usually a vein), representing the more valuable part of the deposit.

organic. Pertaining or relating to a compound containing carbon as an essential constituent arranged in chains and/or rings; also containing hydrogen with or without oxygen, nitrogen, or other elements.

organic reef. A sedimentary rock structural of significant dimensions, erected by, and composed almost exclusively of the remains of corals, algae, bryozons, sponges, and other sedimentary or colonial organisms.

organic rock. A sedimentary rock consisting primarily of the remains of organisms (plants and/or animals), such as of material that originally formed part of the skeleton or tissues of an living system.

organic soil. Any soil or soil.

organic texture. A sedimentary texture resulting from the activity of organisms.

organogenic. Said of a rock or sediment made up of the products of organic activity; *e.g.*, crinotal limestone.

organolite. A rock consisting mainly of organic material, especially derived from plants; *e.g.*, coal, bitumen, and resin.

organosedimentary. Pertaining to sedimentation as affected by organisms.

organotropic. Relating to the development and nourishment of living organs.

oriental amethyst. A violet to purple variety of sapphire.

oriental topaz. A yellow variety of corundum, used as a gem.

orientation diagram. A diagram which relates the crystallographic data to the optical data for a mineral.

origin. Said of a crystal, the point at which the crystallorgraphic axes intersect; the geometric centre of the crystal.

original horizontality. The state of strata being horizontal or nearly horizontal at the time of their deposition.

original valley. A valley formed by hypogene action or by epigene action other than that of running water.

origo facies. Facies of the primary sedimentary environment.

ornoite. A variety of hornablendediorite.

orocline. An orogenic belt with an imposed curvature or shors bend.

oroclinotath. An orogenic belt that has been subjected both to substantial horizontal bending and to stretaching along strike.

orocratic. Pertaining to a period of time in which there is much diastrophism.

orogensis. The process of mountain building, leading to the formation of the intensely deformed belts which constitutes mountain ranges. Most orogenic belts arise on the sites of geosynclines and the resulting mountains therefore consist of sdiments and volcanic rock deformed and metamophosed to a greater or lesser extent according to their position and depth in the orogenic belt. Regional metamorphism and granite emplacement are always found in the deeper levels of an orogenic belt.

Correlation of orogenies in the absence of reliable radioactive dating is a difficult process. On the whole orogenies appear to be more or less world-wide events, although individual maxima are often highly localized. There appears to be an average interval of 200-300 million year between orogenies, but there is little evidence for a marked periodicity in their occurrence. Owing to erosion processes the older an orogenic belt is, the deeper the level which is exposed at the surface at the present day.

orogeosyncline. A geosyncline that later become an area of orogeny.

orography. The branch of physical geography that deals with the disposition, character, formation, and structure of mountains and of chains, ranges, and systems of mountains.

orohydrography. The scientific study dealing with the relations of mountains to drainage.

oroite. Igenous rock of the diorite clan. A medium-grained granular plutonic rock; contains biotite, diopside, hornblende, microcline, plagioclase, and quartz.

orology. The scientific study of mountain building and mountain formation.

orotath. An orogenic belt that has substantially stretched in the direction of its length.

orotvite. A diorite composed of hornblende, biotite, plagioclase, nepheline, and cancrinite, with accessory sphere, ilmenite and apatite.

orpiment. Ag_2S_3. A yellowish mineral crystallizing in monoclinic system; occurs in foliated or columnar masses in veins and in hot spring deposits. Its hardness on Mohs scale is 1.5-2.0.

arthite. Allanite in the form of slender prismatic or acicular crystals.

orthoaxis. The diagonal or lateral axis perpendicular to the vertical axis in the monoclinic system.

orthoclase. $KAlSi_3O_8$. A white, grey, or pink mineral of alkali feldspar group that usually contains some sodium feldspar, either as albite or analbite or in some intermediate state. It is stable at higher temperatures is a common rock forming mineral; occurs in granites and igneous rocks. Also known as orthose; pegmatolite.

orthoclasite. An igneous rock of the syenite clan. A medium to findgrained granular hypabyssal rock; contains orthoclase and minerals.

orthoconglomerate. A conglomerate with intact gravel framework, characterized by a mineral cement, and deposited by ordinary but highly turbulent water currents.

orthodolomite. A primary dolomite formed by sedimentation.

orthoferrosilite. $FeSiO_3$. An orthopyroxene consisting of othorhombic silicate of iron.

orthogneiss. A gniess presumed to have been formed from an original igneous rock.

orthogonal. A curve that is everywhere

perpendicular to the wave crests on a refraction diagram.

orthograde. Pertaining to uniform distribution of dissolved oxygen in hypolimnion of a lake, dependent only on conditions at circulation and on subsequent physical events.

orthomagmatic. The name given to the stage in the crystallization of an igneous mass during which the main mass of silicates crystallies. It is followed by the postmagmatic phase, during which processes such as pneumatoysis, hydrothermy, and pegmatite formation occur. The orthomagmatic phase is sometimes divided into a first or early magmatic phase, during which anhydroussilicatescrystallise,andasecond or late magmatic phase, during which both anhydrous and hydroxyl bearing silicatescrystallise. The temperature and other conditions at which the orthomagmatic stage passes into the postmagmatic stage are definable accurately and main depend upon the amount volatic substances present.

orthomimic feldspar. A group of feldspars that by repeated twining simulateahigherdegreeofsymmstrywithrectangular cleavages.

orthoquarzitic sandstone. *See* orthoquartzie.

orthorhombic lattice. A crystal lattice in which the three axes of a unit cell are mutually perpendicular, and no two have the same length.

orthoschist. A schist derived from igneous rocks.

orthosparite. A sparite cement developed by physiochemical precipitat in open voids.

orthotill. A till formed by immediate release from transporting ice, as by ablation and melting.

osarite. (Os,Ru)AsS. A monoclinic mineral.

osbornite. TiN. A mineral found only in meterorites.

ossipite. An igneous rock of the gabbro clan. A coarse-grained granular plutonic rock; contains labradorite, magnetite, olivine, and pyroxene,. Also known as ossuypite.

ossypite. *See* ossipite.

osteolite. Massive impure variety of apatite.

osumilite. [(K,Na,Ca) (Mg, Fe(II))$_2$-(Al,Fe(II),(Fe(III)$_3$(SiAl)$_{12}$O$_{30}$.-H$_2$O.] Mineral that crysatallizes in the hexagonal system and is commonly mistaken for cordierite.

otavite. $CdCO_3$. A mineral that crystallizes in the hexagonal system and is isostructural with calcite.

ottajanite. An igneous rock of the diorite clan. A fine-grained porphyritic extrusive rock; contains phenocrysts of augite and leucite in a groundmass of calcic plagioclase, leucite, augite, nepheline, and sanidine.

ottrelite. A grey to black variety of chloritod containing manganese.

ouachite. An igneous rock of the ultrabasic clan. a very fine- grained porphyritic hypabyssal rock contains phenocrysts of biotite and augite in a groundmass of analcime glass and autgite.

ouenite. An igneous rock of the gabbro clan. A fine grained granular hypabyssal rock; contains anothite, bronizite. diopside and olivine.

outcrop. The total area over which a particular rock unit occurs at the surface-whether visibly exposed or not. The term is especially used for the delineation of such an area on a geological map.

outcrop water. Rain and surface water that seeps downward through outcropping porous and fissured rock, fault planes old shafts and/or surface drifts.

outer bar. A bar founded at the mouth of an ebb channel of an estuary.

outer core. The outer or upper zone of the earth's core, extending from a depth of 2900 km to 5100 km and including the transition zone. It is equivalent to the *E* layer and the *F* layer. It is presumed to be liquid because it sharply reduces compressional-wave velocities and does not transmit shear waves.

outer reef. One of the reefs comprising the seawards or basinward part of a reef complex or reef tract.

outflow cave. A cave from which a stream issuesout.

outgassing. The removal of occluded gases,generallybyheating.Out-gassing, in which gases and water vapor were released from molten rocks was responsible for the formation of earth's atmosphere.

outlier. An area of rock surrounded by rocks of older age.

outwash. Soil menerial washed down a hill by rainwater and deposited upon gently sloping region.

outwash cone. A cone-shaped deposit consisting chiefly of sand and gravel found at the edge of srinking glaciers and ice sheets.

outwash drift. Sand and gravol transported away from a glacier by streams of meltwater and either deposited as flood plain along a pre-existing valley bottom or broadcast over a preexisitn plain in a form similar to an alluvial fan. Also known as outwash; glacial outwash; overash.

outwash plain. A broad, outspread flat or gently sloping alluvial deposit of outwash in front of or beyond the terminal moraine of a glacier.

overage. Increase in the volume of a liquid with the rise in temperature.

over burden. Waste overlying the stone inaquarryoropenpitmine; worthlesssurface material covering a body of useful mineral.

overfit stream. A misfit stream that is too large to have eroded the valley in which its flows, or whose flood plain is two small for the size of the stream.

overgrowth. The deposit of one mineral growing in oriented crystallographic directions on the surface of another mineral.

overhang. The part of a salt plug that projects from the top.

overite. $Ca_3Al(PO_4)(OH)_6.-15H_2O$. A mineral composed of hydrous basic calcium aluminium phosphate.

overaland flow. That part of surface runoff flowing overland surfaces toward stream channels.

overlap fault. A fault structure in which the displaced strata are doubled back upon themselves.

overload. The amount of sediment that exceedstheabilityofastreamtotransport it, which is therefore deposited.

overloaded stream. A stream that is so heavily loaded with sediment that is forced to deposit a part of its load.

oversatuarted. A term used for igneous rocks or magma that contains silica in excess of the amount required to form saturated minerals from the basic constitutments.present.

oversteepened valley. An ice-free valley in which one side is higher and steeper thens the other.

P

paar. A depression formed as a result of themorning-apartofcrustalblocksrather than by subsidence within a block.

pabstite. $Ba(Sn,Ti)Si_3O_9$.. A rare silicate mineral.

pachnolite. $NaCaAlF_6.H_2O$. A colourless to white mineral composed of hydrous sodium calcium aluminium fluoride; occurs as monoclinic crystals.

pachoidal structure. A metamorphic structure in which small lenses and layers of granular material are surrounded by a matrix of sheared crushed material resembling a crude flow structure. Also known as flaser structure.

pacific suite. A large group of igneous rocks characterized by calcic and calc-alkalic rocks, especially in the region of the circum- pacific orogenic belt.

packed biomicrite. A biomicvite in which skeletal graine make up over 50% of the rock.

packet texture. An obsobeli term for the close grouping of quartz in pegmatite.

packsand. A fine grained sandstore that is so loosely consolidated by calcareous cement so as to be readily cut by a knife.

packstone. A sedimentary carbonate rock whose granular material is arranged in a self- supporting frame work, yet also contains matrix of calcareous mud.

padparadschah. Orange coloured gem variety of corundum.

pahoehoe. A Hawaiin term for basaltic lava flows typified by a smooth or ropy surface; varieties include corded elaphent-hide, entrial, festooned shelly and sleb pahoehoe.

paigeite. '$(FeMg)FeBO_5$. A. black mineral composed of iron magnesium of iron magnesium borate occurs as fibrous aggregates.

pariedterrace. One of the two stream terraces that face each other at the same elevation from opposite sides of the stream valley and that sepresent the remnants of the same flood plain or valley floor.

paisanite. An igneous rock of the granite clan. A fine-grained porphyritic hypybassalrock; contains phenocrysts of quartz and riebeckite in a groundmass of microperthite and quartz.

palaco. A prefix indicating great geologic age or existence early in a geologic time.

palaeobiology. A branch of palaeontology dealing with the study of fossil as organisms.

palaeocene. Lowermost division of the Tertiary System.

paleoclimatology. The study of climate through geologic time, extending back to the Precambrian. Basic data are derived from the analysis of sedimentary rocks, their contained fossils, and other sedimentary features.

palaeoecology. The study of the relationships between ancient organisms and their environment.

palaeothnobotany. The study of plante occurring in archeological sites.

palactiology. Explanation of past changes in the earth's condition as being governed by the laws of cause and effect.

palaegonite. A brown to yellow altered basaltic glass found as interstitial material or amygdules in pillow laves.

palagonite tuff. Igneous rock of the gabbro clan. A brown, fine- grained sandstone like tuff; contains augite, ovivine, and palagonite fragments and feldspar microliths.

palasite. The most abundant of the intermediate typed of meteorites consisting of olivine enclosed in a nickel-iron matrix.

palatinite. An obsolete term for basults and diorites containing orthopyroxenes.

paleozoic. An era of geologic time, from the end of precambrian to the beginning of mesozoic, aabout 570 to 225 million years ago.

palette. A broad sheet of calcite representing a solutional remnnant in a cave. Also known as shield.

palimpasest. The fabric or texture of a metamorphic rock in which remnants of a pre-existing fabric or texture are still preserved.

plingenesis. Formation of a new magma by in situ melting of pre- existing magnetic rocks.

palisade. An extended rock cliff or line of cliffs rising sharply from the border of a river or lake.

pallasites. Stony iron meteorites consisting of grains of ovlivine embedded in a matrix of nickel-iron [10to20%Ni]. The nickel-iron coherent mass enclosing separated stony parts.

palmierite. $(K,Na)_2Pb(SO_4)_2$. A white mineral composed of potassium sodium lead sulphate; crystallizes in hexagonal system.

palygorskite. (1) A chain structure type of clay mineral. (2) A group of light tough fibrous clay minerals showing extrusive substitution of aluminium for magnesium.

palynofacles. A term used for an assemblage of palynomorphs in a portion of a sediment, representing local environmental conditions and not typical of the regional palynoflora.

palynology. The study of pollen and spores, and their dissemination. Ancient pollens have proved most helpful in certain stratigraphic analyses and in paleoecology.

palynomorph. A microscopic resistant, walled organic body found in palynologic maceration residues. Palynomorphs include pollen, spores, acritarchs, chitinaozans, dinoflgellate thecae and cysts, and many acid-in soluble microfossils.

pampa. A vast treeless grassy plaim of temperate regions.

pan. (1) A natural shallow depression often occupied buy a lake. if the lake waters evaporate in time the depression will be lined with salt crystals and irt is called salt pan. (2) The upper portion of the bozen of soils in temporate climates may be hardened or cemented by precipitaion of salts previously leached from a-zonc. The hardened zone is called pan or hard pan.

pancake bomb. A type of volcanic bomb whose flattened shape is due to impact.

pandaite. $(Ba,Sr)_2(Ti,Nb)_2(O,OH)_7$. A mineral of pyrochlore group.

pandemic. A term used for conditions that occur over a broad geographic area and affect a major part of the population.

pandiomorphic. A texture term applied to a rock made up of well- development crystals.

panfan. A graded bedrock surface consisting of a series of coalescing pediments and representing the penultimate stage of an arid cycle of erosion.

pangaea. A supercontinent that existed about 300 to 200 million years ago during Mesozoic time when all continents were joined. The northern portion, named Laurasia, was sometimes partially and sometimes completely separated from the southern portion, named Gondwanaland, by a large sea called Tethys.

pan lake. A lake occupying a shallow natural depression or pan.

panning. A technique of prospecting for heavy metal by rinsing placer of crushed vein material in a pan. Lighter material is washed away, and the heavy metals remain in the pan.

pannonian. A stratigraphic stage name for the Europeran Upper Miocene [Tertiary System].

pantellerite. A green to black extrusive rock characterized by acmite-augite or diopside, anorthoclase, and cossyrite phenocrysts in acmite or feldspar matrix that is either pumiceous, partly glassy, fine-grained holocrystalline trachytic, or microlitic. It may be a rhyolite containing a soda-potash feldspar, quartz, aegirine, and sometimes sodic amphibole. Some pantellerites are low enough in quartz content to be considered varieties of trachyte.

panthalassa The ocean that surrounded Pangaea before its fragmentation.

paper clay. A fine-grained, white, kaolinic clay with high retention and suspending properties, containing very low amount of silica; it is used as filler in paper industry.

paper shale. A highly carbonaceous variety of shale that easily separates on weathering into thin layers.

paperspar. A crystallized variety of calcite, occurring in thin lamellae.

par-,para. A prefix denoting besides. when used with name of a clastic sedimentary rock, it signifies matrix-rich, when it is used with a metamorphic rock name, it indicates that the rock was derived from a sediment.

parabituminous. A term used for bituminous coal containing 84-87% carbon.

parabolic dune. A long, scoop-shaped sand dune having a ground plan approximating the form of parabola, with the horns pointing windward.

paraclase. An obsolate term for a fault.

paraclinal. A term used for a stream valley that is oriented in a direction parallel to the fold axis of a region.

paraconformity. A term used for an obscure or uncertain unconformity in which no erosion surface is discernible or in which the contact is a simple bedding plane, and in which the beds above and below the break are parallel.

paraconglomerate. A term used for a conglomerate characterized by a disrupted gravel framework, is often unstratified and is natable for containing more matrix than gravel-sized fragments; deposited by such modes of mass transport as subaqueous turbidity slides and glacier ice.

paracoquimbite. $Fe_2(SO_4)_39H_2O$. A pale-voilet mineral composed of hydrous ferric sulphate; crystallizes in rhomohedral system; dimorphous with coquimbite.

paradamite. $Zn_2(AsO_4)(OH)$. A monoclinic mineral.

paradiagenatic. A term denoting a close relation with sedimentary diagenesis.

paraffin-base crude. Crude oil that will yield large quantities of paraffin during distillation.

paraffin coal. A variety of light-coloured bitumionus coal from which oil and paraffin are produced.

paragenesis. A characteristic association or occurrence of minerals or mineral assemblages in ore deposits, connoting contemporaneous formation.

parageosyncline. An epeirogenic geosyncline basin located within a carton or stable area.

paraglomerate. A conglomerate which contains more matrix than gravel-sized framents and was deposited by subaqueous turbidity flows and glacier ice rather than normal aqueous flow.

paragneiss. A gneiss presumed to have

formed from an original sedimentary rock.

paragonite. $NaAl_2(AlSi_3)O_{10}(OH)_2$. A yellowish or greenish variety of mica that contains sodium anf usually occurs in metamorphic rock. Also known as soda mica.

parahilgardite. $Ca_8(B_6O_{11})Cl\text{-}4H_2O$. A triclinic mineral composed of hydrous chloride and borate of calcium; dimorphous with hilgardite.

parahopeite. $Zn_3(PO_4)_2.4H_2O$. A colourless mineral composed of hydrous zinc phosphate; crystallizes in triclinic system; dimorphous with hopeite.

paralaurionite. Pb(Cl)OH. A white mineral composed of basic lead chloride; dimorphous with laurionite.

paralic. (1) A term applied to environments of marine margins such as littoral basins and lagoons. (2) Said of coal deposits formed along sea margins, as contrasted with limnic deposits.

paralic coal basin. Coal deposits formed along the margin of the sea.

parallel cleavage. An obsolete term for bedding-plane cleavage.

parallel displacement fault. A term rarely used, for a fault on which the linear features that were parallel before displacement are still parallel afterwards.

parallelepiped. A crystal form bounded by three pairs of parallelograms.

parallel evolution. Similar development of organisms with related, yet different, ancestries under similar environmental conditions.

parallel fold. A fold in which the thickness of the layers is almost constant.

parallel growth. The intergrowth of two or more crystals in which one or more axes in each crystal are almost parallel. Also known as parallel intergrowth.

parallel intergrowth. See parallel growth.

parallel twin. A twinned crystal, the twin axis of which is parallel to the composition surface.

paramagnetism. A property exhibited by substance which, when placed in a magnetic field, are magnetized parallel to the field to an extent proportional to the field [except at very] low temperatures or in extremely large magnetic fields. Substance possing a magnetic permeability slightly greater than unity, *i. e.*, possessing a small positive magnetic susceptilbility, are said to be paramagnetic. The atoms of a paramagnetic substance possess a permanent magnetic moment due to unbalanced electron spins or unbalanced orbital motions of the electrons around the nucleus.

paramater. (1) Any of the axil lenghts or interaxial angles that define a unitcell. (2) On a crystal face, the rational multiple of the rational length interceped by a plane, which determines the position of the plane relative to the crystal lattice. (3) In two- dimensional analytical geometry it is often convenient to express the variables (x,y) each in terms of a third variable t, such that x and y are functions of t; x=f(t),y=f(t). The equations are termed parametric equations, and t is a parameter. A variable that can be kept constant while the effect of other variables is investigated.

paramorph. A pseudomorph having the same composition as the orignal crystal.

paramorphism. The propery of a mineral whose internal structure has changed without change in composition or external form. Formerly known as allomorphism.

paraschist. A schist derived from a sedimentary rock.

parasitic cone. A cinder cone on the side of a large volcano.

parasitic fold. A fold on the limp or hinge

of a larger fold with which it is congruent.

paratacamite. $Cu_2(OH)_3Cl$. A rhombohedral mineral composed of basic cupric chloride; dimorphous with tacamite.

parataxitie. A term used to describe extremely well layered entaxitic texture.

paratectonic. Pertaining to orogenic belts characterized by steep cleavages in low-grade metamorphic rocks.

paratill. A till formed by ice-rafting in a marine or lacustrine environment.

parautochthonous. A term used for a rock unit that is intermediate in tectonic character between antochthonous and allochthonous.

parawollastonite. $CaSiO_3$. Monoclinic phase of the mineral wollastonite.

parental magma. The magma from which a particular igneous rock solidified or from which another magma was derived.

parent rock. The rock mass from which parent materical is derived.

pargasite. $NaCa_2Mg_4Al_3Si_6O_{22}(OH)_2$. A monoclinic mineral of amphibole group; shows unever fracture and good cleavage in one direction. It occurs in igneous and metamorphic rocks.

parisite. $(Ce,La)_2Ca(CO_3)_3F_2$. A brownish-yellow secondary mineral; composed of carbonate and fluoride of calcium, cerium, and lanthanum.

parkerite. $Ni_3(Bi,Pb)_2S_2$. A bright, bronze coloured mineral composed of sulphide of nickel bismuth and lead.

parsec. An astronomical unit of distance, corresponding to a parallax of one second of arc and equal to 3.258 light-years, or $3.0857+10^{13}$ kilo-meters.

parsettensite. $Mn_5Si_6O_{13}(OH)_8$. A red coloured mineral composed of basic silicate of manganese.

parsonite. $Pb_2(UO_2)(PO_4)_2.2H_2O$. A yellow to brown mineral composed of hydrous lead uranyal phosphate.

partial dislocation. The line at the edge of an extended dislocation where a slip through a fraction of a lattice constant has occurred.

partial pluton. The part of a composite intrusion representing a single intrusive episode.

particle diamete. The diameter of a sedimentary particle considered as a sphere.

partical shape. The spatial or geometric form of particles in a sediment or rock. It depends upon the sphericity or/and roundness of the particle.

particle size. The general didmensions of particles or mineral grains in a rock or sediment based on the premise that the particles are spheres; commonly measured by sieving, by calculating setting velocities, or by determining areas of macroscopic images.

parting. (1) Separation of crystals along planes that are not ture cleavage planes. (2) A band or bed of waste material between veins or beds of ores. (3) A joint or fissure, specifically a plane or surface along which a rock is readily reparated or is naturally divided into layers.

parting cost. A sand-filled tension crack produced by creep along the sea floor.

parting lineation. A small scale primary sedimentary structure made up of a series of parallel ridges and grooves formed parallel to the current. Also known as current lineation.

partiversal. Pertaining to formations that dip in different directions roughly as far as a semicircle.

parvafacies. A body of rocks constituting the part of any magnafacies that occur between designated time-stratigraphic planes or key beds traced along the magnafacies.

pascoite. $Ca_2V_6O_{17}.11H_2O$. A red-orange to yellow-orange mineral composed of hydrous calcium vanadate.

passage bed. A stream marking a transition from rocks of one geological system to those of another.

patch reef. (1) A moundlike or flat-topped organic reef, generally less than a kilometre across, frequently forming a part of a longer reef complex. (2) A small, thick, generally unbedded, lens of limestone or dolomite, more or less isolated and surrounded by rocks of unlike facies.

paternoster lakes A chain of small lakes occupying a rock basin in a glacial valley. Their name derives from the similarity of their arrangement to beads on a rosary.

patina. A thin, coloured film produced on a rock surface by weathering.

patronite. VS_4. A dark green mineral composed of vanadium sulphide; extracted from vanadium bearing layer interbeded with shales.

paulingite. An isometric zeolite mineral consisting of an aluminosilicate of sodium, potassium, and calcium.

pavement. A closely packed, smooth natural bare-rock surface that resembles a paved road.

pay streak. The portion of a vein which carries a profitable ore.

Peacock copper. See peacock ore.

peacock ore. Informal name for an iridescent copper mineral having a lustrous, tarnished surface, exhibiting variegated colours, such as bornite and chalcopyrite. Also known as peacock copper.

pearceite. $Ag_{16}As_2S_{11}$. A black mineral composed of silver arsenic sulphide.

pearlite. See perlite.

pearlspar. $(Ca,Mg,Fe(II))CO_3$. A name given to the rhombohedral form of white, grey, pale or brownish coloured mineral dolomite or ankerite.

peat. A dark-brown or black, partially decomposed mass of vegetation which has grown in a shallow lake or marsh. Peat contains recognisable vegetable fragments, but very little mineral material.

peat ball. A lake ball containing an abundance of peaty fragments.

peat bed. *See* peat bog.

peat bog. A bog in which peat has formed under acidic conditions. Also known as peat bed; peat moor.

peat breccia. Peat that has been broken up and then redeposited by water.

peat coal. A variety of coal transition between peat and brown coal or lignite.

peat flow. The flow of peat produced in a peat bog by a bog burst.

peat formation. Decomposition of vegetation in stagnant water with small amounts of oxygen, under conditions intermediate between those of putrefaction and those of moldering.

peat moor. *See* peat bog.

peat soil. An organic soil largely consisting of peat.

peat tar. Tar obtained from the distillation of peat.

pebble. (1) A rock fragment with particle size between 4-65mm. (2) A general term for a small, roundish, especially water-borne stone. (3) A rough gem ocurring in the form of a pebble, as in a stream.

pebble armor. A desert armor made up of rounded pebbles, as on a serir.

ebble bed. Any pebble conglomerate, especially one in which the pebbles weather and fall loose.

pabble coal. Coal that is transitional between peat and brown coal.

pebbledike. (1) A clastic dike composed largely of pebbles. (2) A tabular body containing sedimentary fragments in an igneous matrix.

pebble phosphate., A secondary phosphate of either residual or transported origin, consisting of pellets, pebbles, and nodules of phosphatic material mixed with sand and clay.

pebby sand. An unconsolidated sediment containing at least 75% sand and the amount of pebbles not exceeding 25% of the total aggreagate.

pecoraite. $Ni_3Si_2O_5(OH)_4$. A nickel-bearing mineral of serpentine group.

pectolite. $NaCa_2Si_3O_8(OH)$. A colourless, white, or grey inosilicate that crystallizes in monoclinic system and has a vitreous to silky lustre; shows uneven fracture and perfect cleavage in two directions. Its hardness on Mohs scale is 5.

pediment. A plain of eroded bedrock (which may or may not be covered by a thin veneer of alluvium) in an arid region developed between mountain and basin areas.

pedimentation. The actions or processes by which pediments are formed.

pedimentation pass. A flat, narrow tongue that extends from a pediment on one side of a mountain to join a pediment on the other side.

pedion. A crystal form consisting of only one face.

pediplain. The end-product of the cycle of erosion, according to the scheme of W.Penck and L.C.King . A rock-cut erosion surface formed in a disert by the coalescence of two or more pediments.

pediplantation. The actions or processes by which pediplanes are formed.

pediplane. Any plante erosion surface formed in the piedmont area of a desert, either bare or covered with a veneer of alluvium.

pedocal. An oboslete term for a soil in which there is an accumulation or concentration of carbonates, usually calcium carbonate.

pedogenesis. Soil formation. The sutdy of the morphology, orgin, and classification of soils is pedology.

pegmatite. An exceptionally coarse-grained igneous rock with interlocking crystals, usually found as irregular dikes, lenses, or veins, especially at the margins of patholiths. Most grains are about more than I am in diameter. Although many granite-pegmatites consists more or less exclusively of the common granite minerals, there are also many examples of pegmatites which contain suites of minerals which are otherwise rare, *e.g.*, minerals containing lithium, beryllium, and rare earths. Although diorite and gabbro pegmatites occur, syenite and granite ores atre the most common pegmatites. Many pegmatites have an intermal zonation of fabric and composition. Most are mineralogically simple, comprising mainly quartz and potassium feldspar, with lesser amounts of muscovite and tourmaline. Other pegmatites have rather high concentrations of gaseous constituents. Pegmatites represent the last portions of magma to crystallize.

pegmatitic stage. A stage in the normal sequence of crystalization of a magma containing volatiles, at which time the residual fluid is sufficiently enriched in volatile materials to permit the formation of coarsegrained rocks. The relative amounts of silicate and volatile materials in the fluid, the temperature range, and the relastionship of these fluids to hydrothermal fluids are in dispute.

pegmatitization. Formation of or replacement by a pegmatite.

pegmatoid. An igeous rock that posses-ses the course-grained texture of a pegmatite but lacks graphic intergrowths and/or typically granitic composition.

pelagic. A geologic term applied to any accumulation of deep sea sediments; also used to designate materials and organisms to the marine environmenta sopposed to materials transported to the marine envirment from terrestrial sources.

pelagic deposit. Marine sediment in which the fraction derived from the continents deposition from a dilute mineral suspension distributed throughout deep-ocean water.

pelean-type eruption. A type of volcanic eruption characterized by gaseous clouds and the development of volcanic domes.

pelagochthonous. A term used for the coal derived from a submerged forest or from driftwood.

peldon. An english for a very hard, smooth, compact sandstone with conchoidal fracture.

pele's tears. Volcanic glass in the form of small, solidified drops which precede pendants Pele's hair.

pelite. A sediment or sedimentary rock, such as mudstone, composed of fine clay or mudsize particles.

Pelitic hornfels. A fine-grained metamorphic rock derived from pelite.

Pelitic schist. A foliated crystalline metamorphic rock derived form pelite.

pelitmorphic. Pertaining to clay-size carbonate particles in a limestone or dolomite rock.

pellet. A small approximately 0.2mm diameter, well rounded, ovoid to irregularly shapod homogeneous aggregate of clay-sized calcarcous material, mierite,.Most pellets are considered to be the feces of worms or mollusks;other types include pseudo-ooliths and aggregates produced by gas building, by algal building phenomena or by other intraformational reworking of lithified or semilithified carbonate mud.

pelmatozoan. Any echinoderm living attached to a substrate, either with or without a stem. The term refers to life habit and is not a true taxonomics identification.

pelmictrite. Limestone composed of a variable proportion of carbonite mud (micrite and small rounded aggregates of sedimentary material).

pelsparite. A limestone containing less than 25% each of intraclasts and ooliths having a volume ratio of pellets to fossils greater than 3 to 1, and with the sparry-calcite cement more abundant than the mictrite matrix.

pencil gneiss. A gneiss that splits into thin, sodalike quartz- feld-spar crystal aggregates.

pencil ore. Hard, fibrous masses of hematite that can be broken into splinters.

penecontemporaneous. Formed during or shortly after the deposition of the containing rock stratum.

peneplain. Almost plane, flat surface; a nearly featureless, gently undelating land surface of considerable area, which presumably has been produced by the processes of long-continued subaerial erosion almost to base level in the penultimate stage

of a humid fluvial geomorphic cycle.

peneplantation. The actions or processes by which peneplains are formed.

penfieldite. $Pb_2(OH)Cl_3$. A white, hexagonal mineral composed of basic lead (II) chloride.

pennantite. $Mn_9Al_6Si_5O_{20}(OH)_{16}$. An orange coloured mineral of the chlorite group; composed of basic manganese aluminium silicate.

pennine. 2. $[(Mg,Fe,Al)_6(Si,Al)_4O_{10}(OH)_8]$. A green coloured monoclinic mineral composed of hydrated magnesium iron aluminosilicate; occurs as thick tabular crystals. It shows pearly lustre and perfect cleavage in one direction. Its hardness on Mohs scale is 2.0-2.5. Also known as penninte.

penninite. *See* pennine.

pennsylvarian. A period of palezoic era, thought to have covered the span of time between 320 and 280 million years ago.

pentagonite. $Ca(VO)Si_4O_{10}.4H_2O$. An orthorhombic mineral.

pentahydrite. $MgSO_45H_2O$. A triclinic mineral containing hydrated magnesium sulphate; isostrucheral with chalcanthite.

pentalith. A cocolith formed by five crystal units.

pentalandite. $(Fe,Ni),S_8$. A bronze-yellow isometric mineral containing iron nickel sulphide; occurs in basic and ultrabasic rocks. It is an important ore of nickel.

peperite. A breccialike material occurring in marine sedimentary rock; mainly consisting of a mixture of lava with sediment.

peralkaline. In the Shand classification of igneous rocks, a division embracing those rocks in which the molecular proportion of alumina is less than that of sodium and potassium-oxides combined.

peraluminous. In the Shade classification of igneous rocks, a division embraacing those rocks in which the molecular proportion of alumina exceeds that of sodium and potassium oxides combined.

perched block. A large, detached rock fregment presemed to have been transported and deposited by a glacier, and preched in a conspicuous and precarious position on the side of a hill.

percylite. $PbCuCl_2(OH)_2$. A blue mineral composed of basic chloride of copper and lead; its hardness on Mohs scale is 2.5.

pereletok. A frozen layer of ground, at the base of the active layer, which may exist for one or more years.

perezone. A zone in which sediments accumulate along coastal lowlands; includes lagoons and brackish-water bays.

perfemic. One of the five classes in the CIPW classification of igneous rocks, in which the ratio of salic to femic minerals is less than 1/7.

perforated crust. A type of snow crust containing pits and hollows produced by ablation.

pergelation. The formation of permafrost in the present or in the past.

perhyaline. In the CIPW classification of igneous rocks, the rocks in which the ratio of crystals to glassy materiaals is greater than 1/7.

peri. A prefix denoting around, beyond.

periclase. MgO. A mineral composed of magnesium oxide; occurs in granular forms or isometric crystals; its hardness on Mohs scale is 6. Also known as periclasi

ericlasite. *See* periclase.

pericline. Opaque white variety of albite found in chloritic schists in the Alps.

periclinol. A term applied collectively to domes and basins, or by some authors, to domes along.

peridot. Olivine of gem quality.

peridotite. A class of ultrabasic rocks consisting predominantly of olivine with or without other ferromagnesian minerals.

peridiotiteshell. The upper mantle of the earth.

Periglacial. Referring to locations, conditions, processes, and topographic features of areas that are adjacent to the borders of a glacier. A periglacial climate is characterized by low temperatures, many fluctuations about the freezing point, and strong wind action, at least during certain times.

peristerite. A gem variety of albite that resembles moonstone and has a blue or bluish-white lustre; characterized by sharp internal reflections of blue, green, and yellow.

perknite. Igneous rock of the ultrabasic clan. A fine to coarsegrained granular plutonic rock; contains monoclinic amphiboles and pyroxenes. A class of ultrabasic rocks consisting predominantly of ferromagnesian minerals other than olivine.

perlite. Lustrous, pearly gray glass with perlitic texture.

perlitic A texture found in glassy and devitrifiedrocks,consistingofclustersof concentric curved fractures.

permafrost . Any soil subsoil that or other susficial depicit is permanently frozen. It occurs in those regions (arctic,subarctic,alpine) where the mean annual temperature of the soil remains below freezing. Permafrost underlies almost 20%of the earth's land area, where its thickness ranges from 30 cm to more than 1000m.Ifpermafrostisabletoreturn toanenduring frozen state underexisting climatic condiutions after it has been thawednaturallyorartificially,itiscalled active permafrost.

permafrost island. A small. shallow, isolated patch of permafrost surrounded by unfrozen ground.

permafrost table. The upper limiit of permafrost. Also known as pergelisol table.

permeability. The capacity of a porous rock, soil, or sediment for transmitting a fluds without damage to the structure of the medium. Permeability may be achieved by rock being either porous or percious. The essential feature of a bed of permeable rock is that the liquid it contains may be extracted by pumping. The opposite of permeable is impermeable. Permeability in a rock is measured in darcies.

permeable bed. A porous reservoir formation through which hydrocarbon fluids or water can flow.

permeaition. Penetration by passing throughtheimterstices,asofarockorsoil, without causing physical change.

perneation gneiss. A gneiss formed as a result of or modified by the passage of geochemicallymobilematerialsthrough or into solid rock.

permian. The last period of palezoic era, thought to have covered the span of time between 280 and 225 million years ago; also the carresponding system of rocks. Permain rocks are generally subdivided intotheLowerandUpperPermainSeries, and the boundary placed at the base or middleoftheGuadalupanstage.ThePermain Period began with glaciation and ended with aridity. Sedimentary rocks of glacial origin vary in the time and environment of their deposition. Late Permian was a time of widespread mountain building and volcanism.

permineralization. The process of forssilzation wherein minerals are deposited in the pore spaces of originally hard mineral parts.

permissive intrusion. Emplacement of magma in spaced created by forces; also the magma or rock body so emplaced.

permo-carboniferous. (1) The Perime andCarboniferousperiodsconsideredas one unit. (2) The Permian and pennsylvanian periods considered as one unit. 3. PermianandPennsylvanianperiodsconsidered as a single unit. (3) The period of geologic time transitional between the UpperPennsylvanianandtheLowerPermian periods.

permo-Triassic. Said of strata not differentiated between the Permian and Triassic systems particularly in regions where the boundary occurs within a nonmarine red-beds succession.

perovskite. 8[$CaTiO_3$]. A black,

brownish-black, greyish-black,or yellow, cubic or monoclinic mineral composed of calcium titanium oxide; shows uneven fractures and indistinct cleavage. Its hardness on Mohs scale is 5.5.

perapatic. In the CIPW classification of igneous rocks, the rocks in which the ratio of groundmass to phenocrysts is greater than 7.

perpendicular slip. The component of the slip of a fault that is measured to the trace of the fault on any intersecting surface.

Perpendicular throw. The distance between two points which were formerly adjacent in a faulty bed vein or other surface; measured at right angles to the surface.

persalice. In the CIPW classification of igneous rocks, the rocks in which the ratio of salic to femic minerls is greater than 7.

Persemic. In the CIPW classification of igneous rocks, the rocks in which the ratio of groundmass to phenoerysts is less then 1/7.

Perthite. A variety of alkali feld-spar consisting of an intergrowth of two feldspar minerals [micro-perthite is properly the term used for intergrowths visible under a microscope];i.e., course intergrowth of orthoclase and albite

perthosite. A light coloured syenite composed entirely of perthile, with less than 3% of mafic minerals.

perxenic. In the CIPW classification of igneous rocks, the rocks in which the ratio of rock crystal to chadacrysts is less than 1/7.

petalite. $LiAlSi_4O_{10}$. A colourless, white, or grey monoclinic mineral composed of silicate of lithium and aluminium; occours in foliated masses or crystals.

petr-petro. A prefix denoting rock.

petrifaction. A fossilization process wherebey inorganic matter dissolved in water replaces the original organic materials converting them to a stony substance.

pertochemistry. An aspect of geochemistry in which the chemical composition of rocks is studied; it is not equivalent to petroleum chemistry.

petrogenesis. A branch of petrology that deals with the origin and formation of rocks especially igneous rocks.

petroglyph. Literally a rock carving, it usually excludes writing and therefore is of prehistoric or protohistoric age.

petrographic facies. Fracies distinguished primarily on the basis of appearance or composition, without respect to form boundaries or mutual relation-ships. They consist of large bodies of rock occurring in certain areas and in more or less restricted parts of the stratigraphic section.

petrographic period. The time represented by a rock association.

petrographic province. A wide area in which most or all of the igneous rocks are believed to have been formed during the same interval of magmatic activity.

petrography. Tha branch of geology involved with the description and systematic classification of rocks, esp. by means of microscopic examination Petrography is more restricted in its scope than is petrology.

petroleum. (1) A naturally occurring complex liquid hydrocarbon which after distillation and removel of impurities yields a range of comustible fuels, petrochemicals and lubricants. (2) A general term for all naturally occurring hydrocarbons in gaseous, liquid as well s solid state.

petroleum geology. The branch of economic geology dealing with the origin,occurrence,accumulation,movement, and exploration of hydrocarbon fuels.

petroleum trap. Stable underground formation of such nature as to trap and hold liquid or gaseous hydrocarbons; usually consists of sand and/or porous rock surrounded by impervious rock or clay formations.

pertologen. See kerogen.

petrology. The general term for the study of rocks in all their aspects, including their mineralogies, textures, and structures, origins, field occurences, alterations and their relationships to other rocks.

petromict. Said of a sediment composed of metastable rock fragments.

petzlte. Ag_3AuTe_2. A grey to black mineral composed of silver gold telluride; its hardness on Mohs scale is 2.5-30

phacloith. A concordant, lenticular igneous intrusion. It is generally assumed that the igneous marterial was intruded during or before folding of the host rock, and was deformed along with it. Phacoliths sometimes occur in vertical groups in the domes of anticlinal folds.

phainos. A suffix denoting showing.

phancrite. An igneous rock having phancritic texture.

phancritic. Said of texture of an igneous rock being visibly crystalline.

phanero. A prefix denoting visible.

phanerogenic. Said of a species that is proved to descend from a known species found in older geologic formation.

phanerozoic. That part of geologic time represented by rocks in which the evidence of life is abundant.

phanoclastic rock. An even-grained clastic rock.

phantom. A bed or member that is absent from a specific stratigraphic section but is usually present in charactcristic position in a sequence of similar geologic eye.

phantom crystal. A crystal possessing an earlier stage of crystallization inclusions, or bubbles. Also known as ghost crystal.

phantomhorizon. In seismic reflection prospecting, a line constructed so that it is parallel to the nearest actual dip segment at all points along a profile.

pharmacolite. $CaH(AsO_4).2H_2O$. A white to greyish mineral composed of hydrous acid arsente of calcium; occurs in fibrous form.

phenacite. *See* phenakite.

phenakite. $Be_2\ SiO_4$.. A colourless, white, pink, yellow, brown, or blue glassy mineral that crystallized in the rhombohedral system; used as a minor gemstone. Also spelled phenacite.

phengite. A variety of muscovite possessing high silica content.

phenoclast. One of the larger fragments in a sediment or sedimentary rock composed of various sized of material such as cobble or pebble, in a fine-textured matrix of a conglomerate, or a fragment of breccia.

phenocryst. A crystal in an igneous rock, larger than other crystals of the matrix and formed during the early stages of crystal development. Phenocrysts will be euhedral if the magma system in which they formed was near equilibrium when solidification took place. An igneous rock characterized by the presence of these large crystal is said to have porphyritic texture. The term inset has been suggested as an altervative.

phenomenalgem. A gemstone exhibiting an optied phenomeron, such as asterism, chatoyancy, or play of colour.

phenoplast. A large rock fragment that was plastic at the time of its incorporation into matrix.

phlebite . Melamorphite or migmatite withroughlybandedveinedappearance.

phoenicite. *See* phoenicochroite.

phoenicochroite. Pb_2CrO_5. A red mineral composed of basic chromate of lead; occursincrystalslineformandmasses. Also known as phoenicite.

phlogopite.$K(Mg,Fe)_3AlSi_3O_{10}(OH,F)_2$. A magnesium rich mineral of the mica group. It is yellowish-brown to brownish-redand usuallyoccursincrystalline limestones as a result of dedolomitization.

pholorite. An obsolete name for a clay mineral identical with kaolimite or one of the other kaolin minerals.

phonolite. Igneous rock of the syenite clan. A fine-grained porphyritic extrusive rock. A light-coloured aphanitic rockofvolcanicorigin,composed largely of alkali feldspar, feldspathoids, and smaller amounts of mafic aminerals.

phorogenesis. Slipping of earth's crust over the mantle.

phosgenite..$Pb_2Cl_2(CO_3)_2$. A white, yellow, or greyish mineral that crystallizesinthetetragonalsystem;itshardness on Mohs scale is 3. Also known as horn lead.

phosphaterock. Any rock thatcontains phosphatic mineral[s] in suffiicient quality to permit its commercial use, as a source of phosphatic compounds.

phosphatic nodule. A dark earthy rounded mass, consisting of corals,shells,coprolites,micaflakes,and sand grains, ranging in size from a few millimeters to greater than 30 cm. Phosphatic nodules occur in marine strata.

phosphaferrite. $(Fe,Mn)_3PO_4.\ 3H_2O$. Awhiteofgreenishmineralcomposedof hydrousphosphateofferrousmanganese phosphate; crystallizes in orthorhombic system and exhibits micalike cleavage.

phosphopyllite.$Zn_2(Fe.Mn)(PO_4)_2.4H_2O$. Acolourlesstopale-bluemineralcomposedofhydrouszincferrousmanganese phosphate; exhibits micalike cleavage.

phosphortie. A sedimentary rock composed of phosphate minerals. Also known as phosphate rock; rock phosphate.

phosphosiderite. $FePO_4.2H_2O$. A pinkish-red monoclinic mineral composed of hydrated ferric phosphate.

photicite. A Carbonate rich variety of rhodonite.

phreatic. Said of a volcanic exploision of material such as steam or mud, not being incandescent.

phreatic crycle. The period of time during which the water table rises and thenfalls.Itmaybeadaily,annualofother cycle.

phreatic explosion. A volcanic eruption orexplosion,ofsteam,mud,orother material that is not incandescent; it is causedbytheheatingandconsequentexpansion of ground water due to an underlying igneous heat source such as lava. Such eruptions are usually not accompanied by the usual products of volcanic eruption.

phreaticWater. Originallyappliedonly to water that ocurs in the upper zone of saturation under water-table conditions. but now applied to all water in the zone ofsaturation.Thus,itissynonymouswith groundwater.

phreatomatic explosion. A volcanic explosion that extrudes both magmatic gases and steam; casued by the contact of magma with groundwater or shallow surface water.

phyllite. A medium grade regional metamorphic rock of light silver-grey colour with a texture grading form granoblastic [quartzose phyllites] to prophyroblastic [albitic phyllite] derived from calyey sedimentary rocks containing a residue of organic material, plus chlorite and mica.

planosol. An intrazonal, hydromorphic soil having a claypan or hardpan covered with a leached surface layer; developed in humid to subhumid climate.

plastic. Capable of being deformed permanently without rupture.

plastic deformation. Permanent change in shape or size of a solid body with fracture resulting from the application of sustained stress beyond the elastic limit.

plasticity index. The water content range of a soil at which it is plastic, equal to the differences of the liquid limit and the plastic limit.

plate. (1) A smooth thin, flat fregment of rock. (2) A large, rigid but mobile block involved in plate tectonics; thickness ranges from 50 to 250 km and includes both curst and a portion of the upper mantle.

plateau basalt. Large and extensive eruptions of basalt lava resulting from fissure eruptions are termed plateau basalts. The term has also been used as a 'magma-type', originally for olivine basalts, later for tholeiites. In view of the confusion the term should be abandoned in its latter sense.

plareau eruption. Successive lava flows that spread in sheats over a large area.

plateau glacier. A highland glacior that overlies a generally flat mountain tract; usually overflows its edges in hanging glaciers.

plate crustal. An ice crystal exhibiting typical hexagonal [rarely triangular] symmetry and having comparatively little thickness parallel to its axis.

plate tectonics. A theory of global tectonics in which the lithosphere is divided into a number of plates whose pattern of horizontal movement is that of torsionally rigid bodies that interact with one another at their boundaries, causing seismic and tectonic activity along their boundaries.

plauenite. Igneous rock of the syenite clan. A coarse-grained granular plutonic rock; contains orthoclase, oligoclase, hornblende, sphene, and apatite.

playa. A small sandy land area at the mouth of a stream or along a bay shore.

playa lake. A shallow, intermittent lake in an arid region, occupying a playa in the wet season but drying up in summer, an ephemeral lake that upon evaporation leaves or forms a playa.

plazolite. A cubic mineral composed of hydrated calcium aluminium silicate and carboinate; shows vitreous lustre and conchoidal fracture and does not show any cleavage. Its hardness on Mohs. scale is 6.5

pliocene. An epoch of the period, after the Miocene and before the Pleistocene; also, the corresponding worldwide series of rocks. It is considered to be a period when the Tertiary is desiganted as an era.

plucking. (1) A process of glacial erosion by which blocks or rocks are loosened, detached, and brone away from bedrock by the freezing of water in fissures. (2) A process of stream erosion by which rock fragments are forciably removed by the impact of water entering cracks in a rock.

plug. (1) A mass of sediment filling the part of a stream channel abandoned by the formation of a chtoff. (2) A vertical, pipelike body of megma that represents the conduit to a former volcanic vent. (3) A crater filing of lava, the surrounding material of which has been removed by erosion.

plug dome. A volcanic dome characterized by an upheaved, consolidated mass filling the conduit.

plug reef. A small triangular reef that

grows with its apex pointing seaword through openings between linear shelf-edge reefs.

plum. A clast embedded in a matrix of different kind.

plateaueruption. Successivelavaflows that spread in sheats over a large area.

plumasite. Igneous rock of the diorite clan. A medium-grained granular hypabyssal rock. An undersaturated oligoclase-corundum diorite.

plumboferrite. $PbFe_4O_7$. A dark hexagonalmineralcomposedofleadiron oxide.

plumbogummite. (1) $PbAl_3(PO_4)_2(OH)_5 H_2O$. A mineral composed of hydrous basic lead aluminium phosphate. (2) A group of isostructural minerals that includes gorceixite, goyazite, detaite, dusteritite, crandallite, florencite, as well as plumbogummite.

plume. A persistent column of magma rising upward from the earth's mantle to the crust; it apperas to be the main contributing factor to hot spots.

plunge. In the geometry of folds, a plunge is the inclination of a fold axis, measured in the vertical plain. Together with the strike of the horizontal projection, the plunge defines the attitude of the hinge of a flod.

plungepool. The water occupying a deep hollow scoured in the bed of a stream at the foot of a waterfall; also the hollow or basin itself.

plungingfole. A fold in which the hinge line is inclined to the horizontal.

pluton. A general term for any large scale mass of matasomatic replacement; However, the term can be used in the sense of a high-level, cylindrical mass of granitic rock which was emplaced at low temperature in a near-solid state. This usage is now rather widespread.

plutonic. A term applied generally to rocks, usually igneous, of deep-seated origin, It also describes a rock formed by any process at great depth.

plutonidc rock. A rock formed at considerable depth by crystallization of magma and/or by chemical alteration. It is characteristically medium-to coarse grained or granitoid texture.

plutonic water. Juvenile water in, or derived from, magma at considerable depth, probably several kilometres.

plutriusm. (1) A general term for the phenomenaasassociated with the formation of plutons. (2) The concept of the formation of earth by solidification of a molten mass.

pluvial. (1) Pertaining to rain or, to precipitation. (2) Said of a climate characterized by relatively high precipitation. (3) Said of a geologic process or feature resulting from rain, *e.g.*, pluvial denudation from land sliding and gully erosion.

pluvial lake. A lake formed in a period of heavy rainfall; specifically a lake formed in the Pleistocene epoch during a time a glacial advance.

pneunatic. Pertaining to or operated by air or other gas.

pneumatogenic. Said of a rock or mineral deposit formed by a gaseous reagent.

pneumatolysis. The alteration of rock or crystallization of minerals by gaseous emissions fron solidifying magma.

pneumotectic. Said of processes and products of magmatic con-solidation affected to some extent gaseous constituents of magma.

pod. A n ore body of an elongate or lenticular shape.

Podzol. A group of zonal soils having an oraganic mat and a very thin organic-mineral layer overlying a grey, leached *A*2 horison and a dark brown, alluvial; *B* horizon enriched in iron oxide, alumina and organic matter. It develops under coniferous or mixed forests or under heath, in a cool to temperate moist climate.

poikilitic. 1. Term applied to the speckled appearance of some minerals within igneous rock. Usually, the host completely encloses the inclusion. 2. Textural term describing a condition in which small granular crystals are variously oriented within larger host crystals [oikoscrysts] of a different mineral.

poikilo. A prefix denoting spoted.

poikiloblastic. Said of a texture of metamorphic rocks in which small grains of one mineral lie within larger metacrysts of another. Also spelled as poeciloblastic.

point bar. One of a series of low, arcuate, ridges of sand and gravel developed on the inside of a growiing meander by the slow addition of individual accretions accompanying migration of the channel toward the outer bank.

point-bar deposit. A deposit consisting of a series of alternating point bars and intervening troughs.

point diagram. A fabric diagram in which poles representing lineations, normals to fabric planes, or crystallographic directions have been plotted.

point group. A group consisting of symmetry elements of an object [crystal] having a singler fixed poinr; 32 such groups are possible.

poised stream. A stream, that is neither eroding nor depositing sediment.

polar axis. The directed straight line relative to which angle is measured for a representation of a point in the plane by polar coordinates.

polar cap. An ice sheet centered at one of the poles of the earth.

polar glacier. A glacier whose temperature is below freezing throughout its mass, and on which there is no melting during any season.

polarity zone. A unit of rock characterized its particular polarity; it is the basic unit of polarity-lithostratigraphic classification.

polarization. The modification of light so that its vibrations are restricted to a single plane.

polar symmetry. A type of crystal symmetry in which the two ends of a central crystallographic axis are not symmetrical. Such a crystal ishemimorpic.

pole-fleeing force. A component of forces resulting from the earth's rotation that is supposed to carry the crust away from the poles, toward the equator. The effectiveness of such a force has been shown to be neligible.

pollenite.. Igeous rock of the diorite clan. A greenish-grey, fine-grained porphyritc extrusive rock; contains phenocrysts of sanidine and andesine in a groundmass of olivine, andesine augite, hornblende, biotite, and glass.

poly. A prefix denoting many.

polycrystal. An assemblage of crystal grains of a mineral of unspecified number, shape, size orientation, or bonding, that together form a solid body.

polydymite. Ni_3S_4. A mineral of the linnaeite group.

polygenic. 1. Resulting from more than one processes of formation or, derived from more more than one source, or originating or developing at various place and times. 2. Consisting of more than one type of material, or having a heterogeneous composition.

polygeosyncline. A geosynclinal-geoanticlinal belt along the continental margin, receiving sediments from a boderland on its oceanic side.

polyhalite. $K_2MgCa_2(SO_4)_2.2H_2O$. A hydrous sulphate mineral usually occurringinfibrousmasses.Itmaybepink,red, or grey coloured.

polymetamorphism. Polyphase or multiplemetamorphism,wherebytwoor more successive metamorphic events have left their imprint upon the same rocks. The superimposed metamorphism may be of a higher or lower grade than the earlier type.

polymictic. (1) Term applied to a clastic sedimentary rock composed of many rock types. (2) Said of a conglomerate composedofavariety of fragment types. (3) Said of a lake whose water is continuously circulating and shows no persistent thermal statification.

polymineralic. Saidofarockcomposed of two or more minerals, especially said ofanigneousrockconsistingofmorethan one essential mineral.

polymorphism. The characteristic of crystallizing in two or more distinct forms; *e.g.*, a crystal that has different crystal structures. Iron sulfide crystallizesas twodistinct forms, marcasiteand pyrite; whilesulpurcrystallizesinrhombic and monoclinic forms.

polynya. Any nonlinear opening enclosedin ice, especiallya large expase of water, surrounded by sea ice, but not large enough to be called open water.

polyquart. A group term for Al(PO_4), Al(AsO_4),B(PO_4) etc.

polyschematic. A term used for mineral deposits having more than one testural element.

polysomatic chondrule. A chondrule consisting of several crystals.

polysynthetic twinning. Repeated twinningofthreeormoreindividualsaccordingtothesametwinlawandonparallel composition planes; e.g., albilte twinning of plagioclase.

polytypic. (1) Said of a taxon that includes several units of the next lower category, *e.g.*, a genus with several species. (2) Referring to a species consisting of subspecies that replace each other geographically.

polyzenite. Feldspathoid igneously rock. A medium-grained porphyritic hypabyssal rock; contains phenorcrysts of anomite in a groundmass of lazurite, melilite, olivine, and phlogopite.

pontain. A stratigraphic stage name for the European Lower Pliocene.

pontic. Pertaining to sediments or facies depositedincomparativelydeepmotionlesswater,suchasanassociationofblack shales dark limestones deposited in a stagnant basin.

ponzite. Igneousrockofthesyeniteclan. A fine-grained porphyritic extrasive rock; contains phenocrysts of diopside and aegirine-augite in a ground-mass of augite, magnetite, and orthoclase.

porcelaneous. Resembling unglazed porcelain.

porcellanite. A hard, dense siliceous rock, such as impure chert or indurated clay or shale.

pore. (1) A small to minute opening or intersticinarockorsoil. (2) Asmallopening in the wall or shell of an invertebrate animal, *e.g.*, from the exterior through theeal plates of a crystoid.

pore ice. Groundice thatfills or partially fills pore spacès in the ground; formed as a result of freezing of pore water in situ, without any further addition of water.

pore space. Openspace or void between individual grains in a rock mass; it varies greatly in size and shape.

porosity. Amount of void space in a formationofarock,usuallyexpressedaspercent voids per bulk volume. Absolute porosity refers to the total amount of pore space in a rock, regardless of whether or not that space is accessible to fluid penetration. Effective porosity refers to the amount of connected pore spaces, as in the space available to fluid penetration.

porpezite. A mineral consisting of a native alloy of gold and palladium.

porphyrite. A typabyssal igneous rock containing *plagioclase* as the dominate mineral.

porphyritic. An igneous rock texture; large crystals which have formed first are later surrounded by a fine grained crystalline mass of minerals as the whole magma consolidates.

porphyroblastic. Pertaining to the textures of a recrystallized metamorphic rock having large idioblasts of minerals in a finer grained crystalloblastic matrix.

porphyroclastic. Said of a hetero granular metamorphic texture, characterized by volumetrically significant amounts of both porphyroclasts and meoblasts; also said of a rock with such texture.

porphyrotope. A large crystal enclosed in a finer-graned matrix in a sedimentary rock showing porphyrotopic fabric; such as a large dolomite crystal in finer grained calcitic matrix.

porphyry. Hypabyssal igneous rock containing orthoclase as the dominat mineral. The use of term 'porphyry', as a synonym of phenocryst is not recommended, however this term can be employed in conjunction with mineral names.

posittive area. An area where deposition is not taking place. The opposite of a basin of deposition or negative area.

positive elongation. As seen in thin section, elongation of an anisotropic crystal that is parallel to the slower of the two planepolarized rays.

postglacial. Reffering to the interval of geologic time since the total disappaerance of continental glaciers in middle latitudes or from a particular area.

posthumous fold. A kind of recurrent folding that occurs in younger sedimentary rocks overlying a buried fold belt.

postmagmetic. An indefinite term applied generally to reactions or events ocurring after crystallization of the bulk of a magma, and usually including the hydrothermal stage.

postmineral. In economic geology, said of a structural or other feature formed after miineralization.

postorogenic. Said of a geologic process or event occurring after a period of orogeny; or said of a rock or feature so formed.

potorogenic phase. The fined phase of an orgenic event following the climatic orogeny.

postorogenic pluton. An igneous intrusion emplaced after a period of orogenic activity.

pot. (1) A sedimentary deposit in the shape of a pot, in distinctively different material, *e.g.*, bodies of sandy slit in the upland graverls of Maryland and virginia. (2) A concretion or other rounded object in the roof of a coal seam, whose bottom resembles that of a pot. 3. A pothole or other natural pit or depression, often containing water.

potash bentonite. *See* pottassium bentonite.

potash feldspar. *See* pottassium feldspar.

potash lake. An alkali lake whose waters are rich in dissolved potassium salts.

pot bottom. A large boulder of concretion in the roof of a coal seam.

potash mica. *See* muscovite.

potassium bentonite. A caly of the illite group that contains potassium and is formed by alteration of the volcanic ash. Also known as potash benttonite.

potassium feldspar. Any alkali feldspar (orthoclase, miocrocline, sanidiane, aduleris) containing the molecule $KAlSi_3O_8$. Also know as potash feldpar.

pot hole. (1) A shaftlike cave opening upward to the surface. (2) A small depression with steep sides in a coastal marsh; contains water at or below low-tide level. Also known as rotten spot. (3) A vertical or nearly vertical shaft in limestone. also known as aven; cenote. (4) Any bowlshaped, circular, or cylindrical hole formed by the grinding action of a stone in the rocky bed of a river of stream. Also known as churn hole; colk; eddy mill; eversion hollow; kettle; pot.

potstone. A massive greyish-green, dark-green, iron-grey, or brown-ish-black variety of talc.

potter's caly. A plastic clay free from iron and devoid of fissility, suitable for modeling or making of pottery or adapted for use on a potter's wheel. It becomes white after firing.

powellite. $Ca(WMo)O_4$. A comercially important tungsten mineral, isomorphous with scheelite ($CaWO_4$); crystalliizes in tetragonal system.

Pozzolan. Siliceous material such as diatomaceous earth, opaline chert, and certain tuffs, which can be finely ground and combined with portland cement. It reacts with calcium hydroxide that is liberated as concerete and hardens, forming compounds with cementitious properties Pozzolans also counteract the adverse effects of certain undesirable aggreagtes that may have to be used in concrete. Portland-pozzolan cements are highly resistant to penetration and corrosion by salt water.

prasinite. High grade regional metamoirphic rock; contains albite, chlorite, epidote, glaucophene, and hornblende.

prasopal. A green variety of common opal containing chromium.

precambrian That period of time during which the earth's crust was formed and the first life appeared. The terms Archaeozoic, Archean, Azonic, Cryptozoic, and Proterozoic have been used as synonyma or partial synonyms for this period. The duration of the Precambrian is probably not less than 4000 million years and covers 90% of geologic time. It is generally divided into the Archheozoic and Proterozoic eras, with subdivisions related to recognizable and datable orogenic episodes.

preciousgarnet. An unusally purple and brolliant varity of almandine.

precious metal. A relatively scarcve, valuble metal, such as gold, silver and members of platinum group; platinum, rhodium, rutheriun osmium, iridium, palladium].

precious stone. (1) Any genuine gemstone. (2) A gemstone of high commercial value because of its rarity, beauty durability, and hardness; *e.g.*, diamond, ruby, emerald, and sapphire.

precipitates. A geochemical division of the sedimentary rtocks including the chemically formed limestones.

preconsolidation pressure. Pressure exerted on unconsolidated sediment by ioverlaying material that resulted in compaction; the overburden may have been removed later by erosion.

predazzite. A crystallized limestone that resembles pencatite, but contains less brucite than calcite.

prerglacial. (1) Pertaining to geologic time immediately preceding the pleistocene epoch. (2) Said of materials, underlying glacial deposits.

prehnite. $Ca_2Al_2Si_3O_{10}(OH)_2$. A white to light-green in reniform and stalactitic agtgregates with crystalline surface; crystallizes in orthorhombic system. It has vitreous lustre and its hardness on

Mohs scale iis 6.0-6.5.

preliminary waves. The body waves of an earthquake, including both p and s waves.

premineral. In economics geology, said of a structural or other feature extant before mineralization.

preorogenic. Said of the initial phase of an orogenic cycle, prior to climatic orogany. It is the time of formation of geosynclines and intrusion of ultramafic plutonic rocks.

pressure figure. A patern resembling a six-rayed star,produced by intersecting lines of parting due to gliding, when certain minerals, especially, mica are compressed by a blunt point.

pressure gradient. 1. The rate of variation of pressure in a given direction at a fixed time, e.g., with depth in the ocean, 2.Loosely, the magniude of the pressure gradient.

pressure shadow. In structural petrology, aggregates of new grains growing on opposite sides of a host porphyroblast or detrital grain, thereby producing an elongate structure. This structure is generally aligned parallel to a foliation and may define a lineation.

pressure ice. A general term for ice, whose surface in places has been deformed by stresses genteratd by wind, current or waves. It may rafted, hummocked or tented.

pressure pan. An induced soil pan having a higer bulk density and a lower total proity than the soil directly, below or above it; formed as a result of pressure applied by moral tillage or by other means.

pressure plateau. An uplifled area of thick, panded lava flow.

prevailing current. The ocean current most frequently observed during a given period.

priabonian. European stage; Upper Cocene.

pricetie. $Ca_4B_{10}O_{19}.7H_2O$. A snow-whity, earthy variety of colemarite.

priderite. $(K,Ba)(Ti,Fe)_8O_{16}$. A red mineral of variable composition.

prill. A term dentoing a running stream.

primary dip. The slight dip of a bedded deposit assumed at its moment of deposition.

primary dolomite. A finely textured well-stratified, unfossiliferous dolomite rock formed in place by direct chemical or biochemical precipitation from lake water or sea water.

primary flowage. Movement within an igneous tock that is still partly fiuid.

primary gneiss. A rock that exhibits planer or linear structures characteristic of metamorphic rocks, but lacks observable granulation or recrystallization and is therefore considered to be of igneous origin.

primary magma. A magma originating below the earth's crust.

primary mineral. A mineral formed at the same time as the rock enclosing it, by igneous, hydrothermal, or pneumatolyticprocesses,andthatretains its original form and composition.

primary phase. The only csrystalline phase capable of existing in equilbrium with a given liquid; it is the first to appear on cooling from a molten state, and the last to appear on heating till it melts.

primary porosity. The porosity that developed during the final stage of sedimentation or that was present within sedimentary particles at the time of deposition.

primeval. Pertaining to the earliest stages of the earth.

primitive cell. A parallelepiped whose edges are defined by the primitive translations of a crystal lattice; it is a unit cell

of minimum volume.
primitive translation. For a space lattice, one of three translation which can be repeatedly applied to generate any translation which leaves the lattive unchanged.
principal axis. *See* optical axis.
principal stress. A stress that is perpendicular to one of three mutually perpendicular planes that intersect at a point in a body on which the shering stress is zero; a stress that is normal to a principal plane of stress. The three principal stress are identified at least, or minimum, intermediate, and greatest or maximum.
prismatic. Said of a crystal with one dimension markedly greater than the other two. May also refer to a metamorphic texture characteriized by such crystals.
prismatic class. The crystal class in the monoclinic system having symmentry 2/m.
primatic cleavage. Mineral cleavages parallel to the faces of a prism, *e.g.*, the {100} cleavage of amphibole.
prismatolith. A cocolith constructed of poygonal prisms.
pro. A prefix denting before.
problematic fossil. A structures making, object or other feature in a rock whose nature presents a problem, such as a dowbtful organic nature or origin.
prodelta. The part of a delta that is below the effective depth of a wave erosion, lying beyond the delta front, and sloping gently down to the floor of the basic into which the delta is advancing and where clastic river sediment ceases to be a significant part of the basin-floor deposits.
profile. A graph or chart that shows the variation of one property with respect to another; *e.g.*, gravity, seismic-wava velocity, and geothermal energy may all be plotted against depth below the surface to obtain desired profiles for seismic or geophysical exploration.
profile of equilibrium. (1) The longitudinal [or long] profile of a graded stream or stream whose gradient at every point is just great enough to enable it to transport its load of sediment. Syn: *graded profile*. (2) The *marine profile of equilibrium* is a smooth, sweeping curve, concave upward. It is steep in the breaker zone and flattens sharply seaward. The slope is just suffcient to keep the material deposited by waves and currents in slow seaward transit.
proglacial. Immediately in front of or just beyond the outer limits of a glacier or ice sheet, gererally at or near its lower end, said of lakes, streams, deposits, and other features produced by or derived from glacier ice.
proglacial lake. A lake formed just beyond the frontal margin of an advancing or retreating glacier, generally in direct contact with the ice.
progradation. The building forward or outward the sea of a shoreline or costline by bearshore deposition of river-borne sediments or by continuous accumulation of beach material thrown up by waves or moved longshore drifting.
promontory. A bulff or prominent hill overlooking or projecting into a lowland.
prong reef. A wall reef that has developed irregular buttresses normal to its axis in leeward and seaward directions.
propylite. Igneous rock of the dioritic clan. A green, fine grained porphyritic extrusive rock, hydrothermally altered to produce a complex suite of minerals; contains phenocrysts of biotite, feldspar, and hornblende partly altered to chlorite, carbonates, epidote, and quartz in a groundmass of a clinolite, calcite, chlorite, epidote, pyrite and serpentine.

propylitization. A decteric or hydrothermal process involving the formation of propylite, by the introduction of, or replacement by, an assemblage of minerals including carbontes, epidote, quartz, and chlorite.

protectite. A rock formed by the crystallization of a primary magma.

proterobase. A diabase in which the mafic mineral is primary hornblende.

proteroglacial. Pertaining to the earlier great ice age.

proterozonic. 1. The later of the two major subdivisions of the Precambrian. The entire Precambrian.

protoclastic. A term used for the igneous rocks in which the earlier formed crystals have been broken or deformed due to differential flow of the magma before complete solidification.

Protodolomite. A crystalline calcium-magnesium carbonate with a disordered structure, in which the metallic ions in the same crystalklographic layers instead of in alternate layers as in the mineral dolomite.

protogine. Igneous rock of the granite clan. A coarse-grained gneissose plutonic rock; contains mica, microcline, oligoclase, orthoclase, and quartz.

protolith. The unmetamorphosed rock from which a given metamorphic rock was formed by metamorphism.

promatrix. A term used for uncrystallized clayegy material in weakly, consolidated graywackes and arkoses.

protomylonite. Low grade dynamic metamorphic rock that is transitional between breccia and mylonite.

protoguartzite. A well-sorted, quartz enriched sandstone, lacking the well rounded grains of orthoquartzite.

protore An ore having material too low in concentration or too poor in grade to be mined profitably.

protolith. The unmetamorphosed rock from which a given metamorphic rock was formed by metamorphism.

promatrix. A term used for uncrystallized clayey material in weakly consolidated graywackes and arkoses.

protoquartizite. A well-sorted, quart-enriched savdstone, lacking the well-rounded grains of orthoquartzite.

protore An ore having material too low in concentration or too poor in grade to be mined profitable.

protozic. That part of precambrian time represented by rocks in which traces of life appear.

proustitte. Ag_3AsS_3. A cochinealred mineral that crystallizes in the rhombohedral system, composed of silver arsenic sulphide; it is isomosphous with pyrargyrite and occurs massively and as crystals.

provenance. A term dentoing the accumulated information pertaining to the origin of the constituent materials of sa sedimentary rock; includes the geographic location and environmental conditions of the sediment source area as well as the rock yypes from which the sediment was derived.

province Part of a region defined by certain peculiarities of climate, geology, flora, fauna, etc.

provitrain. Vitrain in which some plant structure can be discerned by microscope. Also known as telain.

proximal. (1) Said of an ore deposit formed adjacent to a volcanic feature to which it is genetically related and from which its constituents have been derived. (2) Said of a sedimentary deposit consisting of coarse clastics, formed nearest ths source area.

psammite. A clastic sediment of or sedimentary rock composed of sand-sized partides; a sndstone.

prammobitic. Said of an organism that lives in sand.

psephite. A sediment or sedimentary rock composed of large fragments set in a matrix varying in kind and amount.

psephyte. A lake-bottom deposit consisting mainly of coarse, fibrous plant remains.

pseudo. A prefix denoting false.

pseudobreecia. A partially dolimitized limestone with a mottled appearance that gives the rock a texture mimicking that of a breccia, or with a weathered surface that appears fragmental. It is produced diagenetically by selective growth of patchy recrystallized masses of coarse calcite in a lighter-coloured and less-altered matrix of calcareous mud.

pseudoconglomerate. A rock that resembles a normal sedimentary conglomerate; or an aggregate of rounded boulders produced in place by spheroidal weathering

pseudogalens. *See* sphalerite.

pseudofossil. A structure natural object, or mineral of inorganic origin that may resemble or be mistaken for a fossil.

pseudogley. A densly packed, stity soil which is alternately water-logged and rapodly dried out.

pseudogritty structure. A variety of mortar strucline in which the larger relics are angular, due to fracture along cleavage planes.

pseudohexagonal. Any crystal habit falsely resembling hexagonal system.

pseudojade. Any mineral falsely resembling jade.

pseudoleucite. *leucite altered* to *orthociase* and *nepheline*.

pseudomorphed. *See* pseudomorphic.

pseudimorphic. Applied to mineral apperinmg in the crystal form of another. due to alteration for example, where serpertines may retain the crystal shape of the original olivines.

pseudophite. Green serpentine-like variety of pennine.

Pseudoporphyritic. Pertaining to a rock that is not a true porphyry, but resembles one because of rapid growth of some of the crystals.

Pseudotanchylite. A black rock that resembles tachylite; carries fragmental enclosures and shows evidence of having been at high temperature.

Pseudovolacno. A large circular hollow or crater believed not to be associated with volcanic activity.

psilomelane. $(Ba,H_2O)_2Mn_5O_{10}$. A manganese mineral, found as a secondary mineral in association with manganese deposits due to weathering.

pterocerian. A stratigraphic stage name for the European Lower Kimmeridgian.

ptygmatic folds. Granitic material within migmatite or gneiss, having the appearance of disharmonic folds. The geneiss of this type of folding is conversial.

pucherite. $BiVO_4$. A redish-brown mineral composed of bismuth vanadate; occurs as small orthorhombic crystals.

pudding stone. In Britain, a conglomerate consisting of rounded pebbles whose colours are in marked contrast with the matrix, giving a section of a rock the appearance of a raism pudding.

puglianite. Igneous rock of the diorite clan. A coarse-grained granular plutonic rock; contains augite, anorthite, biotite, hornblende, leucite, and sanidine.

pulaskite. A light coloured, feldspathoid -bearing, granular or trachytoid alkali; syenite composed mainly of alkali feldspar, sodic pyroxene, arfedsonite, and nephaline.

pull-apart. A early stage during the formation of sedimentary collapse structure.

pumica. A light coloured cellular glassy rock commonly having the composition of rhyolite. It is often suffciently buoyant to float on water and is economically useful as light-weight aggregate and as an abrasive.

pumica flow. A type of pyroclastic flow in which a large proportion of the fragments are of pumice.

pumicite. An accumulation of angular shards and cell walls of rhyolitic glass, with a particle diameter of less than 4 mm. It differs from pumice only in grain size.

pumpelly's rule. The generalization that the axes and axial surfaces of minor folds of an area are congruent with those of the major fold structures of the same phase of deformation.

punky. Said of a semi-indurated rock, such as leached limestone; also said of a tuff that is weakly welded.

pureckian. European stage; Uppermost Jurassic.

purl A surirling or eddying stream or sill, moving swiftly around obstructions.

puy. A samll, servant volcanic come.

pycnite. A variety of topaz occurring in massive columnor aggregations.

p-wava. A type of seismic wave that involves particle motion in the direction of propagation. It is the fastest of seismic waves, travelling 5.5-7.2 km/s in the crust and 7.8/8.5 km/s in the upper mantle.

pyenocline. A layer of water is ocean, charactcrized by a ripid change in density with the change in depth.

pyreneite. A black or greyish-black variety of andradite.

pyriboles. A group mane for pyroxenes and/or amphibolcs.

pyrargyrite. Ag_3SbS_3. A deep ruby-red to black mineral, crystallizing in the hexagonal system, occurring in massive form and in dissiminated grains. Its hardness on Mohs scale is 2.5.

pyrite. FeS_2. A common yellow sulfide mineral, dimorphous with marcasite. Purite forms under a wide range of pressure- temperature, conditions, and so is found in many geollgical environments. It occurs in masses, associated with chalopyrite, among magmatic segregation deposits in mafic rocks.

pyritization. A common process of hydrothermal alteration involving introduction of or replacement by pyrite.

pyritobitumen. Any of various dark-coloured rlatively hard, nonvolatile hydrocarbon substance, generally associated with mineral matter, which decomposes on heating to yield bitumens.

pyrochlore. $(Na,Ca)_2(Nb,Ta)_2O_6(OH,F)$. A complex oxide containing essentially calcium, sodium, miobium and tantalum. It is morphous with microlite and usually occurs in pegmatites derived from alkalic igneous rocks.

pyrochroite. $Mn(OH)_2$. A hexagonal mineral of manganese; similar to brucite in appearance.

pyroclast. An individual particle ejected during volcanic eruption.

pyroclastic. Designating detrital volcanic matereals explosively erupted; formed by fragmentation as a result of vlocanic action; ejected, transported aerially and deposited upon land surfaces, in lanes. Or in marine waters. Pyroclastic material may be calssified according to size. Orgin or petrograhic composition.

1. Bombs and blocks-fragments between 64 mm and a few meters.

2. Laplll-fragments between 64 and 2 mm.

3. Volcanic ash-fragments between 2 and 0.25 mm.

4. Volcanic dust. fragments less than 0.25.

5. Pumice-a light-colored, highly vesicular glassy material; commonly rhyolitic in composition.
6. Scoria- a vesicular material that is usually heavier, darker, and more crystalline than pumice. Both pumice and scoria may exist as pyroclastic material or as froth or lava flows. Ashes, pumice scoriae, and bombs are formed from ejected fragments of fluid lava. Other ejecta, including lapilli and blocks, consist of fragments of all sizes thrown out in a completely soild or semisolid state.

pyroconductivity. Electrical conductivity which develops in a material only at high temperature, mainly at fusion, in solids that are practiucally nonconductive at normal temperatures.

pyroelectric crystal. A crystal that exhibits pyroelectricity. such as tourmaline, lithium sulphate monohydrate, cane sugar, and ferroelectric barium titante.

pyroelectricity. The property of certain crystals to produce a state of electrical polarity by a change of temperatuyre.

pyrolusite. MnO_2. A black mineral that mainly consists of manganese dioxide; it crystallizes in the tetragonal system and its hardness on Mohs scale is 1.2.

pyromagma. A highly mobile lava, oversaturated with gases that exists at shallower depths than hypomagma.

pyrometamorphism. Metamorphic changes occurring without the aid of pressure or water vapour, at temperatures near the melting points of the component minerals; it is a kind of local, intense thermal matauorphism.

pyromatasomatic. Formed as a result of metasomatic changes in rocks, mainly in linestone, at or near intrusive contacts, under the influence of inagmatic emanations and high pressure and temperature.

pyromorphite. $(PbCl)Pb_4(PO_4)_3$. A leadmineral, found in the oxidised zones of lead deposits; it crystallizes in the hexagonal system. Also known as green lead ore.

pyrope. $Mg_3Al_2(SiO_4)_3$. A mineral species of the garnet group characterizesdbyadeepfiery-redcolour; occurs in basic and ultrabasic igneous rocks.

purophanite. $MnTiO_3$. A bloodred, rhombohedral mineral consisting of manganese titante; isomorphous with ilmenite.

pyrophyllite. $Al_2Si_4O_{10}(OH)_2$. A white, grey or brown mineral consisting of basic aluminium silicate; it shows variable streak, uneven fracture and perfect basal cleavage. It resembles talc and occurs in foliated form or i compact masses in quartz veins, granites and inelamorphic rocks.

pyrosehist. A schist or shale that has sufficiently high carbon content. so that it burns with a bright flame.

pyrosmalits. $(Fe(II),Mn)_8Si_6O_{15}(OH,Cl)_{10}$. A colour-less or light-coloured mineral.

pyrosphere. The zone of the earth below the lithosphere.

pyrostilpmite. Ag_3SbS_3. A hyacinth-red mineral composed of silver antimony sulphide, occurring in monoclinic csrystal tufts; it is polymorphous with pyrargerite.

pyroxene. A group of common rock forming minerals, with gereral formula $ABSi_2O_6$, where A is Mg, Fe^{+2}, Ca or Na, and B is Mg, Fe^{+2}, or Al. The minerals cruystallize in both orthorhombic and monoclinic systems, and all species have similar cleavage. The principal orthorhobic members of the group are enstatite and hypersthene ; the monoclinic members are diopside, augite, and jadeite.

pyroxene-hornfels facies. A metamorphic rock facies characteristically formes near larger granitic or gabbroic bodies at depths of a few kilometers; the mineral assemblages reflect lower temperatures. In pelitic rocks, minerals such as quartz, orthoclase, sillimanite, hypersthene, and plagioclase occur. Minerals found in calcareous rocks include plagiocalse, diopside, and grossularite.

pyrixene monzonite. A quartz-poor member containing approximately equal amounts of microperthite and plagioclase.

pyroxene loerthite. Lamellar intergrowths of any of several pyroxenes, as with feldspars.

pyroxenide. An inforal term for any holocryastalline, medium to coaarse-grained igneous rock composed mainly of pyroxene, *e.g.*, pyroxenite.

pyroxenite. A dark-coloured heavy, phaneritic igneous rock composed mainly of pyroxene with smaller amounts of olivine and hornblende; it is formed by crystallization of gabbraic magma.

pyroxenoids. A mineral group (including wollastonite and rhodonite) having composition similar to pyroxene, but SiO_4 tetrahedrons are connected in rings rather than chhains. A more restricted range of cations enters the structure and replacement of sillcon by aluminium is very limited. The following are the commonest pyroxonoids;

Bustamite	$MnCa(SiO_3)_2$
Pectolite	$Ca_2NaH(SiO_3)_3$
Rhodonite	$MnSiO_3$
Wollastonite	$CaSiO_3$.

Pyrrhotite. $Fe_{1-x}S$. A common reddish-brown to bronze mineral that occurs as rounded grains to large masses, more rarely as tabular pseudo-hexagonal crystals and rosettes. It has a defect structure in which some of the ferrous ions are lacking. The mineral is darker and softer than pyrite; often containing 5% nickel, om which case it is mined as a nickel-ore. Its hardness on Mohs scale is 4.0.

pythmic. Pertaining to the bottom of aa lake.

plevial lake. A lake formed in a period of heavy ramfall; specifically a lake formed in the pleistone epoch during a time of glacial advcance.

porphyroclastic. Said of a heterogranular metamorphic texture, characterized by volumetrically signigicant amount of both porphyroclasts and neoblasts, also said of a rock with such texture.

porphyrotope. A large crystal enclosed in a finer-grained matrix in a sedimentary rock showing porphyrotopic fabric; such as a large dolomite crystal in finer grained calcitic matrix.

porphyry. Hypabyssal igneous rock containing orthoclase as the dominat mineral. The use of term 'pornocryst',as a synonym of phenocryst is not recommended, however this term can be employed in conjunction with mineral manes.

posthumous fold. A kind of recurrent folding that occurs in younger sedimentary rocks overlying a buried fold bolt.

postmagmatic. An indefinite term applied generally to reactions or events occurring agter crystallizatrion of the bulk of a magma, and usually including thhe ghdrothermed stage.

Postorgernic phase. The find phase of an orogenic event following the climatiuc orogeny.

postorogenic pluton. An igneous intrsion emplaced after a period of orogenic activity.

potashlake. An alkali lake whose waters are richh in dissolved potassium salts.

pot bottom. A large boulder of concration in the roof of a coal seem.

prasopal. A green variety of common

opal containin chromiium.
praya. A beachh or waterfront.
precious garnet. An unusually purple and brilliant variety of almeandine.
pressure ice. A general term for ice, whosesurfaceinplacehasbeendeformed by stresses gentralid by wind, current or waves.. It may rafled, hummocked or tented.
pressure pan. An induces soil pan having a higher bulk density and a lower totalporositythem thesoildirectlybelow or above it; formed as a result of pressuse applied by normal tillage or bny other means.
pressure plateau. An uplifked area of thick, ponded lava flow.
Prevaritingcrurent. The ocean current most fraquently observed during a given period.
priaborian. Europeam stage; Upper cocene.
Pricete. $CaB_{10}O_{19}.7H_2O$. A snow-white, early variety of colemanite.
priderite. $[K,Ba][Ti,Fe]_8O_{16}$. A red mineral of variable composition.
prill. A term denoting a running stream.
prismaticclass Thecrystalclassinthhhe monoclinic system having symmetry 2/m.
prismatic cleavage. Mineral cleavage porallel is the faces of a prosm, *e.g.*, the [100] cleavage of ainphibole.
prismatolith. A coculith constructed of polygonal prisms.
pro-. A prefix denoting before.
problematic fossil. A structure marking, objectorotherfeatureinarock whose nature presents a problem, such as a donbtful organic nature or origin.
prodelta. The part of a delta that is below the effective depthh of a wave erosion lying, beuond the delta front, and sloping genttly down to the floor of the basin into whichh the delta is advancing anfd where clastic river sediment ceases to be a significant part of he basin-floor deposite.
proglacia lake. A lake formed just beyond the frontal margin of an advancingorreteatingglacier, generallyindirect contact whth the ice.
progradation. The buiilding forward or outword the sea o a shoreline or coastline by mearzhore deposition of river-borne sediments or by continuous aecumulation of beach materiao thrown up by waves or moved longshore drigting.
promontory. A bluff or prominent hill overlookiing or projeeting into a lowland.
prong reef. A wall reeg that has developed viregular buttresses normal ti its axis in leeward and seaward dirctions.
propylitization. A denteric or bydrothermal process involving the formation of propylite, by the introduction of, or replacement by, an assemplage of minerals including carbonates, epidote, quartz, and chylorite.
proteclite. A rock formed by the crystallization of a promary magma.
proterbase. A diabase in whic the mafic mineral is primary hornblensde.
proteroglacial. Peraining to the earlier great iuce age.
protoclastic. A term used for the ignour rocks in which the earlier formed crhstals have been broken or deformed due ti differentical flow of the before complete solidification.
protolith. The unmetamorphosed rock from which a given metamorphic rock was formed by metamorphism.
promatrix. A term used for uncrystallized clayey material in weakly consolidated graywackes and arkoses.
protoguattizite. A well-sorted quartz-enrichrd sandstone, lacking the well-rounded grains of orthhoquartzite.
protozoic. That part of preCambrian

time represented by rocks in which traces of life appeared.

provenance. A term denoting the accumulated information pertaining to the origin of the constituent materials of a sedimentary rock; includes the geographic location and envirtonmental conditions of the sediment source area as well as the rock types from which the sediment was derived.

psammite. A clastic sediment of, or sedimentary rock composed of sand-sized particles; a sandstone.

prammobiotic. Said of an organism that lives in sand.

psephite. A sediment or sedimentary rock composed of large fragments set in a matrix varying in kind and amount.

psephyte. A lake-bottom deposit consisting manily of coarse, fibrous plant remains.

pseudogley. A densly packed, silty soil which is alternately water-logged and rapidly dried out.

pseudogritty structure. A variety of mortar structure in which the larger selics are angular, due to fracture along cleavage planes.

pulaskite. A light coloured, feldspathoid-bearing, granular or trachytoid alkali syenite composed mainly of alkali feldspar, sodic pyroxene, argedsonite, and nepheline.

punky. Said of a semi-indurated rock such as leached limestone; also said of a tuff that is weakly welded.

Purbeckian. European stage; uppermost Jurassic.

purl. A swirrling or eddying stream moving swiftly around obstructions.

puy. A small, remnamt volcanic cone.

pyenite. A variety of topaz occurring in massive columnar aggregations.

pyrochlore. $(Na,Ca)_2[Nb,Ta]_2O_6[OH,F]$. A brown, yellowish-brown, or black mineral composed of hydrated fluoride of calcium, sodium, niobium and tantalum. Most pyrochlore occurs associated with zircon, apatite, and other rare-earth minerals in pegmatites derived from alkalic rocks. It also occurs as an accessory mineral in nepheline syenite and in carbonatites associated with alkalic intrusives.

pyrochroite. $Mn[OH]_2$. A hexagonal mineral of manganese; isostructural with brucite.

pyroclast. An individual particle ejected during volcanic eruption.

pyroclastic. Designating detrital volcanic materials explosively erupted; formed by fragmentation as a result of volcanic action; ejected, transported aerially and deposited upon land surgaces, in lakes, or in marine waters Pyroclatic material may be classified according to size, origin or petrographic compostion.

(i) Bombs and blocks-fragments between 64 mm and a few meters.

(ii) Lapilli fragments of dimensions between 64 and 2mm.

(iii) Volcanic ash-fragments of dimensions between 2 and 0.25 mm.

(iv). Volcanic dust- fragments of diameter less than 0.25 mm.

(v) Pumice-a light-coloured, highly vesicular glassy material; commonly rhyolitic in composition.

(vi) Scoria- a vesicular material that is usually heavier, darker, and more crystalline than pumice. Both pumice and scoria may exist as pyroclastic material or as froth or lava flows. Ashes, pumice, scoriae, and bombs are formed from ejected fragments of fluid lava. Other ejecta, including lapilli and blocks, consist of fragments of all sizes thrown out in a completely solid or semi-solid state.

pyromagma. A highly mobile lava,

oversaturated with gases that exist at shallower depths than hypomagma.

pyrometamorphism. Metamorphic changes occurring without the aid of pressure or water vapour, at temperatures near the melting points of the compoment minerals; it is a kind of local, intense thermal metamorphism.

pyrometasomatic. Formed as a result of metasomatic changes in rocks, mainly in limestone, at or near intrusive contacts, under the influence of inagmatic emanations at high pressure and temperature.

pyroschist. A schist or shale that has sufficiently high carbon content, so that it burns with a bright flame.

pyrosmalite. $[Fe,Mn]_8Si_6O_{15}[OH,Cl]_{10}$. A colourless or light-colured mineral.

pyrosphere. the zone of the earth below the lithosphere.

pyroxene monzonite. A quartz-poor member containing approximately equal amounts of microperthite and plagioclase.

pyroxene perthite. Lamellar intergrowths of any of several pyroxenes, as with feldspars.

pyroxenide. An informal term for any holocrystalline, medium to coarse-grained igneous rock composed mainly of pyroxene, *e.g.*, pyroxenite.

pythmic. Pertaining to the bottom of a lake.

Q

q. A measure of the ability of a system with periodic behaviour to store energy equal to 2 π times the average energy stored in the system divided by the energy dissipated per cycle.

quaking bog. A peat bog floating or growing over water-sataturated land which shakes or trembles when walked on.

quaquaversal. Said of strata and structures that dip outward in all directions from a central point.

quarry. Open workings, usually for the extraction of stones.

quartz. SiO_2 A colourless, transparent, crystalline, rock-forming mineral composed of silicon dioxide. It is next to feldspar, the commonest mineral, occuring either in transparent hexagonal crystals or in crystalline or cryptocrystalline masses. It is the commonest gangue mineral of ore deposits and forms the major proportion of most sands, and has a wide-spread distribution in igneous, metamorphic, and sedimentary rocks. Its hardness on Mohs scale is 7.0.

quartz anorthosite. In IUGS classification, a plutonic rock with *Q* between 5 and 20, *P/(A+P)* greater than 90 and colour index less than 10.

quartzarenite. A sandstone that is composed primarily of quartz. The term is essentially equivalent to orthoquartzite.

quartz diorite. A group of plutouic rocks having the composition of diorite but with an appreciable amount of quartz, *i.e.*, 5-20% of the light coloured constituents; also, any rock of that group. Quartz diorite grades into granodiorite as the alkali feldspar content increases.

quartz-flooded limestone. A varity of limestone characterized by an abundance of quartz particles that were imported suddenly from a nearby source by wind or water currents.

quartz-free wacke. A wacke with more than 90% unstable materials.

quartz gabbro. In IUGS classification, a plutonic rock with *Q* between 5 and 20, *P/(A+P)* greater than 90 and plagioclase more calcic than An_{50}.

quartz index.. (1) A measure of the mineralogical maturity of a sandstone. It is expressed as the ratio of quartz plus chert to the combined percentage of feldspar, rock fragment, and clay matrix. It is used as a basis of evaluating the degree of weathering of the source rock and the degree to which the sediment has been transported. (2) A derived quantity in the Niggli system of rock classification, which may be either positive or negative, and is an indicator of a rock's degree of silica saturation.

quartzine. Chalcedony characterized by fibres having positive crystallographic elongation.

quartzite. (1) A metamorphic rock consisting of quartz grains, formed by the recrystallization of sandstone by thermal or regional metamorphism; a metaquartzite. (2) A sandstone composed of quartz grains cemented by silica; an orthoquartzite.

quartzitic sandstone. Sandstone consisting of 100% quartz grains cemented with silica.

quartz monozite. Granite rock in which 10-50% of the felsic constituents are quartz, and in which the ratio of alkali feldspar to total feldspar is 35-65%.

quartzose. Containing quartz as a principal constituent; especially applied to sediments and sedimentary rocks consisting chiefly of quartz.

quartz porphyry. A porphyritic extrusive or hypabyssal rock containing quartz and alkali feldspar phenocrysts embedded in a microcrystalline matrix. Also known as granite porphyry.

quartz syenite. A group of plutonic rocks having the characteristics of syenite but with a greater amount of quartz (5-20%).

quartz topaz. A frequently used but incorrect term for citrine.

quartz trachyte. A fine-grained igneous rock consisting mostly of alkali feldspar, with normative quartz between 5-20%; the volcanic equivalent of quartz syenite.

quartz wacke. A fine-grained sandstone containing upto 90% quartz and chert, with more than 10% argillaceous matric, less than 10% feldspar, and less than 10% rock fragments.

quarternary. The second period of the cenozoic era, following the Tertiary. It began about two to three million years ago and extends to the present.

quay. A wharf of solid construction, built roughly parallel to the shoreline.

quenselite. $4[PbMnO_2.OH]$. A pitchblack, monoclinic mineral composed of basic lead manganese oxide; shows perfect basal cleavage. Its hardness on Mohs scale is 2.5.

quick clay. Clay that loses its shear strength after being disturbed.

quicksand. A highly mobile mass of fine sand consisting of smooth rounded grains with little tendency to mutual adherence usually throughly saturated with upward flowing water; tends to yield under pressure and to readily swallow heavy objects on surface.

quickstone. A consolidated rock that had flowed under the influence of gravity before lithification.

quick water. The part of a stream characterized by strong current.

quilted surface. A land surface characterized by broad, rounded, uniformly convex hills separating valleys which are relatively narrow.

race. (1) A group of organisms with similar characteristics but not distinctive enough to be classified as a species or subspecies. (2) A strong current of water flowing through a narrow channel, *e.g.*, a tide race; also the constricted channel in which such a current flows.

radial drainage pattern. A drainage pattern characterized by radiating streams diverging from a high central area. Also known as centrifugal drainage pattern.

radial fault. One of a group of faults arranged like the spokes of a wheel, radiating from a central point.

radial flow. Pertaining to spokelike flow of reservoir fluids radially pointing inward toward a wellbore area.

radial symmetry. An arrangement of usually similar parts in a regular pattern around a central axis.

radiation source. Usually, an artificial sealed source of radioactivity used in teletherapy, radiography, as a power source for space batteries, or various types of industrial operations; accelerators, and radioisotopic generators and natural radionuclides may also be considered as radiation sources.

radiation standards. Exposure standards, permissible concentrations, rules for safe handling, regulations for transportation, regulations for industrial control of radiation, and control of radiation exposure by legislative means.

radiation units. Units of measurement used to express the activity of a radionuclide and the dose of ionizing radiation. The units curie, roentgen, rad, and rem are not coherent with SI units but their temporary use with SI units has been approved while the derived SI units becquerel, gray, and sievert become familiar.

The becquerel (Bq), the SI unit of activity, is the activity of a radionuclide decaying at a rate, on average, of one spontaneous nuclear transition per second.

1 Bq $=1\ s^{-1}$. The former unit, the curie [Ci], is equal to 3.7×10^{10} Bq.

The gray (Gy), the SI unit of absorbed dose, is the absorbed dose when the energy per unit mass imparted to matter by ionizing radiation is 1 joule per kilogram. 1 rad = 10^{-2} Gy.

radiative capture. A nuclear capture process whose prompt result is the emission of electomagnetic radiation only.

radioactive. Possessing, or pertaining to, radioactivity. Sometimes only the prefix 'radio-' is used to describe rodioactive nuclides or the substances containing them, *e.g.*, radiocarbon is an abbreviation for radioacative carbon.

radioactive age The age of a mineral, fossil, or wooden object as estimated from its content of radioisotopes. This method assumes that the content of radioisotopes has remained unchanged except for radioactive decay.

radioactive clock. A radioactive isotope such as potassium-40 which spontaneously decays to a stable end product at a constant rate, allowing absolute geological age to be determined.

radioactive dating. A technique for measuring the age of an object or sample of material by determining the ratio of the concentration of a radioisotope to that of a stable isotope in it; for example the ratio of carbon-14 to carbon-12 reveals the approximate age of pieces of wood and other archeological specimens. Most widely used radioactive dating techniques are: carbon-14 dating; lead-lead dating; potassium-argon dating; rubedium strontium dating; thorium-lead dating; and uranium-lead dating. The absolute age of a sample determined by different techniques show some descrepancies. The radioactive dating techniques are also known as isotopic age determination methods; radiometric dating methods; nuclear age determination; radiogenic dating methods.

radioactive debris. Radioactive material which is carried through the air from site of a nuclear explosion.

radioactive decay. The spontanous transformation of a nuclide into one or more nuclides, accompanied by either the emission of particles from the nucleus, nuclear capture or ejection , of orbital electrons, or fission. Also known as decay; nuclear spontaneous reaction; radioactive disintegration; radioactive transformation.

radioactive element. An element capable of changing spontaneously into another element by the emission of charged particles from the nuclei of its atoms. In case of some of elements, such as uranium, all known isotopes are radioactive, while for others, *e.g.*, potassium, one of several isotopes is radioactive.

radioactive emanation. A radioactive gas given off by certain radioactive elements; all of these gases are isotopes of the element radon. Also known as emanation.

radioactive mineral. Any mineral species that contains uranium or thorium, as an essential component of the chemical composition; for example, uraninite, pitchblende, carnotite, coffnite, and autunite.

radioactive equilibrium A state ultimately reached when a radioactive substance of slow decay yields a radioactive product may also decay to give a further radioactive substance, and so on. The amount of any of the daughter radioactive products present after the equilibrium has been reached, remains constant and the loss due to decay being counter-balanced by gain.

radioactive series A series of radioactive nuclides in which each member of the series is formed by the decay of the nuclide before it. The series ends with a stable nuclide. Three radioactive series occur naturally, those headed by throirm-232 [thorium series], uranium-235 (actinium series), and uranium-238 [uranium series]. All three series end with an isotope of lead. The neptunium series starts with the artificial isotope plutonioum-241, which decays to neptunium-237, and ends with thallium-81.

radiocarbon dating. A method of estimating the age of carbon- containing substances which have formed within the last 70,000 years and which have utilized carbon dioxied from the atmosphere or some other portion of earth's dynamic reservoir, based on the decay of carbon-14. The impact of cosmic rays on the earth's atmosphere causes a very small proportion of nitrogen atoms to transform into carbon-14 atoms. Some of these radioactive carbon atoms find their way, *via* carbon dioxide and photosynthesis, into living trees.

radiogenic. Said of a product of a radioactive process.

radiogenic isotope. An isotope that is produced by radioactive decay, but which itself may or may not be radioactive.

radiolarian ooze. A deep-sea pelage sediment containing at least 30%

opaline silica tests of radiolarians.

radiolarite. The hard fine-grained chert-like homogeneous consolidated equivalent of radiolarian earth.

radioluminescence. Luminescence resulting form *X*-rays, β-rays, or β-particles emitted in radioactive decay.

radiolysis. The chemical decomposition of substances as a result of irradiation with the help of *X*-rays, β-rays, or β-particles.

rafaelite. A napheline-free orthoclase-bearing hypabyssal syenite, also containing analcime and calcic plagioclase.

raft. A rock fragment trapped by magma and drifting freely.

rain crust. A type of snow crust formed by freezing of surface snow that had been melted by liquid precipitation.

raindrop erosion. The displacement of soil particles by the impact of large raindrops, particularly under conditions of bare soil affected by intense conventional rainfall. The raindrops have the effect of detaching individual particles, which are then affected by a preferred movement down slope, and helping to seal the soil surface by forcing the finest particles into the voids, reducing infiltration, and increasing the surface runoff.

rain shadow. A dry region on the lee side of a topographical obstacle, general ly a mountain range, where rainfall is significantly less than on the windward side.

rainwash. A very thin sheet of water, at least a few mm in depth, running over a slope, and with the capacity to transport fine soil particles. It may be generated as infiltration excess flow or as saturation overland flow.

raised beach. A deposit of shingle, sand and broken shells, the product of past wave action, standing above the highest level of present-day spring tides. The raised beach, which often rests on a raised beach platform resulting from wave erosion during a period of higher sea level in the past, is frequently preserved either by cementation , or by an overlying layer of head. Many raised beaches are of Pleistocene age, and are related to the fluctuations of the sea level resulting of the waxing and waning of continental ice-sheets and from related isostatic crustal movements.

raised bog. A region of acid, peaty soil, especially that developed from moss, in which the centre is relatively higher than the margins.

raised reef. An organic reef standing above the sea level.

ralstonite. $NaMgAl_5F_{12}[OH]_6.3H_2O$. A colourless, white or yellowish mineral composed of hydrous basic sodium magnesium aluminium fluoride.

ramdohrite. $Pb_3Ag_2Sb_6S_{13}$. A dark grey mineral composed of the complex sulphide of lead, silver and antimony.

rammelsbergite. $NiAs_2$. A grey mineral composed of nickel- diarsenide; it is isomorphous with pararemmelsbergite; shows greyish-black streak, uneven fracture, and does not show any cleavage. Its hardness on Mohs scale is 5.5-6.5. Also known as white nickel.

ramp valley. A trough between faults, forced downward by lateral pressure.

ramsdellite. MnO_2. An orthorhombic mineral composed of manganese dioxide, it is dimorphous with pyrolusite; shows black streak and good cleavage in two directions. Its hardness on Mohs scale is 3.0.

rang. A unit of subdivision in C.I.P.W. classification of igneous rocks.

range zone. Formal biostratigraphic zone made up of a body of strata comprising the total horizontal (geographic) and vertical (stratigraphic) range of occurrence of a specified taxon of a group of taxa.

rank. A coal classification based on degree of metamorphism. The percentage of carbon in dry mineral-free coal.

rankinite. $Ca_3Si_2O_7$. A mineral containing monoclinic calcium silicate.

.ransomite. $Cu(Fe,Al)_2(SO_4)_7$-$7H_2O$. A light-blue mineral composed of hydrated sulphate of copper, iron and aluminium.

rapakivi. A variety of granite or quartz monozite, çharacterized by orthoclase phenocrysts mantled with plagioclase. Also known as wiborgite.

rapids. A part of stream where the current is moving swiftly and where the water surface is broken by obstructions, as where the stream descends over a series of small steps.

rare-earth element. The name given to any of the group chemical elements with atomic numbers 58 to 71; the name is misnomer since they are neither rare nor earths.

rare-earth mineral. A mineral having the high concentration of rare-earths; examples are monozite, xenotime, and bastnaesite.

rashleighite. The name given to light bluish-green variety of turquoise obtained from St. Austell in Cornwell, England.

rasorite. *See* kernite.

raspite. $PbWO_4$. A yellow or brownish-yellow mineral composed of lead tungstate, occuring as monoclinic crystals.

rathite. $Pb_{13}As_{18}S_{40}$. A dark grey mineral with metallic lustre; composed of sulphides of lead and arsenic; occurs as orthorhombic crystals.

rational face. A crystal face peculiar to the internal atomic structure of the mineral type to which the crystal belongs.

Rauracian. A synonym of Argovian (Jurassic System).

ravine. A small and narrow valley with steeply sloping sides.

rayleigh wave. In seismology, a surface wave with a retrograde elliptical motion at the free surface. Also known as *R*-wave.

reaction pair. Any of the two minerals, one of which is formed at the expense of the other by the reaction with liquid.

reaction point. A point on a melting relations diagram in which it is impossible to state the composition of the liquid in terms of all the solid phases in equilibrium at this point.

reaction rim. A peripheral zone around mineral grains that are typically anhedral corroded grains of the unstable high temperature phase of a mineral. The rim is composed of another minerals species, and represents the reaction of the earlier solidified mineral with the surrounding magma.

reaction series. A sequence in which early-formed minerals react with the melt to form new minerals that are further down in the series. A continuous reaction series is one in which there are no abrupt phase changes during the reaction of early-formed crystals with later melts. A discontinuous reaction series is one in which reactions of early-formed crystals with later melts represent a shape phase change.

realgar. AsS. A bright-red to orange-red monoclinic mineral consisting of sulphide of arsenic. It is deposited in hot springs and low-temperature hydrothermal veins, and is frequently associated with orpiment. Also known as red arsenic; sandarac; red orpiment.

rebound. The isostatic readjustment upward of a landmass depressed by glacial loading.

recent. A synonym of Holocene.

recessional moraine. (1) A moraine formed during a minor readvance of the ice front during a period of glacial recession. Also known as stadial moraine. (2) An end moraine formed during a temporary halt in the final retreat of a glacier.

reciprocal lattice. A lattice array of points formed by drawing perpendiculars to each plane [*hkl*] in a crystal lattice through a common point as origin, the distance from each point to the origin is inversely proportional to spacing of the specific lattice planes.; the axes of the reciprocal lattice are perpendicular to those of the crystal lattice.

recombent fold. An overturned fold in which the axial surface is more or less horizontal.

recomposed granite. An arkose composed of consolidated feldspathic residue that has been reworked and decomposed so slightly that upon cementation the rock resembles granite except that its grain is less even and it contains a greater percentage of quartz. Also known as reconstructed granite.

reconstructed granite. *See* recomposed granite.

recrystallization. The formation, essentially while in the solid state, of new crystalline mineral grains in a rock. The new grains are usually larger than the primary grains, and their mineral grains, and their mineral composition may differ or be the same.

rectangular drainage. A pattern of drainage in which *(i)* sections of individual streams are rectilinear in plan, and *(ii)* stream junctions are at right angles. Rectangular drainage is characteristic of granite terrains, in which two sets of joints and shatter zones meet at right angles to each other, and is sometimes is developed in limestone and chalk on local scale.

rectilinear slope. A slope segment which is straight in profile. The rectilinearity may result as a small part of a slope profile, or in some circumstances may dominate the profile.

red beds. Group name for red coloured clastic sediments. Normally sandstones, shales or marls, coloured by hematite coating the constituent grains of the rock.

recurved split. A split whose outer end is turned landward by current deflection, by the opposing action of two or more currents, or by wave refraction.

red clay. A fine-grained, reddish-brown pelagic deposits of windblown particles, meteorite and volcanic dust, pumice shark teeth, manganese nodules. and debris transported by ice. Also known as brown clay.

redingtonite.$(Fe,Mg,Ni)(Cr,Al)_2(SO_4)_4 \cdot 22H_2O$. A purple mineral composed of hydrous sulphate of iron, magstate the compositionnesium, nickel, chromium, and aluminium.

red mud. (1) A type of marine mud that is land-derived and contains as much as 25% calcium carbonate. The red colour in the mud arises due to the presence of ferric oxide. (2) The residue in waste ponds and pits from bauxite processing for aluminium production.

red ocher. A red, earthy variety of hematite, used as a pigment.

red snow. A snow surface of reddish

colour, caused by the presence within it of certain microscopic algae or\and particles of red dust.

reduzates. Sediments that have accumulated under reducing conditions, and which are therefore typically rich in organic carbon and iron sulphide; black shale and coal are examples.

reedmergnerite. $NaBSi_3O_8$. A colourless, triclinc borate mineral that represents the boron analong of albite.

reef. (1) A rigid topographic feature of the near shore marine environment, that tends to be parallel shore and is produced by actively growing sediment binding organisms under high energy conditions. Reefs are generally wave-resistant and serve as protective barriers between the open sea on one side and lagoons,on the shoreward side. (2) Thick lenticular masses of carbonate rock exhibiting limited lateral extent in the stratigraphic record. Some uncertainty exits whether all stratigraphic reef structures were actually formed under the same conditions of modern day reef development.

reef conglomerate. *See* reef talus.

reef debris. *See* reef detritus.

reef detritus. Fragmented material derived from erosion of an organic reef. Also known as reef debris.

reef flat. A flat expanse of dead reef rock which partly or entirely dry at low tide, shallow pools, pothole, gullies and patches of coral debris and sand are features of the reef flat.

reef limestone. Limestone composed of the remains of sedentary organisms such as sponges, and of sediment binding organic constituents such as calcareous algae. Also known as coral rock.

reef patch. A single large colony of coral formed independently on a shelf at a depth less than 70 metres in the lagoon of a barrier reef or of an atoll. Also known as patch reef.

reef pinnacle. A small, isolated spire of rock or coral, especially a small reef patch.

reef rock. A hard unstratified rock composed of sand, shale and the calcareous remains of sedentary organisms; cemented by calcium carbonate.

reef talus. Massive inclined strata composed of reef detritus deposited along the seaward margin of an organic reef. Also known as reef conglomerate.

reflection, seismic. Energy which is propagated down into the earth and which returns after having undergone a single reflection from a subsurface velocity [density] contrast.

refolding. A process by which folds of one generation are subjected to and stressed by a force of different orientation.

refoliation. A foliation that is subsequent to and oriented differently from an earlier foliation.

regime. The existance in a stream channel of a balance betweebn erosion and deposition over a period of years.

regional dip. The nealy unifrom and generally low-angle inclination of strata over a wide area. Also known as normal dip.

regional metamorphism. Heat and high pressure process always associated with orogenesis and igneous intrusions by which rocks in the crust change in nature.

regional slope. The generally uniform dip of rock strata or land surface over a wide area.

regional slope deposit. A sedimentary deposit widely distributed as a thin sheet over a regional slope.

regional unconformity. A continuous unconformity extending throughout a wide region, which may be nearly continent wide, and usually represents a long time period.

regression conglomerate. Type of conglomerate resulting from a relative fall in sea level and a rejuvenation of fluvial activity.

regressive sediments. Sediments deposited during the retreat or withdrawal of water emergence of the land; characterized by an offlap arrangement.

rejuvenate. The act of stimulating a stream to renewed erosive activity either by tectonic uplift or a fall in sea level.

rejuvgenated fault scrap. A fault scrap revived by renewed movement along an old fault line after partial dissection or erosion of the initial scrap.

rejuvenation. The reversal or renewal of some landcape feature or geological form to a former degree of effectiveness, or to a condition leading to a new cycle of erosion.

relative age. The geological age of a rock, fossil, or event, as defined relative to other rocks, fossils, or events rather than in terms of years. The chronological arrangement of fossils, features, or events with respect to the geologic time scale, without reference to their absolute ages.

relative geologic time. Non-absolute geologic time in which events may be placed relatively to one another.

relative relief. The vertical distance between the highest and the lowest points in a landscape. In an area dissected by numerous valleys, relative relief is defined by the interval between the interfluve crests and the valley floors; it is therefore in effect an expression for slope height, and may therefore help to determine the maximum angles at which slopes can develop.

relict structure. Relics of the original rock texture or structure which may still be visible in a metamorphic rock and may have modified the superimposed texture or structure.

renardite $Pb[UO@]_4[PO_4]_2\ [OH]_4\cdot 7H_2O$. A yellowish mineral composed of hydrous basic lead uranyl phosphate.

rendzina. One of an intrazonal calcimorphic group of soils characterized by a brown to black, friable surface horizon and light-grey or yellow, soft underlying horizon; found under forests of grasses in humid to semiarid climates.

reniform. Bean- or kidney-shaped as describing the structure of a crystal in which rounded masses occur at the ends of radiating crystals, or certain structures in animals, and plants.

replacement dike. A dike which is made by gradual transformation of wall rock by solutions along fractures or permeable zones.

replacement texture. The texture exhibited due to the replacement of mineral by the other.

replacement vein. A mineral vein formed by the gradual transformation of an original vein by secondary fluids.

resequent stream. A strem whose direction follows an original consequent stream but is generally lower; resequent streams are generally tributary to a subsequent stream.

reservoir fluid. The subterranean fluid trapped by a reservoir formation; may include natural gas, liquid and vapours of petroleum, and interstitial water.

reservoir rock. Any porous and permeable rock that yields oil or gas; *e.g.*, dolomite, limestone, and sandstone.

residual. (1) Pertaining to residuel left in place after weathering of rock. (2) Said of a mineral deposit, formed by either mechanical or chemical concentration. (3) Said of a topographic feature respresenting the remains of a formerly great mass or area, and rising above the surrouding surface.

residual anticline. In soil tectonics, a relative structural high resulting from the depression of two adjacent rim synclines. Also known as residul dome.

residual clay. Clay material formed in place by the weathering of rock, derived either from the chemical decay of feldspar, and other rock minerals or from the removal of nonclay-mineral constituents by solution from a caly-bearing rock.

residual debris. A layer of highly decomposed granite, resulting from prolonged chemical weathering in a warm and humid climate. Breakdown of the ferromagnesian minerals and feldspar crystals by percolating water gradually produces a structureless mass of clay (mainly kaolinite) and stable quartz particles.

residual dome. *See* residual anticline.

residual liquid. *See* residual magma.

residual magma. The still-molten part of a magma that remains in the magma chamber after partial crystallization has taken place during a series of differentiations. Also known as residual liquid; rest magma; ichor.

residual soil. Soil formed in place by the decomposition of rocks like those on which it lies.

resin. Any of various hard, brittle, transparent or translucent materials formed, especially in plant secretions and obtained as exdates of recent or fossil origin by the condensation of fluids agter the evaporation of volatile oils. Resins are yellowish to brown in colour, fusible and flannable, soluble in most of the oragnic colvents but insoluble in water.

resistates. A geochemical division of the sedimentary rocks, including the arenaceous and redaceous rocks.

resurgence. The emergence of an underground river from a cavern, usually at the base of a scrap face or cliff in limestone (spring or seepage , in which a smaller quantity of water is involved).

reticulate. (1) Said of a vein or load with netlike structure. (2) Said of a rock structure in which crystals are partially altered to a seconday mineral, forming a network that encloses remnants of the original mineral. 3. Said of a metted pattern of an invertebrate.

retinalite. Massive yellowish-green, waxy variety of chrysotile.

retrograde metamorphism. A type of poymetamorphism by which minerals of lower grade are formed at the expense of minerals characteristic of a higher grade, a readjustment necessiated by a change in physical conditions *e.g.,* lowering of temperature.

retrograde shoreline. A shoreline that is being moved backward as a result of wave attack. Also known as abrasion shoreline.

retzian. $Mn_2Y[AsO_4][OH]_4$. A brown orthorhombic mineral containing manganese, ytterium and arsenic.

reversed drainage.. A type of drainage pattern in which, as a result of earth-movements, rivers are forced to flow in the opposite direction to their original courses One feature of reversed drainage is the presence of tributaries which join the main stream at an angle pointing upstream.

reverse fault A fault along which the hanging wall is displaced upward with respect to the footwall.

reverse saddle. A mineral deposit associated with the trough of a synclinal fold and follwing the bedding plane.

r_g wave. A slow, short-period, Rayleigh wave that travels only along a nonoceanic path.

rhegmagnesis. Orogeny characterized by the development of large-scale strike slip faults.

rheid. A substance below its melting points, which deforms by visous flow during applied stress at an order of magnitude at least three times of elastic deformation under similar circumstance.

rheid fold. A fold whose strata deform by visous as if thay were fluid.

rheidity. Relaxation time of a substance, divided by 1000.

rheo-. A prefix denoting streming or flowing.

rheomorphism .The process by means of which a rock becomes, mobile as a result of fusion; the change is ofter accompained by the addition of new material during diffusion.

rhodesite. $[Ca_2,Na_2,K]Si_{16}O_{40}.11H_2O$. An orthorhombic mineral composed of hyprous sodium calcium potassium silicate; shows cleavage visible in one direction. Its hardness on Mohs scale is 4.0.

rhodizite. $NaKLi_4Be_3Al_4B_{10}O_{27}$. A colourless or white complex beryllum aluminium alkali boratc mincral occurring as dodechedral crystals; shows conchoidal fracture and indistinct cleavage. Its hardness on Mohs scale is 8.0.

rhodocrosite. $MnCO_3$. A hexagonal carbonate mineral, found in lead and silver-lead ore veins and in metasomatic deposits. Its hardness ion Mohs scale is 3.5-4.0.

rhodolyte. Rose-red variety of pyrope.

rhodonite. $MnSiO_3$. A mineral containing manganese silicate; a pyroxenoid, sometimes found in hydrothermal veins or metasomatic deposits. Its hardness on Mohs scale is 5.5-60.

rhodusite. Fibrous variety of glaucophane.

rhmboclase. $HFe[III][SO_4]_2.\ 4H_2O$. A colourless mineral composed of hydrous acid ferric sulphate, occurring in rhombic plates.

rhombohedral. 1. Said of or pertaining to rhombohedral system. 2. Said of or pertaining to crystal cleavage in or a centered lattice of the hexagonal system.

rhombohedral lattice. A crystal lattice in which the three axes of a unit cell are of equal length and three angles between them are the same and are not right angles. Also known as trigonal lattice.

rhombohedron. 1. A trigonal crystal form that is a parallelepiped, the six idential faces being rhombs, 2. A prism with six parallelogram faces.

rhombiod ripple-mark. An aqueous current ripple mark characterized by a reticular arrangement of diamond shaped tongues of sand, with each tongue having two acute angles, one pointing upcurrent and the other pointing down-current.

rhyodacite. A group of extrusive porphyritic igneous rocks containing quartz, plagioclase, and biotite phenocrysts in a finegrained to glassy groundmass; composed of alkali feldspar and silica minerals. Also known as dellenite.

rhylite. A light coloured, aphanitic volcanic rock composed largely of alkali feldspar and free silica with minor amounts of mafic minerals. Mineralogically they are similar to granites and microgranites, although chemically they appear somewhat richer in SiO_2. Occasionally quartz in rhyolites is replaced by the high- temperature beta-form, and very rarely by tridymite or cristobalite. Ferromagnesian minerals are less obvious than in the corresponding plutonic rocks. Rhyolites in the strict sense are divided into sodic and potassic forms, according to the type of feldsper present.

rhyolitic glass. Volcanic glass that is chemically equivalent to rhyolite.

rhyolitic lava. A highly viscous, silica-rich lava.

rhyolitic magma. A type of magma formed by differentiation from basaltic magma in combination with assimilation of siliceous material, or by melting of portions of the earth's sialic layer.

rhyolitic tuff. A tuff composed of fragments of rhyolitic lava.

rhythmic accumulations. Regular patterns of ripples and cusps in sediment on the beach or the sea floor, formed by currents and waves.

ria (1) A long narrow arm of the sea, produced by the partial submergence of an area dissected by subaerial erosion; it is shallower and shorter than a fjord. (2) A long, narrow river inlet that gradually decreases in depth from mouth to head.

ribbon. One of set of parallel bands in a rock or mineral.

ribbon diagram. A continuous geologic cross-section that is drawn in perspective along a curved or sinuous line.

ribbon rock. A rock showing a succession of thin layers of differing composition or appearance.

ribbon structure. A succession of thin layers of different mineralogy and texture often contorted and deformed.

ribbon slate. Slate produced as a result of incomplete metamorphism, with still visible bedding that ends across the cleavage surface.

richterite. $[Na,K]_2[Mg,Mn,Ca]_6Si_8O_{22}[OH]_2$. A yellow, red or brown monoclinic mineral composed of basic silicate of sodium, potassium, magnesium, manganese, and calcium; a member of amphibole group.

rickardite. Cu_4Te_3. A dark purple coloured mineral composed of copper telluride, occurring in masses.

ricolite. Trade name for green serpentine from New Mexico.

rideau. A small ridge or mount of earth, or a slightly elevated piece of ground.

ridge. 1. A general term for a long narrow, elevation, usually sharp-crested and with steep sides. 2. An elongate. steep-sided elevation of ocean floor, having rough topography. 3. An irregular wall of broken floating ice, buckled upward by wind or current pressure.

ridge line. A slope line linking peaks via passes or pales.

ridging. A kind of deformation of floating ice, caused by lateral pressure, whereby ice is forced randomly one piece over the other to form ridged ice.

riebeckite. $Na_2[Fe,Mg]_5Si_8O_{22}[OH]_2$. A blue or black mineral of amphibole group, occurs as a primary constituent in some acid or sodium-rich igneous rocks.

riedel fracture. A branching fault developed in the cover as a result of movement in the basement.

riegel. A rocky bar or mound, forming an irregularity in the long-profile of a glaciated upland valley. Riegels are usually separated by near-level 'treads' or rock basins containing alluvial flats or lakes, and coincide with outcrops of particularly hard rocks crossing the valley floor; the later are abraded on their up-valley slopes and plucked on their down-valley faces.

rift. 1. A narrow opening in a rock caused by cracking or splitting. 2. A high and narrow passage in a cave.

rift block mountain. A mountain range which is horst block bounded by normal faults.

rift valley. A valley that has developed along a tectonic rift resulting from plate separation. A long narrow valley running along the crust of midocean ridges. the valley wall represent the edges of two oceanic plates diverging or spreading away from each other. The valley floor consists of the youngest exposed oceanic crust and is the frequent site of submarine volcanic activity.

rift zone. A system of crustal fractures associated with plate boundaries.

right lateral fault. A strike-slip fault where the ground opposite to the observer has moved to the right.

rill. (1) A small, transient unnel. (2) A small brook or stream.

rill erosion. The formation of numerous, closely spaced rills due to uneven removal of surface soil by streamlets running water.

rill mark. A small, dendritic channel formed on beach mud or sand by a rill, especially if on the lee side of a partially buried obstruction.

rim cement. A thin layer of calcium carbonate, or silica developed on the surface of detrital grains during diagensis.

rime. A white or milky and opaque granular deposit of ice formed by the rapid freezing of detrital grains during diagensis.

rime. A white or milky and opaque granular deposit of ice, formed by the rapid freezing of supercooled water drops as they impinge upon an exposed object.

rim rock. A top layer of resistant rock on a plateau outcropping with nearly vertical walls.

ring complex . A composite igneous intrusion, the individual components of which appear as circular or arcuate outcrop. Ring dikes and cone sheets are the two main members of ring complexes.

ring dike. An almost circular dike that is vertical or inclined away from the centre of the arc.

ring fault. A steep-sided fault patterm which is cylindrical in outline and is associated with cauldron subsidence.

ringite. An igneous rock formed due to mixing of silicate and carbonatite magmas.

rinkite. *See* mosandrite.

rinkelite. *See* mosandrite.

rinneite. NaK_3FeCl_6. A colourless, pink, yellow, or violet mineral; mainly composed of sodium potassium ferrous chloride; occurs in granular masses.

ripple drift. A pattern of cross lamination formed by sedimentray deposits on both sides of a migrating ripple.

riparian. Pertaining to or situated on the bank of a body of water.

ripidolite. $2[Mg,Fe,Al]_6[Si,Al]_4O_{10}[OH]_8$. A mineral of chlorite group; consists of basic magnesium ferrous aluminium silicate. It shows colourless streak amd good cleavage in one direction. Its hardness on Mohs scale is 1.0-2.0. Also known as aphrosiderite.

ripple. A small curling or undulating wave controlled to a significant degree by both surface tension and gravity.

ripple drift. A pattern of cross lamination formed by sedimentary deposits on both sides of a migrating ripple.

ripple index. On a ripple surface, the ratio of the crest--to- crest distance to crest-to-trough distance.

ripple load cast. A load cast of a ripple mark showing evidence of penecontemporaneous deformation in the accumulation of its trough and crest and in the oversteepening of the component laminae.

riple mark. A series of subparallel ridges and troughs of small scale, rarely exceeding a few centimetres from ridge to ridge. Ridges may be asymmtrical current ripple marks or symmetrical oscilatory ripple marks. These are created by bottom currents in streams, lakes, estuaries, lagoons or the marine environment. They are useful in reconstructing paleoenvironment, in general and palcocurrent strengths and directions, in particular.

rips. A turbulent agitation of water, generally causued by the interaction of currents and wind; in nearshore regions; they may be currents flowing swiftly over an ittegular bottom; sometimes referred to erroneously as tide rips.

rip up. A sedimentary structure formed when currents lift plates or discs of finely laminated and partially consolidated sediment and transport them to a new site of deposition. The individual plates or clasts may themselves by referred to as rip-ups.

rise. A long, broad elevation which rises gently from its surroundings, such as the seafloor.

riser. A steplike topograaphic feature, such as a steep slope between terraces.

rising limb. The rising part of the hydrograph resulting from runoff, or rainfall, or snowmelt.

Riss. (1) A European stage of geologic time pleistocene. (2) The thrid stage of glaciation of Pleistocence in the Alps.

Riss-wurm. The third interglacial stage of pliestocene in the Alps, following the Riss glaciation and preceding the Wurm glaciation.

river. A channelled flow of water, which moves downhill because of gravity. Most rivers flow in the sea, although there are some which flow into lakes, and some flow into deserts, where they evaporate to leave salt flats or very saline lakes. Most rivers flow throughout the year, although the amount of water flowing through the river varies depending upon the climatic factors, especially the rainfall, and the season. The source of a river is generally in an area of high ground, it may be a spring, a lake, a marshy patch, or a place where the water table reaches the surface. From its source, it generally flows through the mountains, then through an upland area, and finally onto a lowland. The upper parts of a river are usually rocky and turbulent, and the narrow valley is V-shaped with steep sides. There may be water falls to interrupt the long profile. In the middle course, the valley becomes wider, and the meanders begin to develop. Bluffs and small flood plains may be forming. The lower course passes through a wide flood plain; the meanders are larger, and some may have been abandoned to form oxbow lakes.

river capture. The capture of a river's water supply by a second river, which works back uphill by headward erosion and intercepts the flow. Also known as river piracy.

river bar. An accumulation of alluvium in the channel, at the mouth, or along the banks of a river or stream. At low water, it is commonly exposed and constitutes a navigational obstruction.

river basin. The area drained by a river and all of its tributaries.

riverbed. The channel which contains, or formerly contained, a river.

river cliff. The steep cutbank formed by the lateral erosion of a river.

river drift. Rock material deposited by a river in place after having been moved from another.

river profile. The longitudinal profile of a river.

river terrace. A near-level surface in a river valley, usually covered by a thin layer of gravel or alluvium. At the rear of the terrace there is a steep bluff leading up to higher ground; at the front, there is a steep drop either to the river or the flood plain. Many river terraces result from lateral plantation at a time when the river was graded to the higher base level. Subsequently, the river under went rapid vertical incision, leaving the former valley floor upstanding as the terrace. Sometimes, the river terraces are formed when the phase of the extensive valley-floor aggradation is followed by a phase of degradation.

river tide. A tide that occurs in rivers emptying, directly into the sea, showing three charcteristic modifications of ocean tides; the speed at which the tide travels upstream depends on the depth of the channel, the further upstream the longer the duration of the falling tide and shorter the duration of the rising tide, and range of the tide decreases with distance up- stream.

river valley. An eloganted depression in the earth's surface, carved by river or stream erosion in conjunction with mass wasting or collapse of river banks.

robinsonite. $Pb_7Sb_{12}S_{25}$. A mineral composed of lead antimony sulphide.

roche mountonnee. A glaciated bedrock surface in the form of asymmetrical mounds of varying shape. The up-ice side of a roche moutonnee has been glacially scoured and is smoothly abraded; the down-ice side has a steeper, jagged slope as a result of glacial plucking.

rock. (1) Any mass of mineral matter, whether consolidated or not, which forms part of the earth's crust is a rock. Rocks may consist of only one mineral species, in which case they are called monomineralic, but they usually consist of an aggregate of mineral species. (2) A hard. consolidated, and/or load-bearing, which where necessary has to be removed by blasting. This concept also accords with the popular idea of the meaning of the word.

rockallite. An igneous rock of the granite clan. A coarse-grained granulat plutonic rock; contains aegirite, albite, and quartz.

rock association. A group of igneous rocks within a petrographic province that are related chemically and petrographically, gererelly in a systematic manner such that chemical data for the rocks plot as smooth curves on variation diagrams.

rock avalanche. A very rapidly moving, turbulent mass of broken-up bedrock. Movement in a rock avalanche is a flowage on a grand scale.

rockbridgetite. $Fe(II)(Fe(III)_8(PO_4)_4(OH)_8$. A basic phosphate mineral of iron ; isomorphous with frondelite.

rock burst. A sudden and violent

breaking of a mass of rock from the walls of a tunnel, mine or deep quarry, caused by the release of accumulated stream energy. It may result in closure of a mine opening, or projecttion of broken rock into it, accompanied by ground tremors, rockfalls, and air concussions.

rock crystal. A clear variety of quartz. Also known as mountain crystal.

.**rock cycle.** The interrelated sequence if events by which rocks are intially formed, altered, destroyed, and reformed as a result of magmatism, erosion, sedimentation and metamorphism.

rock-defended terrace. (1) A river terrace protected from later undetermining by a projecting ledge or outcrop of resistant rock at its base. (2) A marine terrace protected from wave erosion by a mass of resistant rock at the base of the wavecut cliff in the overlying coastalplain sediments.

rock desert. An upland desert in which the bedrock is either exposed or is covered with a thin veneer of coarse rock fragments.

rock drumlin. A hill having the form of a drumlin but consisting of bedrock veneered with hill.

rock element. The coharent intact piece of rock that is the basic constituent of the rock system and which has physical, mechanical, and petrographic properties that cannot be described or measured by ordinary laboratory tests.

rock failure. Fracture of a rock that has been stressed beyond its ultimate strength.

rock fan. A fan-shaped rock surface whose apex is where a mountain stream deboundches upon a piedmount slope, and which occupies an area where a pediment meets the mountain slope.

rock flour. Chemically unweathered material consisting of slit and clay-sized angular particles of rock-forming minerals, mainly quartz; formed when rock fragments are pulverized while being transported or are curshed.

rock forming. Referring to those minerals that are part of the composition of rocks and determine their classification. Among the more important rock-forming minerals are quartz, feldspars, micas, amphiboles, and pyroxenes.

rock glacier. A mass of angular boulders and finer material, with interstitial ice or an ice core. It occurs in high mountains in permafrost area, and is derived from a cirque wall or other.

rock gypsum. Massive, coarsely crystalline to earthy, finely granular type of gypsum found in rock.

rock island. *See* meander ore.

rock pediment. A pediment formed on the surface of bedrock.

rock permeability. The ability of a rock receive, hold or pass fluid materials by mature of the interconnections of its internal porosity.

rock pillar. A column of rock formed by differential weathering or erosion, as along a joint plane. Also, a columnlike structure that is residual bedrock.

rock pressure. (1) The presure exerted by surrounding solids on the support system of underground openings, including that caused by the weight of the overlying material, residual unrelieved stresses, and pressure associated with swelling clays. (2) The compressive stress within the solid body of underground geologic material.

rock salt. (1) Coarsely crystalline halite occurring as a massive or granular aggregate, and constituting a nearly pure sedimentary rock that may occur in domes or plugs as extensive beds resulting from evaporation of saline water. (2) Artificially prepared salt in the form of large crystals or masses.

rockslide. The sudden, rapid, downward movement of newly detached bedrock segments over a surface of weakness. Once sliding begins, a rock slab usually breaks into rubble. Like rockfalls, rockslides can be caused by under cutting at the base of the slope, due to erosion or construction activities. Also known as rock slip.

rock slip. *See* rockslide.

rock stream. Rocks moving in a mass down a slope under the influence of gravity.

rock system. In rock mechanics, all natural environmental factors that can influence the behaviour of that portion of earth's crust which will become part of an engineering structure.

rock terrace. A stream terrace on the side of a valley composed of resistant bedrock which remains during erosion of weaker overlying and underlying beds.

rock type. (1) One of three major rock groups : igneous, sedimentary, metamorphic, (2) A rock having a unique identifiable set of characters, such as basalt.

roedderite. $[Na,K]_2[Mg,Fe]_5Si_{12}O_{30}$. A meteorite mineral composed of silicates of sodium, potassium, magnesium and iron.

roemerite. $Fe(II)[Fe(III)]_2[SO_4]_4\text{-}14H_2O$. A yellow to brownish mineral composed of hydrous ferrous and ferric sulphate.

roesslerite. $MgH[AsO_4].7H_2O$. A monoclinic mineral composed of hydrous acid arsenate of magnesium; it is isomorphous with phosphorroesslerite.

roof. The rock lying above a coal bed; also, the country rock bordering the upper surface of an igneous intrusion.

roof foundering. Collapse of overlying rock into a magma chamber following excavation of a large quantity of magma.

ropey lava. A synonym of pahoehoe.

rose muscovite. Manganese rich variety of muscovite.

rose quartz. A pink variety of quartz; commonly massive and used as a gemstone. Also known as Bohemian ruby.

rosolite. A synonym of landerite.

rosterite. A synonym of vorobyevite.

rougemonitite. An igneous rock of gabbro clan. A medium to finegrained granular plutonic rock; contains anorthite, augite, biotite, hornblende, and olivine.

roumanite. A brittle, brownish-yellow massive variety of amber occurring in Rumania.

rounding. The grinding away of sharp edges and corners of rock fragments during transportation. This process is particularly effective in sand and gravel as rivers, glaciers, or waves cause particles to hit and scrape one another or against a rock surface, such as a rocky stream bed.

routivarlte. An igneous rock of the gabbro clan. A fine-grained granular plutonic rock; contains garnet, orthopyroxene, and plagioclase.

rouvillite. An igneous rock of the gabbro clan. A mottled, medium to coarse-grained plutonic rock; contains augite, hornblende, labradorite, and nepheline.

roweite. $[Mn,Mg,Zn]Ca[BO_3]_2[OH]_2$. A light-brown mineral composed of basic borate of manganese, magnesium, zinc and calcium.

rtectonite. A tectonite in which the

fabric is believed to have resulted from rotation.

rubber ice. Newly formed sea ice which is clastic and weak.

rubellite. The red to violet coloured mineral tourmaline that is rich in lithium; its hardness on Mohs scale is 7.0-7.5.

rubicelle. Yellow or orange-red gem variety of spinel.

ruby. The red variety of corundum, containing small amounts of chromium; in its fine quality, it is used as gemstone.

ruby copper. A crystallized variety of cuprite.

ruby mica. The first grade of Indian mica; used for electrical capacitors.

ruby silver. Either of two red silver sulphide mineral; pyrorgyrite (dark ruby silver) and proustite (light ruby silver).

rudaceous. Of or pertaining to a sedimentary rock composed of large quantities of fragments that are larger than sand grains (diameter greater than 2 mm).

rudite. Any sedimentary rock composed of coarse fragments larger than sand grains (2mm) in diameter. Inluded are both conglomerate (rounded pebbles and grains) and breccia([angular pieces).

rudus. A prefix denoting pebbly.

rumanite. A synonyrn of roumanite.

run off. The part of the precipitation appearing in surface streams.

rupelian. A stratigraphilc stage name for the European Middle Oligocene.

S

saddle. A low point sag, or depression along the surface axis or axial trend of an anticline.

saccharoidal. Having a granular texture resembling that of a loaf of sugar. The term is used to describe some marbles and sandstone.

saddle back. A hill or ridge with a concave outline along its crest.

saddle fold. A flexuaral fold perpendicular to the parent fold and having an additional flcxurc at its crest.

saddle reef. A mineral deposit associated with the crest of an anticlinal fold and following the bedding planes, usually found in vertical succession, especially the gold-bearing quartz veins of Australia.

safforite. $CoAs_2$. A mineral composed of cobalt arsenide; occurs in tin-white masses and is dimorphous with smaltite.

sagenite. A variety of rutile that occurs in groups of needle-like crystals crossings at 60°, often enclosed in quartz or alike minerals.

sagenitic. A term denoting containing acicular minerals.

sagittal. Pertaining to or situated in the median anterior plane of a body having bilateral symmetry, or in any plane parallel thereto.

sagittate. Like an arrowhead in form; triangular, with the basal lobes pointing downward or concavely toward the stalk.

sag pond. A small body of water occupying an enclosed depression of sag formed where active or recent fault movement has impounded.

sahlite. *See* salite.

salar. A salt flat in a region of arid climate and usually associated with a saltlake or playa lake.

Sakmarian. A stratigraphic stage name for the base of the East European and Russian Permian.

saleeite. Mg $(UO_2)_2(PO)4$. 10 H_2O. A yellow mineral composed of hydrous phosphate of magnesium and uranium.

saleite. $Cu(IO_3)(OH)$. A bluish-green mineral composed of hydroxy-iodate of copper(II).

salic. A term derived from silica and aluminum and applied to the group of standard normative minerals in which one or both of these elements occur in large amounts. These include quartz, feldspars, and feldspathoids. The corresponding term for the slicic and aluminous minerals actually present in a rock is felsic.

salina. An area such as a salt flat, in which deposits of crystalline salts are containing high salt concentration.

salite. $(Mg,Fe)_2Si_2O_6$. A greenish to black mineral variety of diopside containing more magnesium than iron; member of clinopyoxene group. Also spelled sahlite.

saltation. Transport of a sediment in which the particles are moved forward in series of short intermittent bounces from a bottom surface. The size of particles can be saltated depends upon the velocity of the current and its density, *e.g.*, water will saltate larger particles than air at the same velocity.

salt dome. A circular piercement structure produced by the upward movement of a pipelike plug of salt. Salt domes occur in certain areas underlain by thick layers of sedimentary rock, and are derived from deposits of deeply buride halite. A cap rock of gypsum, anhydrate, and varying amounts of sulphur and calcite are found on top of many salt plugs, and some contain petroleum deposits.

salt lake. A body of water in and or semi-arid region, having no outlier to the sea and containing a high concentration of salt and Louisiana sulphur.

salt marsh. An area of coastal marsh which is flooded by the sea from time to time. Special salt-resistant plants colonize the areas; they trap silt and help to build up the land. Many of the coastal marshes can be reclaimed once the land has become well vegetated.

salt tectonics. The study of the formation and mechanics of emplacement of salt domes and other salt-related structures.

samarskite. $(Y,Ca,U,CaFe,Pb,Th)\text{-}(Nb,Ta,Ti,Sn)_2O_6$. A brown to blackish metamict orthorhombic mineral with splendent vitreous to resinous lustre; occurs in granite pegmatites.

sampleite. $NaCaCu_5(PO_4)_4Cl.5H_2O$. A blue mineral composed of hudrous chloride and phosphate of sodium, calcium and copper.

samsonite. $Ag_4MnSb_2S_6$. A black mineral composed of sulphide of silver, manganese, and antimony; occurring in monoclinic prismatic crystals.

sanbornite. $BaSi_2O_5$. A white triclinic mineral composed of barium silicate.

sand. (1) A detrital particle smaller than a granule and larger than a silt grain, having a diameter in the range of 1/16 to 2 mm. (2) A loose aggregate of such particles, most commonly of quartz.

sand bar. A bar or ridge of sand built up to or near to the surface of the water by currents in coastal waters.

sand crystal. A large euhedral or subhedral loaded with detrital sand inclusions upto 60%; level developed by growth in a sandstone during cemantation.

sand drift. (1) An accumualtion of sand aganist the leeward side of a fixed obstruction. (2) Movement of windblown sand along the surface of a desert of shore.

sand dune. A mound of loose windblown sand commonly fouud along low-lying seashores above high-tide level.

sandkey. A small sandy island parallel to the shore.

sand ridge. *See* sand wave.

sand stone. A clastic sedimentary rock composed of grains of sand size set in a matrix of slit of clay and more or less firmly united by a cementing material (such as silica, iron oxide, calcium carbonate); the consolidated equivalent of sand. The sand particles usually consist of quartz, and the term 'sandstone' when used without qualification, indicates a rock containing about 85-90% quartz.

sandstone dike. A clastic dike composed of lithified sand.

sand wave. A small, ridgelike primary stucture resembling a water wave on the upper surface of a sedimentary bed, that is formed by high-velocity air or water currents. Also known as sand ridge.

sand wedge. A wedge-shaped accumulation of sand with the apex downward formed by the filling in of winter contraction cracks.

sanidine. $KAlSi_3O_8$. A silicate mineral having glassy appearance; occurs in volcanics and near-surface dike rocks such as rhyolite and trachyte. It can be distinguished from plagioclase microscopically, by the absence of twinning striations.

sanmartinite. $ZnWO_4$. A mineral composed of zine tungstate.

sannoisian. A stratigraphic stage name for the European Lower Oligocene.

santa catherina. A stratigraphic name for the Brazillian equivalent of the Karroo.

santonian. A stratigraphic stage name for the European mid Upper Cretaceous.

santorinite. An igneous rock of the diorite clan. A fine-grained porphyritic extrusive rock; contains phenocrysts of labradorite with oligoclase rims in a groundmass of hypersthene and oligoclase.

saponite. A trioctahedral, magnesiumrich clay mineral of the montmorillonite group. It represents an end-member in which magnesium almost completely replaces aluminum.

sapphire. Any pure, gem-quality corundum other than ruby. Its varieties other than the well-known blue sapphire include green sapphires and yellow sapphires.

saprolite. A soft, earthy-brown or red, decomposed igneous or metamorphic rock that is rich in clay and formed in place by chemical weathering. Also known as saprolith.

saprolith. *See* saprolite.

sard. Light to dark brown translucent semiprecious variety of chalcedony.

sardonyx. A gem variety of chalcedony having straight, parallel bands of sard alternating with another chalcedony.

Sarmatian. A stratigraphic stage name for the European Upper Miocene.

sarmientite. $Fe_2(AsO_4)(SO_4)(OH).5H_2O$. A yellowish mineral composed of hydrous basic sulphate and arsenate of iron.

satin spar. Silky fibrous variety of aragonite or calcite.

sauconite. The zinc bearing end member of the montmorillonite group; a trioctahedral clay mineral.

saussurite. A compact mineral aggregate consisting of albite and zoisite or epidote; it is an alteration product of calcic plagioclase.

saxonite. An igneous sock of the ultrabasic clan. A coarse grained granular plutonic rock; contains bronzite, enstatite, hypersthene, and olivine.

scacchite. $MnCl_2$. A mineral composed of manganese(II) chloride; found in volcanic regions.

scaglia. A dark, very fine-grained, somewhat calcareous shale usually developed in the Upper Cretaceous and Lower Tertiary periods of the Northern Apennines.

scalar. Said of physical features of a rock fabric that are non-directional ; *e.g.*, a grain shape, crystal habit, or porosity.

scalenohedron. A closed crystal form whose faces are scalene triangles; the hexagonal scalenohedron has twelve faces, and the tetragonal scalenohedron has eight faces.

scapolite. A white or light coloured complex aluminosilicate of sodium and calcium belonging to tectosilicate group of silicate minerals having general formula $(Na,K,Ca)[Al_3(Al,Si)_3Si_6O_{24}](F,Cl,OH,CO_3,SO_4)$. It crystallizes in the tetragonal system; occurs in metamorphosed limestones, locally in considerable amounts; also occurs in schists and gneisses, commonly replacing plagioclase. Its hardness on Mohs scale is 5-6.

scar. A short steep slope, which is generally bare rock. The term is associated with areas of carboniferous limestone.

schairerite. $Na_3(SO_4)(F,Cl)$. A colourless rhombohedral mineral composed of sodium sulphate, chloride and fluoride.

schalstein. A slaty rock formed by shearing basaltic or andesitic tuff or lava.

scheelite. $CaWO_4$. A mineral composed of calcium tungstate; found in pneumatolytic and hydrothermal veins and in contact metamorphic deposits. In scheelite, sometimes Mo may subtitute for W, and a partial series extends toward powellite ($CaMoO_4$). Most scheelite, however contains only minor amounts of Mo. The colour of the fluorescence of scheelite varies with Mo content; whereas with X-rays, shortwave ultraviolet, and cathode rays, more-or-less pure scheelite fluoresces bright blueish white; with increasing Mo content, it tends to fluoresce white toyellowish white. It commonly occurs with wolframite in deposits of apparently high-temperature origin.

schefferite. $(Ca,Mn)(Mg,Fe,Mn)Si_2O_6$. A brown to black variety of pyroxence that crystallizes in the monoclinic system.

scheteligite. $(Ca,Y,Sb,Mn)_2(Ti,Nb,Ta,W)_2O_6(O,OH)$. A mineral composed of oxides of titanium, niobium, tantalum, tungstan, and calcium.

schist. A regionally metamorphosed rock characterized by presence of quartz, feldspar, mica, chlorite, talc or hornblende, schistose and lineated texture.

schistose. Pertaining to rocks exhibiting schistosity.

schistosity. A type of cleavage characteristic of metamorphic rocks notably schists and phyllites, in which rocks tend to split along parallel planes defined by the distribution and parallel arrangement of platy mineral crystals.

schlieren. Elongated segregations of light/dark minerals in igneous rocks, presumably caused by magma or lava flow prior to immobilization by crystallization.

schlieren dome. A term used for an intrusive igneous body, more or less completely outlined by flow layers which culminate in one central area.

schmidt net. A coordinate of net system used to plot a schmidt projection.

schmidt projection. A Lambert azimuthal equal-area projection of the lower hemisphere of a sphere on to the plane of a meridian; used in structural geology.

schoepite. $UO_3.2H_2O$. A yellow mineral composed of hydrous uranium (VI) oxide.

schonfelsite. A type of basalt containing embeded crystals of olivine and augite in a complex, dense fine-grained groundmass.

schorl. *See* schorlite.

schorlite. The black, iron-rich, opaque variety of tourmaline; occurs in the form of radiating clusters of needled. Also known as schorl.

schorlomite. $Ca_3(Fe,Ti)_2(Si,Ti)_3O_{12}$. A black mineral of garnet group that has vitreous lustre and occurs in masses. Its hardness on Mohs scale is 7.0-7.5.

schreibersite. $(Fe,Ni)_3P$. A silvery-white magnetic meteoritic mineral that crystallizes in tetragonal system.

Schrund line. The base of the bergschrud, of deep crevasse, at a late stage in the excavation of a cirque; the schrund line separates the steep slope of the cirque wall from the gentle slope below.

schungite. An amorphous carbon rich material occurring in Precambrian schists.

scolecite. $CaAl_2Si_3O_{10}$. A zeolite mineral that occurs in radiating group of white fibrous or acicular crystals.

scolite. Any of small tubes in rock belived to be the fossilized burrows of forms.

scopulite. A crystallite in the form of a rod with terminal brush or plume.

scoria. A vesicular cindery crust on the surface of lava flows, the cellular nature of which is due to the escape of volcanic gases before solidification. It is heavier, darker and more crystalline than pumice.

scorodite. $FeAsO_4.2H_2O$. A mineral consisting of hydrated iron arsenate.

scouring. *See* glaical scour.

scree. A heap of rock debris produced by eathering at the base of a cliff, or a sheet of coarse waste covering a mountain slope. 'Scree' is frequently considered to be a synonym of 'talus' but is a more inclusive term. Whereas talus is an accumulation of debris at a cliff base, scree also includes loose debris lying on slopes without cliffs.

screw axis. A symmetry element of some lattices, in which the lattice in unaltered by a rotation about the axis combined with a translation parallel to the axis and equal to a fraction of the unit lattice distance in this direction.

screw dislocation. A dislocation in which atomic planes form a spiral ramp widing around the line of dislocation.

scroll. One of a series of crescent-shaped sediments on the inner bank of a moving channel, deposited there by the stream.

scroll meander. A type of forced cut meander, in which the scrolls built on the inner bank cause erosion of the outer bank.

scyelite. An igneous rock of the ultrabasic clan. A hornblende biotite peridotite in which the biotite poikitically encloses the olivine.

sea cave. A split, or hollow opening, usually at sea level, in the base of a sea cliff, formed by waves acting on weaker parts of the weathered rock. Also known as marine cave; sea chasm.

sea chasm. *See* sea cave.

sea cliff. An erosional landform, produced by wave action, which is either at the seaward edge of the coast or at the landward side of a wave-cut plateform and which denotes the inner limit of the beach erosion.

seamanite. $Mn_3(PO_4)(BO_3).3H_2O$. A yellowish orthorhombic mineral composed of hydrous phosphate and borate of manganese.

seascrap. A submarine cliff that is relatively long, high and straight.

sea surge. A large movement of oceanic water, which is likely to cause large waves or high tides when it reaches land. In many parts of the world, the coastal land is affected ocasionally by a sea surge, when a large volume of water is moved southwards through the sea. If this coincides with high tide, it can cause considerable flooding.

seat clay. *See* under clay.

seat earth. A british term for a bed of rock underlying a coal seam, representing an old soil that supported an old soil thaqt supported the vegetation from which the coal was formed. Also known as underclay.

sea terrace. *See* marine terrace.

sebastianite. An igneous rock of the diorite clan. A coarse-grained granular plutonic rock; contains anorthite, apatite, augite, and biotite.

secondary. (1) A term applied to rocks and minerals formed by the alteration of pre-existing minerals. Secondary minerals may be found at the site as paramoprhs or pseudomorphs, or may be deposited from solution in rock interstices through which the solution is percolating; *e.g.*, amygdules. (2) Formed of material derived from the disintengration or erosion of other rocks; *e.g.*, clastic rocks.

secondary depression. A small area of low pressure which is found on the edge of a major depression. It may just be a bulge in the shape of the isobars, or it may be separate system with its own set of roughly circular isobars. Secondary depressions often become deeper and more intense than the primary depression, and they often move around the primary in an anti-clockewise direction. Like all depressions, they gradually fill in and become weaker.

secondary interstices. Openings in a rock that formed after the enclosing rock was formed.

secondary mineral. A mineral produced in an enclosing rock after the rock was formed as a result of weathering or metamorphic or solution activity, and usually at the expense primary material that came into existence earlier.

secondary stratification. The layering that occurs when sediments that were at one time deposited are resuspended and redeposited.

secondary structure. A structure that originated after the deposition or emplacement of the rock in which it was found, *e.g.*, a fault; especially, an epigentic sedimentary structure, such as concretion or sedimentary dike.

secondary wave. *See S* wave.

secular movement. Systematic, persistent movements of the earth's crust, either upward or dowdward, that takes places slowly and imperceptibly over long periods of geologic time.

sedentary. Said of a sediment or soil that is formed in place, with-out transportation, by disintegration of underlying rock or by accumulation of organic matter.

sediflucion. The subaqueous or subaerial movement of material in unconsolidated sediments, occurring in the primary stages of diagenesis.

sediment. (1) Solid material that has settled down from a state of supension in a liquid. (2) A mass of organic or inorganic solid fragmental material transported and deposited by wind, water, or ice, chemically precipitated from solution, or secreted by organisms, and that forms in layers in loose unconsolidated form, such as mud and sand.

sedimentary. Pertaining to or containing sediment, or formed by its deposition.

sedimentary dike. A tabular mass of sedimentary material that cuts across the structure or bedding of pre-existing rock in the manner of an igneous dike. It is formed by the filling of a crack or fissure by forcible injection of sediments, under abnormal pressure, or by simple infillings.

sedimentary facies. A stratigraphic facies differing from another part or parts of the same unit in both lithologic and paleontologic characters.

sedimentary ore. A sedimentary rock of ore grade; an ore deposit formed by sedimentary processes; *e.g.*, saline residues and phosphatic deposits.

sedimentary quartzite. *See* orthoquartzite.

sedimentary rock. Any of layered rocks resulting from the consolidation of sediment, *e.g.*, a clastic rock as sandstone, a chemical rock such as rock salt, or an organic rock such as coal. Clastic or detrital sedimentary rocks originate with the accumulation of discrete mineral or rock particles derived from weathering and the erosion of pre-existing rocks. Nonclastic varieties originate from molecular quantities of minerals of chemical precipitation, or from the biogenic action of organisms . Texture of sedimantary rock is described as rudaceous (coarse), arenaceous (medium) or lutaceous (fine). Rocks characterized by these textures are referred to as rudite, arenite, and lutite, respecetively. These are purely textural terms, independent of composition. The composition of sedimentary rocks is determined by conditions prevailing at the source of the sediments and at the site of deposition, as well as by the nature of transport. The rocks may contain fragments of any pre-existing rock or mineral.

sedimentary tectonics. Buckling and folding of strata in geosynclinal basins produced by subsidence of the geosyncline.

sedimentary trap. An area between a high energy and low-energy environment in which sedimentary marerials accumulate.

sedimentary tuff. A tuff containing a small amount of nonvolcanic detrial material.

sedimentary volcanism. The expelling. extruding or breaking through of overlying formations by a mixture of sediment, water, and gas, driven by the gas under pressure.

sedimentation. (1) The process of forming sediments in layers, including the separation of rock particles from the parent material, transportation of these particles to the site of deposition the actual deposition or settling the diagenetic changes occurring in the sediment, and its ultimate consolidation into rock. (2) Less broadly, the process of deposition of sediment, especially, by mechanical means from a strata of suspension in air or water.

sedimentation basin. A depression in the ocean floor with a flat and wide bottom in which sediment accumulates.

sedimentation trough. A depression in the ocean floore with a narrow *U*- or *V*- shaped bottom in which sediment accumulates.

sediment discharge. The amount of sediment moved by a stream in a given time, measured by dry weight or by volume; the rate at which sediment passes a section of a stream.

sediment load.. The solid marerial transported by a stream, expressed as dry weight of all the sediment that passes a given point in a given period of time.

seep. A spot where water or petroleum oozes from the earth, often forming the source of a small tricking stream.

segregation. (1) *Magmatic segregation.* (2) A secondary feature, such as a nodule of iron sulfide, that is farmed by the chemical rearrangement of minor components within a sediment subsequent to its deposition.

segregation banding. A compositional band in gneisses that is the result of segregation of material from an originally homegenous rock.

seif. A very large, sharp-crested tapering longitudinal dune or chains of duner found in the Sahara desert. Its crest in profile consists of a series of peaks and cols, and it bears on one side a succession of curved slip faces produced by strong but infrequent cross winds that tend to increase its width and height.

seismic. Pertaining to an earthquake or earth vibration, including those that are artificially induced.

seismic wave. A general term for all elastic waves produced by earthquakes or generated artificially by explosions. It includes both body waves and surface waves. The use of seismic waves has been invaluable on determining physical properties of the earth and the structure of its interior. Arrival times are used to locate reflecting discontinuities with in the earth, and the velocities of *P* and *S* waves give information about the ratio of any two of the parameters incompressibilly, elastic rigidity, and density of the material through which the waves pass.

selagite. An igneous rock of the syenite clan; a fine grained porphyritic extrusive rock; contains phenocrysts of biotite in a ground mass of orthoclase, diopside, oligoclase and apatite.

selective fusion. The fusion of only a portion of a mixture, such as a rock.

selective replacement. The replacement or one mineral by another, preferrentially within an altered rock mass.

selvage. A marginal zone of a rock form, showing some distinguishing peculiarity, such as in composition or fabric.

seracs. A very irregular surface of ice, generally found at the foot of an icefall on a glacier. It is the result of a crevassed section of a glacier becoming compressed when the gradient decreases, or the speed of ice movement is reduced. Pillars and pinnacles of ice stand up several metres above the general level of the ice, and make an almost impenetrable barrier.

sere. A sequence of ecological communities that succeed one another in development from pioneer age to climax community.

sericite. A fine-grained mica, generally occurring in small flakes. It is particularly common in schists, but is also found as an alteration of wall- rock material.

serpentinite. An altered igneous rock of the ultrahasic clan, normally compact rock with a banded or streaked texture; original minerals replaced by varieties of serpentine.

serpentine. $(Mg,Fe)_3Si_2O_5(OH)_4$. A group of common, rock forming minerals containing many polytypes, almost always occurring as fibrous yellowish-white or green aggregates.particularly in altered gabbros and ultrabasic igneous rocks. Its important varieties are bastite, bowenite, ophiotite, antigorite, chrysotile, lizardite, deweylite. Serpentine decomposes at high temperatures to forsterite, talc, and water; its presence in a rock may, there fore, indicate an origin below this temperature. The structures of the diverse polytypes are similar to those of minerals of the kaolinite group. In atleast some serpentines, however, the layers are curved to form tabular structures. The nickel-bearing variety of serpentine, with an apple-green colour, is sometimes reffered to by name garnierite. Massive serpentine is sometimes cut and polished as an ornamental stone.

serpierite. $(Cu,Zn,Ca)_5(SO_4)_2(OH)_6.3H_2O$. A bluish-green mineral composed of hydrous basic sulphate of calcium, zinc, and copper; occurs in tabular crystal and tufts.

serrate. Said of topographic feature that are notched or have a saw-toothed profile, *e.g.*, a serrate divide.

seybertite. A synonym of clintonite.

shadow zone. A region 100°-140° from

the epicentre of an earthquake in which, owing to refraction from low- velocity zone inside the core boundary, there is no direct penetration of seismic waves.

shale. A fine-grained laminated or fissile sedimentary rock nade up of slit-or clay-size particles or mud ; it is well indurated but not as hard as argilite or slate. It may be brown, re, grey, or black in colour. It generally consists of about one-third quartz, one-third clay materials, and one-third miscellaneous minerals, including carbonates, iron oxides, feldspars, and organic matter.

shale clay. A clay made from ground slate.

shale oil. A crude oil obtained from oil shale after the destructive distillation of oil shale; further processing is done to convert shale oil into products similar to petroleum oils.

shallow-pool test. A well located within the known limits of an oil or gas pool and drilled with the object of searching new producing zones above the producing zone of the pool.

shallow-focus earthquake. An earthquake with a focus at a depth of less than 70 km.

shaly. Pertaining to or having the characteristics of shale.

shandite. $Ni_3Pb_2S_2$. A rhomobohedral mineral composed of sulphide of nickel and lead; occurs in crystals.

shard. A curved, spiculelike fragment of pumice.

sharpite. $(UO_2)(CO_3).H_2O$. A greenish-yellow mineral composed of hydrous uranyl carbonate.

sharpsand. An angular-grain sand free of clay,loam,and other impurities.

sharpstone. Any rock fragment larger than a sand grain(diamater greater than 2mm) having angular edges and corners.

shatter cone. A striated conical rock fragment along which fracturing has taken place.

shattuckite. $Cu_5(SiO_3)_4.H_2O$. A blue mineral occurring in fibrous masses; composed of hydrous basic cupric silicate.

shear. (1) A deformation resulting from stresses that cause contiguous part of a body to slide relatively to each other in a direction parallel to their plane of contact. It is the mode of failure in which the portion of a mass on one side of a plane or surface slides past the portion on the opposite side. In geology the term refers in variably to strain rather than to stress: It is also used to refer to surfaces and zones of failure by shear, and to surfaces along which differential movement has taken place. (2) To make vertical cuts in a coal seam that has been undercut.

shear fold. A fold model of which the mechanism is shearing or slipping along closely spaced planes parallel to the fold's axial surface. The resulting structure is a similar fold.

shear fracture. A fracture that results from stresses that tend to shear one part of a rock past the adjacent part.

shear joint. A joint that is formed as a shear fracture.

shear strain. A measure of the amount by which parallel lines have been sheared past one another by deformation.

shear structure. A local structure in which earth stresses have been relieved by many small, closely spaced fractures.

shear zone. A layer or slab-like portion of a rock mass traversed by closely, spaced surfaces along which shearing has taken place. Many faults are charaeterized by closely spaced fractures among which movements have been distributed. The shear zones of some faults are silicified to varying degrees or intruded by a network of quartz veins which fill the fractures.

sheet erosion. A form of erosion in which the surface materials are all removed. It may be the result of wind blowing away all the fine particles, or rain wash carrying away the particles after sheet flooding. Sheet erosion occurs only if there is no vegetation cover. When it occurs, it removes the fine particles of soil, leaving the coarse less fertile parts behind. Once the land has been denuded by sheet erosion, it is likely that future rainstorms will start to cause gully erosion as well.

sheeting A form of rupture, similar to jointing, that occurs in massive rock bodies. It is characterized by tabular surfaces that are somewhat curved and essentially parallel to the topographic surface, except in regions of recent rapid erosion.

shield basalt. A basaltic lava flow from a group of small, close spaced shiled valcano bents that coalesced to form a single unit.

shield cone. A cone or dome-shaped volcano built up successive outpourings of lava.

shield volcano. A broad, gently sloping volcanic cone of flat domical shape, built chiefly of overlapping and interfingering basaltic lava flows. Also known as basaltic dome; lava dome.

shift. The relative displacement of the units affected by a fault but outside the fault zone itself.

shingle barchan. A dune like ridge formed of shingle perpendicular to the beach in shallow water.

shingle beach. A narrow beach composed of shingle and generally having a steep slope on both its landward and seaward sides. Also known as cobble beach.

shoal. A submerged elevation rising from the bed of a shallow body of water and consisting of or covered by unconsolidated meterial, and may be exposed at low water.

shoaling. The bottom effect which influences the height of waves moving from deep to shallow water.

shoal reef. A reef formed in irregular masses amid submerged shoals of calcareous reef detritus.

shock breccia. A fragmental rock formed by the action of shock waves.

shock metamorphism. The complete and permanent changes including physical, chemical, mineralogic, and metamorphic changes in rocks caused by transient high pressure shockwaves that act for short-time interval, ranging from a few microseconds to a fraction of a minute.

shock wave. A fully developed compressional wave of large amplitude, across which density pressure, and particle velocity change drastically.

shock zone. A volume of rock surrouding an explosion or impact crater in which the efforts of shock metamorphism are present.

shonkinite. An igneous rock of the syenite clan; a mottled. coarse-grained granular plutonic rock. It contains augite, orthoclase, plagioclase, olivine and nepheline.

shore. The narrow strip of land bordering any water-body; the most seaward part of the coast.

shore face. The narrow zone seaward

from the low-tide shoreline permanently covered by water, over which beach sands and gravels actively oscillate with varying wave conditions.

shoreline. The intersection of the sea or a lake with the shore or beach; migrates with changes of the tide or of the water level. The term is frequently used in the sense of high-water-shoreline or the landward limit of the intermittently exposed shore. Also known as waterline.

shore platform. The horizontal or slightly sloping surface produced along a shore by wave erosion.

shore terrace. *See* marine terrace.

shortite. $Na_2Ca_2(CO_3)_3$. A mineral composed of sodium and calcium carbonate.

shungite. A black, hard, coal-like amorphous material, composed of more than 98% carbon.

si. In structural petrology, a fabric defined by the preferred orientation of grains within or internal to a porphyroblast. It may or may not be parallet to the perferred orientation of grains outside the porphyroblast.

sial. A petrologic name for the upper layer of earth's crust, composed of rocks rich in silica and alumina. It is an acronym for silica and alumina.

sialma. A layer of the earth's crust that is intermediate in dapth as well as composition between sial and sima.

sicklerite. $(Li,Mn)PO_4$. A dark brown mineral composed of phosphate of lithium and manganese; occurs in cleavable masses.

siderite. (1) $FeCO_3$. A brownish, greenish, or grey rhombohedral mineral composed of ferrous carbonate. It is widespread as a bedded deposit in sedimentary rock sequences consisting largely of shale and/or coal seams. It is present in many hydrothermal veins as a gangue mineral of primary origin; also occurs as a hydrothermal replacement of limestone by iron-bearing solutions. It alters readily to hematite or limonite.

siderite mudstone. A variety of mudstone rich in chalybite , normally found associated with coal deposits.

sideritic limestone. A variety of limestone enriched in chalybite by diagenetic processes.

siederolites A name given to stony iron meterorites,

sienna. Any of various brownish-yellow earthy limonitic pigments for oil stains and paints. Its colour becomes dark when burnt in air, and even more darker and more transparent in oils than others.

sierra. A high range of hills and/ or mountains, especially one having jagged or irregular peaks that when projected against the sky resemble the teeth of a saw.

silerete. A conglomerate consisting of surficial sand and gravel cemented into a hard mass by silica.

silexite. (1) Chert occurring in calcareous beds. (2) Igneous rock composed mainly of primary quartz.

silica. SiO_2. Naturally occurring silicon dioxidw; occurs in five crystalline polymorphs; quartz, tridymite, cristobalite, coesite, and stishovite. It occurs in cryptocrystalline form (as chalcedony), in amorphous and hydrated forms (as opal), and combined form in silicates. The structures of quartz, tridymite, cristobalite, consist of SiO_4 tetrahedra linked by bridging oxygen atoms, but the arrangement of tetrahedra is different for each form. In the low-temperature form of SiO_2, quartz, the tetrahedra are arranged so as to form helical chains running through the crystal. In a given crystal, these chains can either be left- or right-handed, so that it is possible to obtain quartz in two optically active forms. Such isomeric crystalline forms are easily identified and can be mechanically separated. The high temperature forms of SiO_2 are less dense than quartz and have more open structures. The cubic crystobalite structure is essentially that of zinc blende. The arrangement of atoms in the tridymite structure is similar to that in the wurtzite structure.

silica sand. An industrial term for a sand or an easily disaggregated sandstone that has a very high percentage of quartz. It is a source of silicon and is also used as a raw material for glass and other industrial products.

silica stone. A sedimentary rock composed of siliceous materials.

silicates. (1) The generic term for a compounds containing silicon, oxygen and one or more metals. (2) large group of minerals whose crystal lattice contain SiO_4 tetrahedra, either isolated or joined through one or more oxygen atoms to form groups, chains, sheets, or three dimensional structures with metallic elements. These are the most important group of compounds occurring in the crust of the earth, and probably make up 95% of the crust. Silicates in which none of the oxygen of the tetrahedra are shared , have an isolated silicate structure. The individual silica tetrahedra are bonded together by positively charged ions. If two oxygen of each tetrahedron are shared with adjacent tetrahedra, chain silicate s are formed. When each tetrahedron shares three oxygens, sheet silllicates such as mica, are formed. When all the four oxygens of the tetrahedron are shared by adjacent tetrahedra, a framework silicate structure, such as quartz is formed.Silicates are classified by crystal structure patterns taken by the silicate tetrahedrons as follows.

Tektosilicates: Framework of tetrahedra extending in three dimensions; *e.g.*, silica group, feldspar group, zeolite group.

Phyllosilicates: Tetrahedra in sheets; *e.g.*, mica group, chlorite series.

Inosilicates: Single and double chains of tetraheadra; *e.g.*, amphibole group, pyroxene group.

Cyclosilicates: Tetrahedra forming rings; *e.g.*, beryl, tourmaline.

Sorosilicates: Tetrahedra linked in pairs; *e.g.*, epidote group, hemimorphite.

Nesosilicates: Single tetrahedron; *e. g.*, garnet group, aluminum silicate group.

silicated. Said of a rock in which the process of silication has taken place.

silication. The process of converting into or replacing by silicates, especially in the formation of skarn minerals in carbonate rocks.

silicatization. The sealing off of water by the injection of calcium silicate under pressure; sometimes used to reduce the leakage of water through defective lengths of tubing in a shaft.

siliceous earth. A loose, soft, porous, friable, fine-grained, light-weight, and usually white siliceous sediment usually derived from the remains of organisms.

siliceous ooze. Any pelagic deep-sea sediment containing at least 30% siliceous skeletal remains, such as radiolarians.

siliceous sediment. A sediment composed of fragmental, concretionary, precipitated siliceous materials.

siliceous shale. A hard, finegrained rock of shaly texture with an exceptional amount of silica (55%). It may be formed by silicification of normal shale, as by precipitation of silica derived from volcanic ash, or by accumulation or organic material.

siliceous sinter. A white, lightweight, porous, opaline variety of silica deposited by hot springs.

silicic. Said of magma or igneous rock rich in silica (usually at least 65%); for example, granite is a silicic rock.

silicification. 1. The introduction of, or replacement by, silica particulary in the form of fine-grained quartz, chalcedony or opal, which may fill cavities and pores and replace existing minerals. 2. Process of fossilizarion wherein the original hard parts of an organism are replaced by quartz. chalcedony or opal posed of basic fluoride and silicate of calcium and magnesium.

sill. (1) A sheet like body of igneous rock which conforms to bedding or other structural planes. Sills may be multiple or composite and are generally medium-grained, although larger ones may be coarse-grained. (2) A submarine ridge, located at a shallow depth, separating one basin from another or from the open sea. A sill is also a tabular intrusive structure, but it is concordant. It differs from dikes in being parallel to any planes or layering present in the country rock. Typically, the country rock bounding a sill is layered sedimentary rock. As magma squeezes into a crack between two layers, it solidifies into a sill. (3) A ridge at shallow depth near the mouth of a fjord. It functions as a barrier between the deep water of the fjord and that of the ocean.

sillar. A pyroclastic igneous rock of the granite clan; white to salmon red unconsolidated rock.

sillenite. Bi_2O_3. A mineral composed of bismuth (III) oxide; it is polymorphous with bismite and occurs as earthy masses.

sillimanite. Al_2SiO_5. A white, pale-green, or brown neosilicate mineral that crystallizes in orthorhombic system and commonly occurs in slender crystals, as well as in fibrus aggregates. It resembles other fibrous silicates, such as wollastonite and tremolite, but can be distinguished by its single direction of cleavage and its insolubility in acids. It is formed in aluminium rich rocks under conditions of high-grade regional metamorphism, and is found in schists and gneisses. Its hardness on Mohs scale is 6-7. Also known as fibrolite.

silt. (1) A detrital particle finer than fine sand and coarser than clay, generally in the diameter range of 1/16 to 1/256 mm. (2) A loose aggregate of rock or mineral particles of slit size, commonly with a high percentage of clay minerals. (3) Mud or fine earth suspended in water.

siltstone. A sedimentary rock composed of indurated silt, having the composition and texture of shale, but lacking its fissility or fine lamination. It occurs in layers, rarely thick enough to be classified as formations. siltstones are intermediate between sandstones and shales but less common than either. They contain less alumina, potash, and water than shales, but more mica.

sima. A petrological name for the lower of the earth's crust, composed of rocks that are rich in silica and magnesia. It is equivalent to oceanic crust and to the lower portion of the continental crust. underly the sial. Also know as intermediate layer.

similar fold. A fold in deformed beds in which the successive folds resemble each other.

simple cubic lattice. A crustal lattice whose unit cell is a cube, and whose lattice points are located at the vertices of a cube.

simple dike. An igneous dike emplaced in a single episode.

sincosite. $Ca(VO)_2(PO_4)_2.5H_2O$. A greenish mineral composed of hydrous calcium vanadyl phosphate; occcurs in tetragonal scales or plates.

sinistral fold. An asymmetric fold with the asymmetry of an *S* as opposed to that of a *Z* when seen in profile. The long limb appears to be offset to the left.

sink. (1) A swallow-hole or doline. (2) A collapse depression on the flank of a volcano. (3) A depression with no outlet, as where a desert stream comes to an end or disppears by evaporation.

sinkhole plain. A regionally extensive plain or plateau characterized by well-developed karst features.

sinter. A sedimentary rock deposited by precipitation from mineral waters, especially siliceous sinter and calcareous sinter. Also *see* siliceous sinter.

sinus. A groove, silt, or notch, as developed in shells or several invertebrates.

sjogrenite. $Mg_6Fe_2(OH)_{16}(CO_3)4H_3O$. A mineral composed of hydrous basic carbonate of magnesium and iron.

skarn. A contact metamorphosed lime stone or dolomite, consisting of calcium, magnesium, and iron silicates that has been derived from nearly pure limestone or dolomite into which abundant amounts of silicon, iron, aluminium, and magnesium were metasomatically introduced.

skeleton texture. Descriptive of the texture of limestone that consists of an in-place accumulation of skeletal material, that is, the hard parts secreted by organisms.

skialith. A vague remnant of country rock assimilated in granite.

skleropelite. A general term used to describe massive dense uncleavable rocks produced by the low grade regional or contact metamorphism of argillaceous rocks.

skutterudite series. The series including two minerals skutterudite ($CoAs_{2-3}$) and nickel skutterudite ($NiAs_{2-3}$). However some mineralogists also include isostructural arsenic-deficient minerals smaltite and chloanthite. The skutterudite minerals typically occur in veins with other cobalt and nickel minerals, especially cobaltite nickeline, but also arsenopyrite, silver, and bismuth.

slack ice. Ice fragments on still or slow-moving water.

slaking. (1) The breaking up of dried clay when saturated with water. (2) Crumbling and disintegration of earth materials when exposed to air and moisture.

slate. A group name for various fine grained rocks derived from mudstone, siltstone, and other clayey sediments as a result of low-degree regional metamorophism. Slate develops under temperatures and pressure only slightly greater than those found in the sedementary realm. The temperatures are not high enough for the rock to crystallize. The important controlling factor is intense shearing or compressive directed pressure. A slate indicates that a relatively cool and brittle rock has been subjected to intense tectonic activity. Because of the ease which it can be split into thin flat sheets, slate finds many specific applications.

slaty cleavage. A parallel foliation of fine-grained, platy minerals, mainly chlorite and cericite, in a direction perpendicular to the direction of maximum finite shortening, developed in slate or other homogeneous rock by deformation and low grade metamorphism. Host slaty cleavage is also called axial plane cleavage.

slick. Area in which capillary waves are absent or suppressed.

slickens. A layer of fine slit deposited by a flooding stream.

slickenside A rock surface that has become polished and striated from the grinding or sliding motion of an adjacent rock mass, commonly found along fault planes or zones.

slide. (1) The movement of descent of a landslide or rock-slides, also the mass of material so moved. (2) The track of bare rock or furrowed earth left by a landslide.

slip. The relative displacement of formerly adjacent points on opposite sides of a fault, measured in the fault surface.

slip face. (1) The leeward surface of a sand wave, exhibiting forest-bedding. (2) The steeply slopping surface on the beside of a dune, standing at or near the angle of repose of loose sand advancing of slides wherever the angle is exceeded.

slip faul[illegible] [illegible] normal fault.

slope failure. Gradual or rapid downslope movement of soil or rock under gravitational stresses, often as a result of factors caused by human-beings such as removal of material from the base of a slope.

slope stability. The resistance of a natural or artificial slope to failure by landsliding.

slope wash. (1) The material that is or has been tansported. (2) The mass wasting process, assisted nonchanneled running water, by which rock and soil is transported down a slope, specifically, sheet erosion.

slough. (1) A marshland or tidal waterway which usually connects other tidal areas; often more or less equivalent to a bayou. (2) Rock material that has crumbled from the sides of a borehole; it may obstruct the hole or be washed out in the drilling mud.

sludge. A soft or muddy bottom deposit as on tideland or in a stream bed.

sludge lump. An irregular mass of sludge formed as a result of strong winds.

sluicing. Concentration of heavy minerals *e.g.*, gold cassiterite, by washing consolidated material through boxes (sluices) equipped with riffles that trap heavier minerals on the floor of the box while allow the lighter particles to pass through.

slump. (1) The downward silding of rock debris as a single mass, usually with a backward rotation relative to the slope along which movement taies place. (2) The slipping or sliding down of a mass of sediment relatively soon after its deposition on a subaqueous slope

slump bedding. A term applied loosely to any disturbed bedding; specifically, deformed bedding produced by subaqueous slumping or lateral movement of newly deposited sediment. Also known as slurry bedding.

slump block. A coherent mass of material form away during the formation of a slump, the slide mass remains virtually intact and moves outward and downward.

slump fault. *See* normal fault.

slump fold. Any intraformational fold produced by slumping of soft sediments.

slurry bedding. *See* slump bedding.

slush. Snow or ice on the grond that has been reduced to a soft, watery mixture by rain, warm temperature, or chemical treatment.

smaltite. $CoAs_2$. A metallic grey mineral composed of arsenide of cobalt; some iron and nickel often occur with cobaltite. It shows metallic lusture, conchoidal fracture, and good cleavage in two directions.

smaragdite. A green amphibole mineral that is pseudomorphous after pyroxene in rock such as eclogite.

smectite. A group of dioctahedral (saponite) clay minerals of general formula $M_{0.33}Al_2Si_4O_{10}(OH)_2.nH_2O$ where M may be Na^+, K^+, Mg^{2+}, Ca^{2+}, and possibly others. These are characterized by three-layer crystal lattice; by deficiencies in charge in the tetrahedral and octahedral positions balanced by the presence of cations subject to base exchange; and by smelling or wetting, due to introduction of inter-layer water in the *c*-axis direction.

smithsonite. $ZnCO_3$. White, yellow, brown, or green secondary carbonate mineral associated with sphalarite, and occurs in oxidised zone of ore deposits, apparently derived by alteration form primary zinc minerals. It is typically associated with hemimorphite and secondary lead and copper minerals such as cerussite, malachite, anglesite, and pyromorphite. Simthsonite is isostructural with calcite, and thus the metal ion Zn^{2+}, is in octahedral coordination, which is unusual for zinc, since it normally occurs in tetrahedral coordination with oxygen. Also known as calamine; zinc spar.

snaking stream. *See* mendering stream.

snout. The downstream end of a glacier. It will be an area of melting for much of the year, and a river is likely to be flowing from underneath the ice, often through a cave. Masses of morainic debries may make the end of the ice look very dirty, and there will be depositional material all around the snout, and also further down the valley if the galcier is in a period of retreat.

snow. A type of precipitation consisting of water vapour which has frozen into ice crystals. Several ice crystals join together to form a snowflake, which will gradually fall down to earth. Some snowflkes may melt as they fall, especially if the temperature is only just down to freezing point. Under such condition sleet occurs, and for the same reason it is possible for clouds to give snowfall on mountains, while producing rainfall on adjacent lowland areas.

snowline. The line on a mountainside above which there is perpetual snow. There may be snow below the line during the winter months but it will not persist throughout the year. The altitude of the snowline varies with latitude.

sodalite. $Na_8(AlSiO_4)_6Cl_2$. A white, grey, or green mineral tectosilicate of feldspathoid group; crystallizes in isometric system. It occurs in both volcanic andplutonic rocks of nepheline syenite family, but more commonly in plutonic types, generally in association with nepheline. Blue sodalite occurs in considerable quantity in pegmatitic nepheline syenite. It is sometimes cut and polished as an ornamental stone. Its hardness on Mohs scale is 6.

soda mica. *See* paragonite.

soggendalite. An igneous rock of the gabbro clan, a fine grained porphyritic hypabyssal rock, contains phenocrysts of pyroxence in a ground mass of labradorite and ilmenite.

soil. Product of weathering of the bedrock before removal by erosion. The upper portion of the regolith capable of supporting plant life.

soil creep. The gradual, steedy downhill movement of soil and loose rock material on a slope.

soil genesis. The mode by which soil originates, with particualr reference to processes of soil-forming factors responsible for the development of true soil from unconsolidated parent material. Also known as pedogenesis; soil formation.

soil horizon. A layer of soil that is distinguishable from adjacent layers by characteristic physical properties such as colour, texture, or structure, or by chemical composition, including content or organic matter and degree of acidity or basicity.

soil profile. A vertical section of a soil, showing horizontal and parent material.

soil series. A family of soils having similar materials under the influence of similar climate and vegetation.

soil stripes. Alternating bands of fine and coarse material in a soil structure.

soil structure. Arrangment of soils into various aggregates, each differing in the characteristics of its particles.

solar faculae. Bright streaks or regions on the surface of the sun, especially near solar sunsport.

solar flares. Short high temperature outbursts seen as bright areas in the chromosphere of the chromosphere of the sun. Jets of particles, known as the solar wind and strong radio frequency electromagnetic radiations are emitted during solar flares; associated with sunspots; cause magnetic and radio disturbance on earth.

solar parallax. The angle subtended by the earth's equatorial radius at the centre of the sun at the mean distance between the earth and the sun (*i.e.* at 1 astronomical unit). It has the value 8.794 148 arc seconds.

solar pond. An experimental means of storing solar energy in either fresh or salt water. The 'pond' is a large, suitably lined open container of varying dimensions.

solar prominence. The large eruptions of luminous gas from the sun's surface ; they appear dark against the sun's disc but bright against the dark sky, and occur only in regions of horizontal magnetic fields.

solar radio noise. Radio noise originating at the sun. Its intensity increases during sunspots and solar flares. It is heard as a hissing noise on short wave radio receivers. Also known as solar noise.

solar still. A device for converting sea water into fresh water by evaporation. The sea water is collected in one or more pools and a transparent cover made of glass or plastic is placed over it. Heat from the sun evaporates the water, leaving behind a residue of salts ; the vapours from the evaporated water condenses on the surface of the receiving tank.

solar system. The sun and the celestial bodies moving about it; includes nine planets : mercury, venus, the earth, mars, jupiter, saturn, uranus, neptune and pluto ; and of the belt of asteroids revolving in elliptical orbits round the sun, along with satellites of the planets, commets, and meteor swarms. Over 99% of the mass of the system is concentrated in the sun. The solar system as a whole moves in an approximately circular orbit about the centre of the Galaxy, taking about 2.2 × 108 years to complete its orbit.

solar wind. The supersonic flow of gas containing streams of electrically charged particles such as ionized hydrogen, electron and helium, which continuously flows from the sun out through the solar system with velocities of 300 to 1000 kilometres per second emitted by the sun, predominantly during solar flares and sunspot activity.

sole. (1) The undersurface of a rock body or a vein, especially the bottom of a sedimetary stratum. (2) The fault plane underly a thrust sheet. (3) The middle and lower parts of the shear surface of a landslide. 4. The basal ice of a glacier.

sole fault. A low-angle thrust fault forming the sole of a nappe; also the basal main fault of an imbricate structure.

sole injection. An igneous intrusion that was put in place along a thrust plane.

sole mark. An irregularity or penetration on the under surface of a sedimentary stratum.

solfatra. A fumarole from which sulphurous gases are emitted.

solfatric. The action of sulphur when dissolved in a sillicate melt, prolonging the period of crystallization.

solifluction lobe. An isolated, tongue-shaped feature, upto 25 metres wide and 150 metres long, formed by more rapid solifluction on certain sections of a slope showing variation in gradient. Generally, it has a steep front and a relatively smooth upper surface.

solifluction stream. A narrow, laterally confined streamlike deposit of solifluction material.

solitary coral. A coral that does not form part of colony; an individual corallite exists unattached to other corallites.

solstice. (1) Either of the two points on the ecliptic midway between the equinoxes, at which the sun is at its greatest angular distance north (summer solstice) or south (winter solstice) of the celestial equator. (2) The time at which the sun reaches either of these points. The summer solstice occurs on June 21 and the winter solstice or December 21 in the northern hemisphere; the dates are reversed in the southern hemisphere.

solution breccia. A collapse breccia formed where soluble material has been removed as solution, allowing the overlying rock to settle and become fragmented.

solvsbergite. An igneous rock of syenite clan. A medium to finegrained granular hyperbyssal rock; contains sodaclase, aegirine, and microline.

soma-. A prefix denoting body.

somma. The rim of a volcano.

sommaorite. An igneous rock of the syenite clan; a medium to coarse-grained porphyritic plutonic rock. It contains phenocrysts of augite and olivine in a groundmass of leucite, bytownite, and orthoclase.

sordawalite. *See* jaspoid.

sorosilicate. A class of structutal type of silicate characterized by the linkage of two SiO_4 tetrahedra by sharing of one oxygen, with a Si : O ratio of 2:7 ; *e.g.*, hemimorphite, $Zn_4Si_2O_7(OH)_2.H_2O$.

sorotite. A type of meteorite similar to pallasites, with troilite substituting for olivine.

sorption. Adsorption (a surface process) or absorption (a volume process). The term is often used when the mechanism of a particular process is not known or is not specified. This term can also be used for ion-exchange, ion-exclusion, ion-retardation, chemisorption, and dialysis.

space. (1) That part of the boundless four-dimensional continuum in which matter can be physically (rather than temporally) extended . (2) The part of the universe that lies beyond the earth's atmosphere, in which the density of matter is very low.

spandite. A term used for the varieties of garnet intermediate in composition between spessartite and andradite.

space-time continuum. A geometry that includes the three dimensions and a fourth dimension of time. In Newtonian physics, space and time are considered as separate entities and whether or not events are simultaneous is a matter that is regarded as obvious to any component observer. In Einstein's concept of the physical universe, space and time are regarded as entwined, so tht two observers in relative motion could disagree regarding the simultaneity of distant events.

space wave. A radiowave that travels between a transmitting and a receiving aerial situated above the ground and that includes the direct wave and the ground-reflected wave. It is the component of the ground wave that does not travel along the surface of the earth.

spallation. A nuclear reaction in which a high energy incident particle, or photon, causes several particles or fragments to be emitted by the target nucleus. The mass number and atomic number of the target nucleus may thus be reduced by several units.

spar. A term imprecisely applied to any transparent or translucent light-coloured mineral that is somewhat lustrous and usually shows easy cleavage.

sparagmite. A groups name for the late pre-Cambrian arkoses of Scandinavia.

sparite. (1) A descriptive term for the crystalline transparent or translucent interstitial component of limestone, consisting of clean, relatively coarse-grained calcite or aragonite that either accumulated during deposition or was introduced later as a cement. (2) A limestone in which the sparite cement is more abundant than the micrite matrix.

spatter cones. Small structures built around unimportant or immature vents. Explosions throw lumps of sticky lava into the air, and on falling, they build cones or domes. Some spatter cones form on top of lava flows, where gas gathers in pockets and makes tiny explosions.

specific. (1) Denoting that an extensive physical quantity so described is expressed per unit mass. For example, the specific latent heat of a body is its latent heat per unit mass. When the extensive physical quantity is denoted by a capital letter (*e.g.*, L for latent heat), the specific quantity is denoted by the corresponding lower-case leter (*e.g.*, for specific latent heat). (2) In some older physical quantities the adjective 'specific' was added for other reasons (*e.g.*, specific gravity, specific resistance). these names are now no longer used.

specularite. Fe_2O_3. The popular name for haematite crystallized in the form of brilliant, black, metallic crystals. Also known as iron hat, grey hamatite.

speloehronology. The dating or chronology of a cave's formation, or of its mineral deposits or filling. The dating may be either relative or absolute.

speleology. The exploration and scientific study of caves, including their genesis, morphology, and mineralogy.

speleothem. A mineral deposit formed in a cave by the action of water.

spergenite. A calcarenite that contains ooliths and fossil debris having that has a quartz content not exceeding 10 per cent.

sperrylite. $4(PtAs_2)$. A mineral composed of platinum arsenide; crystallizes in cubic system, shows poor cleavage. Its hardness on Mohs scale is 6.0-7.0.

spesasartite. $8(Mn_3Al_2Si_3O_{12})$. A brownish-red mineral composed of manganese aluminium silicate, occurs as dodecahedral crystals, shows subconchoidal fracture. Its hardness on Mohs is 7.0-7.5.

sphaerosiderite. Globular concretions of fibrous chalybite.

sphalerite. (Zn,Fe)S. An isometric sulfide mineral; its colour varies from yellow or reddish brown to blackish; occurs in aggregates of distorted crystals and banded concretionary masses. Sphhalerite is found mainly in pegmatitic pneumatolytic veins, and in hydrothermal veins associated with galena, argentite, and chalcopyrite jack; pseudo-galana; zinc blende. The structure of sphalerite is analogous to that of diamond, with four zinc atoms at the points of a face-centered cubic lattice, and four sulphur atoms at the points of a face-centered cubic lattice that is displaced one-quarter along the body diagonal of the first cube. Each zinc atom is coordinated to four sulphur atoms and each sulphur atom is coordinated to four zinc atoms. The sphalerite structure is characteristic of a large number of compounds of zinc, cadmium, and mercury with sulphur.

sphene. $CaTiSiO_5$. An accessory mineral found in acid igneous rocks and in metamorphosed limestones shows; good cleavage in two directions. Its harness on Mohs scale is 5.0-5.5.

sphenoid. A tterm denoting wedge-shaped.

spicule. A minute slender pointed hard body; one of the minute calcareous or siliceous bodies supporting the tissues of a number of inverteb rates.

spike. The known amount of an isotope added to a sample to determine the unknown amount present in analysis by isotope dilution.

spilite. An igneous rock of the gabbro clan; a fine grained extrusive rock displaying pillowlava structure. It contains albite, chlorite, epidote and calcite. It is thought that spilites form from sea-floor basalts near oceanic rifts, where a metasomatic interchange with heated sea water occurs.

spilosite. *See* spotted slate.

spinel. $8(MgAl_2O_4)$. A colourless, greenish, yellow, or black mineral, usually forming octahedral crystals, also occurs as granular masses. Its hardness on Mohs scale is 7.5-8.0. It occurs in basic igneous rocks, metramorphosed aluminous rocks. Metamorphosed limestones, granite pegmatites high temperature ore veins and placer deposits. Because of its resistance to both physical and chemical weathering and erosion processes, spinel also occurs in gravels and sands.

***S-P* interval.** In earthquake seismology, the time interval between the first arrivals of longitudinal and transverse waves, which is a measure of the distance from the earth-quake source.

spit. A small point of sand or gravel projecting from the shore into a body of water; a finger-like extension of the beach. Spits are formed by the action of waves and currents, and are frequently found on irregular coastlines.

***s*-plane.** In structural geology, a nongenetic term for any planer fabric element, *e.g.*, foliation or bedding.

splash errosion. Erosion caused as a result of the impact of falling raindrops.

splent coal. *See* splint coal.

spliced. Relating to veins that pinch out and are overlapped at that point by another vein.

splint coal. A type of banded coal that is dull, blocky, and greyish-black in colour, with uneven fracture and granular texture. Quantitatively, it is defined as having more than 5 per cent anthraxylon and more than 30 per cent opaque attritus. Also known as splent coal.

split. A coal seam that cannot be mined a single unit because it is separated by a parting of other sedimentary rock. Also known as split coal.

split spread. A type of seismic spread in which the shot point is at the centre of the arrangement of geophones. It is used for continuous profiling and for dip shooting.

spodsol. A soil order characterized by a spodic or placic horizon overlying a frage.

spodumene. $4(LiAlSi_2O_6)$. A greenish-white, yellowish-green, purple, or green clinopyroxene mineral composed of lithium aluninium silicate; occurs prismatic crystals; shows uneven to subconchoidal fracture and perfect cleavage in one diredction. Its hardness on Mohs scale is 6.5-7.0. Spodumene is found in lithium-bearing pegmatites associated with quartz, feldspars, beryl, and tourmaline. It is a source of lithium and its salts.

spongework. A pattern of small irregular interconnecting cavities on walls of limestone caves.

spongolite. A rock or sediment composed mainly of the remains of sponges. Also known as spongolith.

spongolith. *See* spongolite.

spot correlation. In seismology, the correlation of reflections on isolated seismograms by observing similarities in character and interval.

spotted slate. Contact metamorphic rock, a fine-grained rock with an impressed cleavage and clotted concentrations of cryptocrystalline iron oxides in a streaked matrix of sericite, chlorite and quartz. Also known as spilosite.

spread. (1) A marsh or shallow-water resulting from the expansion in width of a stream, as where a natural obstruction impedes flow. (2) In seismology, the layout of geophone groups from which the data from a single shot are recorded simultaneously.

spring. Water from an underground source issuing at the surface.

spring line. A series of springs along the foot of a chalk or limestone hill, which may give rise to farms or even villages. The spring line generally follows the junction of the limestone or chalk and an impervious rock which makes up the lowland areas adjacent to hills.

spur. An area of high ground which protrudes into an area of lower ground such as a river valley. Spurs alongside valleys are gradually eroded by the river but in the upper and middle courses especially, they often form interlocking spurs. In galcial valley many of the spurs will have been truncated.

spurrite. $Ca_5Si_2O_8CO_3$. A white mineral composed of calcium silicate and carbonate, occurs as granular masses. It possesses vitreous lustre and shows uneven to splintery fracture and good cleavage in one direction.

stack. Erosional remnant in the form of a small bedrock island a short distance off shore. stacks are pillar-like formations detached from a headland by marine erosion.

stadial moraine. *See* recessional moraine.

staffelite. Green fibrous variety of apatite ; also synonym of francolite.

stage. (1) A phase in the development of cycle of erosion ; *i.e.,* the stages of youth, maturity, and old age. (2) A chronostratigraphic unit next in rank below series and above substage, based on biostratigraphic zones considered to approximate time-equivalent deposits ; the rocks formed during and age of geologic time. (3) A term for a major subdivision of a glacier epoch; it includes glacial stage and interglacial stage. 4. The height of a water surface above an established datum plane.

stalactite. A massive concretionery accumulation of culcium carbonate found on the roofs of limestone caves.

stalagmite. Massive concretionary mineral deposit of calcium carbonate formed on the floors of limestone caves by dripping solutions.

staurolite. $(FeMg)_4Al_{18}Si_8O_{46}(OH)_2$. A dark-reddish to brownish-black mineral composed of silicate of iron, magnesium, and aluminium, occurs as prismatic crystals; shows good cleavage in one direction. Its hardness on Mohs scale is 7.0-7.5.

steatite. 1. A compact, massive rock consisting mainly of talc but usually containing large quantities of other materials ; an impure talc-rich rock. 2. A term originally used as an alternative mineral name for talc, oftern restricted to grey-green or brown massive talc that can be easily carved.

steatite talc. A high-grade variety of talc, suitable for use in electronic insulators. It is the purest commercial form of talc.

***S*-tectonite.** A tectonite whose fabric is dominated by planer fabric elements caused by deformation, *e.g.,* slate.

steinkern. Consolidated mud or sediment that filled the hollow interior of a fossil shell or other organic structure. Also, the fossil thus formed after dissolution of the mold.

stellate. Said of an aggregate of crystals in a starlike arrangement, *e.g.,* wavellite.

stenohaline. Said of a marine organism that tolerates only a narrow range of salinities.

step fault. *See* fault.

stephanite. $4(Ag_5SbS_4)$. An iron-black mineral composed of silver antimony sulphide ; occurs as prismatic or tabular crystals; also massive. It shows subconchoidal to uneven fracture with imperfect cleavage. Its hardness on Mohs scale is 2.0-2.5.

sternbergite. $AgFe_2S_3$. A dark brown or black mineral composed of silver iron sulphide ; occurs as tabular crystals or flexible laminae.

sternberg's law. The wearing away of transported particles is proportional to their weight in water and distance travelled.

stewartite. (1) Variety of diamond. A steel-grey, iron-containing variety of bort that has magnetic properties. The high percentage of ash is due to the presence of oxide of iron. (2) A brownish-yellow mineral composed of hydrous mangenese phosephate occurring in minute crystals or fibrous tufts in pegmatites.

stibiopalladinite. Pd_3Sb. A mineral composed of palladium antimonide.

stibnite. Sb_2S_3. An antimony ore mineral, found in low-temperature hydrothermal veins. It occurs as prismatic or acicular crystals, or columnar masses, shows subconchoidal fracture with perfect cleavage in one direction. Its hardness on Mohs scale is 2.0.

stichtite.$3(Mg_6Cr_2(CO_3)(OH)_{16}.4H_2O)$. A pale-lilac to rose-pink coloured rhombohedral mineral composed of hydrous basic carbonate of magnesium and chromium; shows perfect cleavage in one direction. Its hardness on Mohs scale is 1.5-2.0.

stilbite. $2[NaCa_2Al_5Si_{13}O_{36}.14H_2O]$. A white, yellow, or brown mineral belonging to zeolite silicates ; crystallizes in monoclinic system ; occurs in sheaf-like aggregates of tabular crystals. It has pearly lustre, shows uneven fracture, and perfect cleavage in one direction. Its hardness on Mohs scale is 3.5-4.0.

stilpnomelane. $K(Fe(II),Fe(III),Al)_{10}Si_{12}O_{30}(O,OH)_{12}$. A black or greenish-black mineral composed of basic hydrous potassium iron aluminium silicate ; occurs as fibres, incrustations, and foliated plates; crystallizes in monoclinic system. It shows perfect cleavage in one direction and its hardness on Mohs scale is 3.0-4.0.

stinkstone. A stone containing decomposing matter and gives off an offensive smell when struck or rubbed.

stipoverite. *See* stishovite.

stishovite. SiO_2. A high-density tetragonal form of silica produced under static conditions of high pressure. It is the only mineral in which coordination number of silicon is six. Also known as stipoverite.

stock. An instrusive body of deep-seated igneous rock, usually discordant and resembling a batholith, except for its size. Contacts with the country rock generally dip outward and may vary from a steep to a shallow angle.

stolzite. $4(PbWO_4)$. A reddish-brown, yellow, or green mineral composed of lead tungstate ; occurs as tabular crystals; shows conchoidal to uneven fracture with imperfect cleavage. Its hardness on Mohs scale is 2.5-3.0.

stone line. A line or angular or subangular rock fragment that parallels a sloping topographic surface at a depth of several metres. It crops out in natural and artificial cuts.

stone stripe. A sorted stripe consisting of coarse rock debris, and occurring between wider stripes of finer material.

stony-iron meteorite. A general name for rare meteorites containing atleast 25 per cent by weight of both nickel-iron and heavy basic silicataes such as pyroxene and olivine, *e.g.*, mesosiderite.

stony meteorite. A general name for meteorites consisting largely or entirely of silicate minerals; *e.g.*, cohondrite and achondrite. These resemble ultramafic rocks in composition, and constitute more than 90 per cent of all meteorites seen to fall.

stope. An underground excavation formed by the extration of ore.

stoping. A process in which magma displaces country rock fragments, which then sink or are assimilated. Piecemeal stoping usually occurs on a minor scale, and is inefficient as an emplacement mechanism for large igneous rock masses. It serves rather to modify contacts of intrusions implaced in other ways.

storm beach. A low, rounded ridge of coarse gravel, cobbles, and boulders, piled up by powerful storm waves behind or at the inner margin of a beach, above the level reached by normal high spring tides or by ordinary waves.

storm berm. A low beach ridge marking the upper limit of wave action during storms.

storm surge. A rise above normal water level on the open coast due only on the water surface; includes the rise on level due to atmospheric pressure reducttion as well as that due to wind stress. Also known as storm wave; surge.

storm wave. *See* storm surge.

stoss. Said of the side of a hill, knob, or prominent rock, facing the upstream side of a glacier.

strain. The resistance to deform, that is, the tendency to resume its original shape, of a material subjected to a static or dynamic force; it increases as a function of stress. When a body is deformed by an applied stress, the strain is the ratio of the dimensional change to the original or unstrained dimension. The strain may be a ratio of length, area, or volume.

strain rosette. A paterrn of intersecting lines on a surface along which linear strains are measured to find stresses at a point.

strain seismometer. A seismometer that measures relative displacement of two points in order to detect deformation of the ground.

strain-slip. A rock fracture resulting in a slight displacement.

strait. (1) A narrow waterway connecting two layer water-bodies. (2) A neck of land.

strand. A beach bordering a sea or an arm of an ocean.

strand flat. Any wave-cut platform, especially a low, flat platform upto 65 km wide, extending for many hundred of kilometres along the rocky coast of westeern Norway, supporting thousands of stacks and other small islands.

strandline. (1) The level at which of a body of standing water meets the land. (2) A beach raised above the present sea-level.

strath. (1) An extensive remmant of a broad, flat vally floor that has undergone degradation following uplift. (2) A broad, elongate depression with steep sides on the continental shelf.

strath terrace. An extensive remnant of a strath from a former erosion cycle.

stratification. (1) The formation of layers in a mass of snow, ice, or firn. (2) An arrangement or deposition of sedimentary material in layers, or of sedimentary rock in strata. (3) The arrangement of a body of water, as a lake into two or more horizontal layers of differing characteristics, especially densities.

stratification plane. A demarcation between two layers of sedimentary rock, often signifying that the layers were deposited under different condition.

stratified drift. Fluvioglacial drift composed of material deposited by a meltwater stream or settled from suspension.

stratified flow. A two-phase folw in which a liquid flows along the bottom of a pipe and gas flows separately above it.

stratified ocean. An ocean where there is a vertical gradient of density.

stratifled rock. *See* sedimentary rock.

stratiform. (1) Descriptive of a layered mineral deposit of either igneous or sedimentary origin. (2) Consisting of parallel bands layers, or sheets.

stratigraphic range. The duration and distribution of any taxonomic group of oraganisms though geologic time, as indicated by their presence in strata whose geologic age is known.

stratigtaphic throw. The thickness of strata which originally separated two beds brought into contact at a fault. Also known as stratigraphic separation.

stratigraphic trap. Sealing of a reservoir bed due to lithologic changes rather than geologic structure.

stratigraphic unit. A stratum of rocks or a body of strata classified as a unit on the basis of character, property, or attribute.

stratigtaphy. (1) The branch of geology concerned with all characteristics and attributes of rocks as they are in strata, and the interpretation of strata in terms of derivation and geological background. (2) That aspect of the geology of an area that pertains to the character of its stratified rock.

stratovolcano. A volcano contructed of lava and pyroclastics, deposited in alternating layers. Also known as cmposite volcano.

stratum. Synonymous with bed, Rocks which display layering or bedding may be described as displaying stratification.

stratum contour. A synonym of strike line.

stray. A lenticular rock formation encountered unexpectedly in drilling an oil or a gas well; it differs from an adjacent persistent formation in lithology and hardness.

stream. A body of running water or any other moving liquid under the influence of gravity to lower levels in a narrow, clearly defined natural channel.

stream capacity. The ability of a stream to carry detritus; expressed as per unit time at a given point.

stream channel. A long, narrow slopping trough-like depression where a natural stream flows or may flow.

streamer. A non-piercing injection structure.

stream erosion. The progressive removal of exposed matter from a surface of a stream channel by a stream.

streamflow. A type of channel flow, applied to surface runoff moving in a stream.

streamline. A line which is every where parallel to the direction of fluid flow ar a given instant.

stream load. All the material that is transported by a stream. The dissolved load consists of material in ionic solution; the ions are part of the fluid and move with it. They are derived from groundwater that has filtered through weathering soil and rocks. The suspended load is comprised of clay, silt, and sand particles so fine that t' remain in suspension almost indefi- ly.

stream terrace. One of a series of level surfaces on a stream valley flanking and parallel to a stream channel and above the stream level, representing the uneroded remnant of an abondened floodplain or stream bed. Also known as river terrace.

stream transport. Movernent of rock material in and by a stream.

strengite. 8($FePO_4.2H_2O$). A colourless, red, or violet mineral composed of hydrous ferric phosphate; occurs oa octahedral or tabular crystals, also occurs as crusts or aggregates. It gives white streak and shows goods cleavange in one direction. Its hardness on Mohs scale is 3.5.

stress. Force set up inside a mass as a result of externally applied strains.

stretch. The increase in length of oilwell casing or tubing when freely suspended in fluid mediums.

stretch fault. *See* stretch thrust.

stretched pebbles. Pebbles in a sedimentary rock which have been elongated from their original shape by deformation.

stretching. In metamorphic rocks, the elongation of mineral grains, or other features; a type of lineation.

stretch thrust. A revgerse fault development as a result of shear in the middle limb of an overturned fold. Also known as stretch fault.

stria. One of a series of parallel staight lines on the surface of crystal, as in pyrite, indicative of an oscillation between two crystal forms; also one of a series of such limes on the cleavage planes of a mineral, as of plagioclase, indicative of polysynthetic twinning.

striation. The process in which scratches and groves (striae) are left by ice scraping over rocks. It is the small rock fragments in the ice which probably do most of the damage. The marks show the direction of the ice movement, and can be seen in many places where there has been recent ice activity.

strike. (1) The direction taken by a structural surface such as a fault of bedding plane as it intersects the horizontal; it is the compass direction of the horizontal line in an inclined plane. (2) To be aligned in a direction at right angles to the direction of dip.

strike fault. A fault that strikes parallel with the strike of the strata involved.

strike joint. A joint that strikes parallel to the strike or lineation of the enclosing rock.

strike separation. The distance of separation on either side of a fault surface of two formerly adjacent beds.

strike-shift fault. A fault whose direction of movement is parallel to the strike of the fault. Also known as strike-slip fault.

strike-slip. The component of the slip of a fault that is parallel to the strike of the fault. Also known as horizontal displacement; horizontal separation.

strike-slip fault. *See* strike-shift fault.

stringer lode. A lode that consists of many narrow veins in a mass of country rock.

striped ground. A pattern of alternating stripes formed by frost action on a sloping surface. Also known as striated ground; striped soil.

striped soil. *See* striped ground.

stripped plane. The upper, exposed surface of resistant stratum that resistant stratum that fro rms a stripped structural surface when extended over a considerable area.

stromobolian eruption. A type of volcanic eruption characterized by jetting of clots or fountains of fluid basaltic lava from a certral crater.

stronatianite. $SrCO_3$. A colourless or grey mineral composed of strontium carbonate, found in veins, as nodules in limestones and as concretionary masses; occurs as prismatic crystals in fibrous masses.

strontium unit. A measure of the concentration of strontium-90 in an organic medium (e.g. milk, bone, soil, etc.) relative to the concentration of calcium in the same medium.

1 SU = 10–12 curie of strontium - 90 per gram of calcium.

strouhal number. A quantity used in the study of vibrations of a body past which a fluid is flowing; it is equal to the characteristic dimension of the body times the frequency of vibrations divided by the fluid velocity relative to the body. It is also called reduced frequency.

structural. Of or pertaining to rock deformation or to features that result from it.

structural contorl. The influence of structurtural features on ore deposition, *e.g.*, ore minerals filling fractures.

sturctural crystallography. Study of the internal arrangement and spacing of atoms and molecules composing crystalline solids.

structural feature. A feature produced as a result of deformation or displacement of rocks, such as fold or fault.

structural relief. The vertical distance between stratigraphically equivalent points at the crest of an anticline and in the trough of an adjacent syncline.

structural terrace. (1) A terrace-like landform conrtolled by the structure of the underlying rocks, especially, one produced by the removal of weaker strata from more resistant rocks in a formation with horizontal bedding. (2) A local shelf or steplike flattening in otherwise uniformly dipping strata.

structural trap. A trap for oil or gas that is the result of folding, faulting, or other deformation.

structure. (1) A general term relating to the terrain underlying a landscape. (2) The arrangment of various constituents present in a mineral. (3) Feature or interrelationships in rocks of a larger scale than texture, as foliation and cross laminarion. (4) Subsuface folds, faults, unconformities or other arrangements of rock exhibiting a potential for containing petsduems or natural gas. (5) Attitude of beds, surface and rock bodies and geometry of changes in these factors as well as their inter-relationships.

structure contour. A synonym of strike line.

stylolite. An irregular surface within a bed, usually calcite developed in marks and limestones; characterized by pits and teethlike projections on one side that fit into counterpart negatives on the other. Stulolite seams are formed of material less soluble than limestone, such as clay or organic matter.

sub-. A prefix denoting under.

subaluminous. In Shand classification of igneous rocks, division embracing those rocks in which there is little or no excess of aluminium oxide over that reqired to from feldspars or feldspathoids.

subangular. Somewhat angular but free from sharp edges and corners.

subaqueous. Pertaining to conditions and processes occurring in, under, or beneath the surface of water, especially fresh water.

subarkose. Sandstone that is intermediate in composition with arkose and pure quartz sandstone; it contains less feldspar than arkose.

subconchoidal. Pertaining to a fracture that is partly or vaguely conchoidal in shape.

subcrop. An occurrence of strata beneath the subsurface of an inclusive stratigraphic unit that succeeds an unconformity on which there is marked overstep.

subduction. Movement of one crustal plate (lithospheric plate) under another. Subduction refers to the process, not the site.

suberinite. A variety of provitrinite composed of corky tissue.

subfeldspathic. Referring to mature lithic wacke or arenite containing an abundance of quartz grains with less than 10 per cent feldspar grains.

subgelisol. Unfrozen ground beneath permafrost.

subglacial. Pertainning to an area in or at the bottom of, or immediately beneath, a glacier.

subgreywacke. An argillaceous sandstone with a composition intermediate between greywacke and orthoquartizte; a clay matrix usually upto 15 per cent is also present.

subhedral. Descriptive of a crystal having partially developed crystal faces. 2. Pertaining to an individual mineral crystal that is partly bound by its own crystal faces and partly bound by surfaces against preexisting crystals.

submarine canyon. A V-shaped subnarine valley resembling land canyons that have been cut by streams. It is characterized by high, steep walls and an irregular floor that slopes continually outwaard.

submarine delta. *See* submarine fan.

submarine eruption. An eruption from a volcano located on the oceanic crust. Because of their location, the magmas of such volcanoes are basaltic.

submarine fan. A shallow marine sediment that is fan- or coneshaped and lies off the seaward opening of large rivers and submarine canyons. Also known as abyssal cave; sea fan; submarine deta.

submarine pit. A cavity on the bottom of the sea. Also known as submarine well.

submarine weathering. A slow alteration of the form, composition, and texture of the sea-floor as a result of chemical, biological, and thermal causes.

submarine well. *See* submarine pit.

subsequent stream. A stream which is tributary to the consequent. The consequent stream could have been the main stream to develop on a slope. Subsequent streams generally flow at right angles to the consequent, and generally follow the strike of the rocks. They often flow along the softer and weaker rocks, and may carve larger valleys than the consequent streams.

subsoil. The layer which is situated below the top soil. It may not contain any organic matter, but is likely to include fragments of rock. It is approximately equivalent to the *C*- horizon.

subsolvus granites. Granites containing two feldspars.

sulphides. The minerals containing sulphur (II) in combined form. The sulfides include about eighty species, but only those of iron, copper, and zinc are abundant. The remaining sulphides are uncommon or rare, particularly in specimens suitable for collections. Sulphides concentrate in hydrothermal ore veins, tending to form small to large masses of several species together and, less commonly, substantial masses in which one sulfide predominates. Mostly, veins are filled wall-to-wall and euhedral crystals are diffcults to obtain despite the abundance of the sulphides concerned. As a class, sulphides are characterized by dark colour, metallic lustre, high specific gravitiy, and low tenacity. Most are therefore easily broken, and readily produce dark streaks on a porcelain test plate. Additionally, many fuse easily on charcoal and otherwise react distinctively to various tests. Practically all are opeque or at best translucent.

sulphur ball. (1) A pyritic impurity in coal, occurring as spheroidal or irregular mass. (2) A sulphurous mud skin that forms on a bubble of hot volcanic gas and becomes firm on contact with air.

sunstone. Reddish-grey gem variety of oligoclase. Also known as heliolite.

supergene. A term applied to ores or ore minerals formed by processes that almost always involve water, with or without dissolved material, descending from the surface.

supermassive star. A star with a mass exceeding about .0 times that of the sun.

supernova. An explosive brightening of an star in which the energy radiated by it increases by a factor of about 1010. It takes several years to fade and while it lasts dominates the whole galaxy in which it lies. It is estimated that there could be a supernova explosion in the Milky Way every 30 years, although only six have actually been observed in the last 1000 years. A supernova explosion occurs when a star has burnt up all its available helium and the core collapses catastrophically.

superstructure. The upper structural layer of an orogenic belt, which is subjected to relatively shallow or nearsurface deformation processes, as distinct from an underlying and more profoundly deformed infrastructure.

surf. Wave activity in the area between the shoreline and the outermost limit of breakers, that is, in the surf zone.

surface creep. A stage of the wind erosion process in which grains of sand move each other along surface.

surface current. Any current whose maximum velocity core is at or near the surface.

surface run-off. The portion of rainfall which runs off either as channel flow or as overland flow, but does not percolate into the soil and rocks, and is not evaporated into the atmosphere. Also called immediate run-off .

surface wave. (1) A seismic wave that travels along the surface of the earth, or along subsurface interface; includes Love wave, Rayleigh wave, and coupled wave. (2) A progressive gravity wave in which the disturbance is of gtreatest amplitude at the air water interface.

surge. *See* storm surge.

sussexite. (1) An igneous rock of the syenite clan. A medium grained porphyritic hypabssal rock; contains phenocrysts of nepheline in a groundness of nepheline, orthoclase, aegirine, and perovskite. (2) $MnBO_2OH$. A white mineral composed of basic manganese borate, occurring in filbrousveins.

svabite. $Ca_5(AsO_4)_3F$. A colourless, yellow or reddish-brown mineral composed of fluoride and arsenate of calcium.

svanbergite. $SrAl_3(PO_4)(SO_4)(OH)_6$. A colourless, white, or yellow mineral composed of basic phosphate and sulphate of strontium and sulphate of strontium and aluminium; isomorphous with corkite and woodhouseite.

swale. (1) A low place in a land area, usually marshier than the adjacent higher land. (2) A depression in an undulation ground moraine, due to uneven glacial deposition. (3) A long, narrow depression between beach ridge.

swallow-hole. A large pothole in a limestone area down which a river is flowing. The hole will be an enlarged joint, which has been weathered by solution and also eroded by running water.

swartzite. $CaMg(UO_2)(CO_3).12H_2O$. A green, monoclinic mineral composed of hydrous carbonate of calcium, magnesium, and uranium.

swash. 1. A bar over which the sea washes. 2. A narrow channel or ground within a sand bank, or between a sand bank and the shore. 3. The rush of water up onto the beach following the breaking of a wave.

swash mark. A thin delicate wava or arcuate line of very small ridge on a beach, marking the farthest advance of wave uprush. It is convex landward and contains of fine sand, mica flakes, bits of seaweed, and other debris.

***S*-wave.** A seismic body wave propagated in the crust or mentle of the earth by a shearing motion of material; its speed is 3-4 km/s in crust and about 4.4-4.6 km/s in the mantle. Also known as distortional wave; rotational wave; secondary wave; shake wave; shear wave; tangential wave; transverse wave.

sweet. Said of crude oil or natural gas that contains very little or no sulphur compounds.

syenite. An igneous rock of the syenite clan. A medium to coarse-grained granular plutonic or hypabyssal rock; contains orthoclase, mirocline, sodaclase, biotite and diopside. Syenite is the intrusive equivalent of trachyte; with an increase in quartz, it grades into granite.

sylvanite. $2[AgAuTe_4]$. A steel grey, silver-white, or brass-yellow mineral that crystallizes in monoclinic system and often occurs in implanted crystals; composed of silver gold telluride. It shows uneven fracture and perfect cleavage in one direction. Its hardness on Mohs Scale is 1.5-2.0.

sylvine. $4[KCl]$. (1) A white or colourless isometric mineral, occurring in cubes or crystalline masses or as saline residue; shows uneven fracture and perfect basal cleavage. Its hardness on Mohs scale is (2) Also known as leapoldite; sylvite.

sylvite. *See* sylvine.

symmetry element. Some combination of rotations and reflections and translations which brings a crystal into a position that cannot be distinguished from its original position. Four elements of symmetry are recognised in crystallography:

symplesite. $Fe_2(AsO_4)_3$. $8H_2O$. A blue or bluish-green triclinic mineral composed of hydrous ferric arsenate.

syn-. A prefix denoting with; at the same time.

synadelphite. $(Mn,Mg,Ca,Pb)(AsO_4)(OH)_5$. A black mineral composed of basic arsenate of manganese, often associated with calcium, magnesium, lead or other metals.

sychusite. $(Ce,La)Ca(CO_3)_2F$. A mineral composed of fluoride and carbonate of cerium, lanthanum and calcium.

synclinal axis. The line which moved parallel to itself, generates the form of a syncline.

synclinal mountain. A mountain whose underlying structure is that of a syncline.

syncline. A fold of which the core contains the stratigraphically younger rocks; it is generally concave upward.

synclinorium. A composite synclinal structure in a region of lesser folds.

syncology. The study of relationships between communities and their environments.

synegabbro. An intrusice rock differing from gabbro by the presence of alkali feldspar.

syneresis The spontaneous throwing-off of a liquid by a gel during aging, resulting in shrinkage and in the formation of cracks.

synform. A *U*-shaped fold whose limbs close downward in strata for which the stratigrsphic sequence is unknown.

syngenesis. The formation of unconsolidated sediments in place; early diageneis.

synegnetic. (1) Said of a primary structure, such as a ripple mark, formed contemporaneously with the deposition of the sediment in which it occurs. (2) Said of a mineral deposit formed at the same time, and by the same processes, as the enclosing rock.

syngenite. $2(K_2Ca(SO_4)_2..H_2O)$.A colourless or milky-white monoclinic mineral composed of hydrous potassium calcium sulphate; shows conchoidal fracture and perfect cleavage in two directions. Its hardness on Mohs scale is 2.5.

synorogenic. Said of a geologic process or event occurring during a period of orogenic activity; or said of a rock or frature so formed.

syntectic. *See* syntexis.

syntectonic. Refers to a geologic process or event occurring during tectonic activity. Also known as synkinematic.

syntexis. Generation of magma by the melting of two or more rock type, and the assimilation of host rocks.

syserskite. A mineral composed of an alloy of osmium (50-80%) and iridium (20-50%).

szechenyiite. A variety of richterite containing iron and aluminium.

szmikite. A monoclinic mineral composed of hydrous manganese sulphate.

szomolnokite. $FeSO_4.H_2O$. A yellow or brown mineral composed of hydrous ferrous sulphate.

tabbyite. A variety of solid asphalt; used as rubber filler with roofing materials.

tableland. A broad, elevated, almost level and extensive region of land that has been deeply cut at intervals by valleys or broken by escrapments. Also known as continental plateau.

table reef. A small isolated organic reef which has a flat top and does not enclose a lagoon.

tabula. A transverse septum that closes off the lower part of the polyp cavity incertain extinct corals and hydroids.

tabular. (1) Said of a feature having two dimensions that are much larger or longer than the third, such as igneous dike or plateau. (2) Said of the shape of a sedimentary body whose width/thickness ratio is greater than 50 to 1 but less that 1000 to 1. 3. Said of a sedimentary particle or a crystal form that shows one dimension markedly smaller that the other two. 4. Said of a metamorphic texture in which a large proportion of grains are tabular and have approximately parallel orientation.

tachydrite. $CaMg_2Cl_6.12H_2O$. A colourless or yellowish, hexagonal mineral composed of hydrous chloride of calcium and magnesium; shows perfect cleavage in one direction. Its hardness on Mohs scale is 2.0.

tachygenesis. The extreme crowding and eventual loss of those primitive phylogenetic stages that are represented early in the life of an individual.

tachylyte. Name given to natural glass in extrusive ignerous rocks. Also known as hyalobasalt; jaspoid; sordawalite; wichtisite.

tachymeter. A surveying instrument designed for use in rapid determination of distance, direction, and difference of elevation from a single observation, using a short base, which may be an integral part of the instrument. Rangefinders with self-contained bases belong to this class.

taconite. The siliceous iron formation from which high grade iron ores of the Lake Superior district have been derived; consists chiefly of fine-grained silica mixed with magnetic and hematic.

tactite. A rock of complex mineralogical composition, formed by contact metamorphism and metasomatism of carbonate rocks. It is typically coarse grained and rich in garnet, ironrich pyroxene, epidote, wollastonite and scapolite.

taenite. Cubic nickel iron mineral; also containing about 12 per cent nickel; found in meteorites.

tahitite. An igneous rock of the diorite clan. A fine-grained porphyritic extrusive rock; contains phenocrysts or hauyne in a groundmass of analcite glass, augite, hauyne, orthoclase and leuctic.

taimyrite. An igneous rock of the syenite clan. A medium-grained granular extrusive rock; contains nosean, anorthoclase, biotite and sanidine.

talasskite. A brown variety of fayalite containing about 12 per cent ferric oxide.

talc. $Mg_3Si_4O_{10}(OH)_2$. (1) A very soft, white, greenish-white, grey or brownish mineral, with a hardness of about 1 on the Moh scale; found in foliated, fibrous, or granular masses as an alteration product of magnesium silicates or ultramafic rocks. It is also formed by metasomatism in impure dolomitic marblels. Its massive varieties are potstone and steatite; while iron rich variety is minnesotaite. (2) In commercial usage, a rock containing talc, chlorite, tremolite and related minerals. Talc has many commercial uses; the most familiar, although one that consumes comparitively very little of the mineral, is as talcum powder. It has very low conductivity for heat and electicity, is fire resistant, hardens when heated to high temperature, and is not attacked by most of the acids, are some of the properties, that are valuable in most of the industrial applications. It is also used as a lubricant, in ceramics, and as a filler in paint, paper, and rubber.

talcose rock. A rock having a soft and soapy feel, that is, resembling talc.

talc schist. A type of schist containing talc as the dominant schistose mineral; variety is dolerine.

talus. Rock fragments usually coarse and angular, lying at the base of a cliff or steep slope from which they have been derived; also the heap or mass of such broken rock, considered as a unit. Also known as scree.

talus creep. The slow downslope movement of talus, either in individual rock fragments or the moss as whole.

tamaraite. An igneous rock of the gabbroclan. A fine-grained granular or porphyric hypabyssal rock; contains augite, biotite, hornblende, nepheline, plagioclase and orthoclase.

tantalite-coulambite series. $(Fe,Mn)(Nb,Ta)_2O_6$. These minerals form an almost continuous series of solid solutions within the range shown in the formula. Iron or manganese may predominate in either mineral, and magnesium may be a major component of coulambite. These minerals occur in granite pegmatites, particularly in those that contain albite, lithium silicates, and Li-Mn-Fe phosphate minerals. The commonly associated minerals are albite, microcline, beryl, lepidolite, muscovite, tourmaline, spodumene, amblygonite, apatite, microlite, and cassiterite.

taphrogeosyncline. A synonym of rift valley.

tapiolite. $2[FeTa_2O_6]$. A tetragonal mineral composed of iron tantalate; occurs as primatic crystals; shows uneven to subconchoidal fracture without any cleavage. Its hardness on Mohs scale is 6.0–6.5.

taramite. $(Ca,Na,K)_3(Fe(II),Fe(III))_5(Si,Al)_8O_{22}(OH)_2$. A monoclinic mineral composed of hydrous calcium sodium potasium and iron aluminosilicate; occurs as prismatic crystals; shows perfect cleavage in one direction. Its hardness on Mohs scale is 5.0.

tarn. A small lake formed in a cirque; also called a cirque lake.

tar pit. An accumulation of natural bitumen that is exposed at the land surface, and becomes a trap into which animals can sink; the hard parts of the carcasses are preserved.

tar sand. A sandy area that is large enough to contain a commercial store of asphalt; it may be an oil sand from which more volatile materials have escaped.

tasmanite. An impure coal transitional between cannal coal and oil shale; rich in sulphur.

tautonym. The name of species in which the term designating the genus is the same as that for species.

tavolatite. An igneous rock of the syenite clan. A fine-grained porphyritic extrusive rock; contains phenocrysts of leuctic,

hauyne, and augite in a groundmass of **leuctie**, hauyne, orthoclase, labradorite, biotite, aegirine- augite and garnet.

tawite. A feldspathoid igneous rock; a coarse-grained granular plutonic rock; contains sodalite and aegirite.

tawmawite. A chromium rich variety of epidote.

taxis. A suffix denoting arrangement.

taylorite. *See* bentonite.

teallite. 2[$PbSnS_2$]. An orthrhombic mineral composed of lead tin sulphide ; occurs as thin tabular crystals and massive foliated aggregates. It shows black streak and perfect basal cleavage. Its hardness on Mohs scale is 1.5.

tear fault. A fault in which the movement is along the horizontal. The amount of movement may only be a few metres, but can also be several kilometres in certain cases. This type of faults are frequent in areas of tectonic activity.

tectogenes. 1.A narrow, elongate unit of downfolding of sialic crust believed to be related to mountain-building processes. 2. Deficiency of gravity measured over the sites of mountain ranges led to an earlier belief in roots of mountains, or crumpled and thickened portions of the outer earth shell that had been compressed and forced down into the sima. This concept of 'sialic roots' was later extended to cover similar gravity deficiencies associated with island arcs and the name 'tectogene' was proposed for the elongated zone of crustal thickening and downfolding thought to be associated with mountain building. Plate tectonics has helped in understanding the relationship between mountain roots and orogeny. Tectogene is little used at present because of its strong association with invalid aspects of the geosynclinal theory of mountain building receding plate tectonics.

tectonic. Relating to structures of or forces associated with tectonics.

tectonic breccia. An aggregation of angular rock fragments that have formed as a result of tectonic movement; a crush breccia.

tectonic style. The total character of a group of related structures that distinguishes them from other groups of structures in the same way that the style of a building or an art object distinguishes it from others of different periods or influences.

tectonics. A branch of geology that is closely related to but not synonymous with structural geology and deals not only with larger features of the earth, but also with the forces and movements that produced them.

tectonite. A rock in which crystallization has taken place under stress, resulting in the mineral constituents taking up a preferred orientation or fabric.

tectonosphere. The zone of layer of the earth above the level of isosatic equilibrium,in which crustal or tectonic movements originates.

tectosilicate. A class of structural type of silicate characterized by the sharing of all four oxygens of the SiO_4 tetrahedra with neighbouring tetrahedra, and with a Si:O ratio of 1:2, *e.g.*, quartz, SiO_2.

tectosome. A body of strata indicative of uniform tectonic condition ; the sedimentary rock record of a tectonic environment or tectotope.

tectotope. An area of uniform tectonic environment, recognition of which depends on interpretation of a tectosome. The term is of very little importance.

teilzone. A time term originally introduced to disignate the local duration of existence of a species.

tektites. Black or green natural glass

containing 73 per cent silica. It may be button discoid-, or teat-shaped. Its fused appearance may be the result of entering the earth's atmosphere; possibly a result of meteorite impact of desert sands.

telain. *See* provitrain.

temperature. The property of a body or region of space that determines whether or not there will be a net flow of heat into it or out of it from a neighbouring body or region and in which direction (if any) the heat will flow. If there is no heat flow the bodies or regions are said to be in thermodynamic equilibrium and at the same temperature. If there is a flow of heat, the direction of the flow is from the body or region of higher temperature. The temperature of a body is a measure of its 'hotness', which can be defined as a property determining the rate at which heat will be transferred to or from it. Temperature is thus a measure of the kinetic energy of the molecules, atoms or ions of which matter is composed. the basic physical quantity, The thermodynamic temperature, is expressed in kelvins.

tender. A barge anchored alongside an offshore drilling platform.

tennantite. 8[Cu_3AsS_3]. A cubic mineral composed of copper arsenic sulphide ; occurs as tetragonal crystala as well as granular masses. It shows black to brown streak and subconchoidal to uneven fracture and does not show cleavage. Its hardness on Mohs scale is 3.7-4.5.

tenorite. CuO. A monoclinic mineral composed of cupric oxide ; occurs as elongated lath-like crystals; shows conchoidal to uneven fracture and poor cleavage. Its hardness on Mohs scale is 3.5.

tensile strength. The tensile (pulling) stress that has to be applied to a material to break it. It is measured as a force per unit area ; *e.g.*, newtons per square metre; dynes per square centimetre ; pounds or tons per square inch.

tephra. All fragmental material ejected from a volcanic vent and transported through the air, *e.g.*, ash, pumice, bombs, etc; a synonym for pyroclasic material. ash, cinders, lapilli, scoriae, pumice bombs *etc*.

tephrite. An igneous rock of the gabbro clan. A fine-grained porphyritic extrusive rock ; contains phenocrysts of augite and labradorite in a groundmass of augite, labradorite, leucite, and nepheline.

tephroite. Mn_2SiO_4. A mineral of the olivine group. It shows subconchoidal fracture and good cleavage in two directions. Its hardness on Mohs scale is 5.5-6.0

terminal moraine. The outermost end moraine of a glacier or ice sheet, marking maximum advance of the ice.

terminus. The outer margin or extremity of a glacier.

terra. An upland or mountainous region on the surface of the moon, characterized by a lighter colour than that of a mare, by relatively high albedo and by a rough texture formed by large overlapping craters.

terrace. (1) A benchlike structure bordering an undersea feature. (2) A horizontal or gently sloping embankment of earth along the contours of a slope to reduce erosion,control run off, or conserve moisture. (3) A long narrow , nearly level surface bounded by a steeper descending slope on one side and by a steeper ascending slope on the other side. (4) A narrow coastal strip sloping toward the water.

terraced pool. One of the shallow, rimmed pools on a reef, produced by the growth of lime and silica-secreting algae, and arranged in successively lower terraces.

terracette. A small steplike form developed on the surface of a slumped soil mass along a steep grassy incline.

terra-cotta. A brownish-orange clay used in the production of high-quality earthenware, vases, and statuettes, and for tile-floors and roofing.

terrain. A region of the earth's surface that is treated as a physical feature or as a type of environment.

terra rosa. A reddish soil developed on limestone in areas of mediterranean climaate; the colour of thesoil is die to the presence of iron hydroxides. Terra rosas largely comprise insoluble particles within the limestone which have been exposed as the surface has been lowered by carbonation. It is thus essentially a residual soil, though there has usually been some lateral movement of the soil particles into valley bottoms and enclosed hollows such as dolines. Terra rosas have been formed over a long period embracing moist conditions than at present (for example , Pleistocene interglacial periods or even a warm humid climate in the lateTertiary.

terrestrial sediment. A sedimentary deposit on land above reach.

terrigeneous. Derived from the land or a continent.

terrigeneous sediment. Shallow marine sedimentary deposits composed of eroded terrestrial material.

tertiary. A geological period that is part of the Cenozoic Era.

teschenite. An igneous rock of the gabbro clan; a fine to coarse grained hypabyssal rock; contains hornblende, augite, labradorite, and analcite. Barkevicite is common, and olivine- teschenites are of frequent occurrence; the presence of alkali feldspar feldspar gives rise to the variety glenmuirite. Bekinkinite is a nepheline-bearing teschenite often containing barkevicite. Crinanite is a term which has been used for rocks so similar to teschenite that it appears to be superfluous.

test. (1) A drill-stem test or other procedure for sampling the content of an oil or gas reservoir. (2) The shell or supporting skeleton of many invertebrates, *e.g.*, of an echinoid or a foraminifer.

test well. A well drilled to determine the presence of petroleum oil and its potential commerical value in terms of abundance and accessibility.

tetradymite. Bi_2Te_2S. A hexagonal mineral composed of bismuth tellurium sulphide; normally occurs as foliated or granular masses, shows perfect cleavage in one direction. Its hardness on Mohs scale is 1.5-2.0.

tetragonal lattice. A crystal lattice in which the axes of a unit cell are perpendicular, and two of them are equal in length to each other, but the third axis is different in length.

tetrahedrite. $(Cu,Fe.Zn,Ag)_{12}Sb_4S_{13}$. A greyish black mineral crystallizing in the isometric system as tetrahedrons, and occurring in massive or granular form; found with other copper ores in hydrothermal veins. It shows subconchoidal to unven fracture and does not show any cleavage. Its hardness on Mohs scale is 3.0-4.0.

tetrahedron. An isometric crystal form in cubic crystals, in the shape of a four-faced polyhedron, each face of which is a triangle.

tetrahexahedron. A form of regular crystal system with for triangular isosceles faces on each side of a cube; there are altogether 24 congruent faces.

tetrakalsilite. A synonym of meionite.

textural maturity. A type of sedimentary maturity in which a sand approaches the textural end product to which it is driven by the formative processes that operate upon it. It is defined in terms of uniformity of particle size and perfection of rounding, and is independent of mineral composition.

thalweg. (1) The line connecting the lowest points along a stream bed or valley; a longitudinal profile. (2) The line of continuous maximum descent from any point on a land surface, *e.g.*, the line crossing all countour lines at right angles. (3) The ground water stream percolating benath and in the same direction as a surface stream. (4) The deepest or best navigable channel used in defining water boundaries between states.

thaw lake. A shallow, rounded depression containing a circular or semi-circular pond in a lowland periglacial region. Thaw lakes are very common features of thermokarst, and result from the surface thawing of frozen ground (for example, at points where he vegetation cover has been disrupted). This produces localized collapse, and the formation of an irregular depression which becomes occupied by meltwater.

thenardite. 8[Na_2SO_4]. A colourless or milky-white orthorhombic mineral composed of sodium sulphate; shows uneven to jagged fracture and perfect cleavage in one direction. Its hardness on Mohs scale is 2.5-3.0.

thermal. (1) Pertaining to or caused by heat. (2) A vertically moving current of air that is caused to rise by differential heating of the ground below it. (3) An interglacial interval.

thermal belt. Any of several possible horizontal belts of a vegelation type found in mountainous terrain, resulting primarily from vertical temperature variation. Also known as thermal zone.

thermal diffusion. A phenomenon in which a temperature gradient in a mixture of fluids gives rise to flow of one constituent relative to the mixture as a whole. Also known as thermodiffusion.

thermal fracture. The cracking of a rock surface owing to alternate heating and cooling; since diurnal heating by the sun's rays is believed to be involved. The process is also referred to as insolation weathering and may lead to exfoliation. It has been suggested that; (i) dark-coloured rocks (which absorb solar energy most effectively), and (ii) heterogeneous crystalline rocks (comprising crystals with different colours and coefficients of expansion are most pron to thermal cracking. However , it is increasingly believed that, even in hot deserts where ranges to temperature from day to night are at a maximaum , the physical beakdown of rock surfaces is due more to chemical processes such as hydration and salt weathering.

thermal gradient. The rate of temperature increase with depth below the surface of the earth.

thermal metamorphism. Metamorphism that results from chemical reconstitution controlled by a temperature increase, and influenced to a lesser extent by confining pressure; there is no requirement of simultaneous deformation.

thermal spring. A spring whose water temperature is higher than the local mean annual temperature of the atmosphere.

thermal stratification. Horizontal layers of differing densities produced in a take by temperature changes at different depths.

thermal structure. A distinct structural

pattern, defined by the arrangement of metamorphic zones of increasing grade.

thermocline. (1) A layer in a thermally stratified body of water in which such a gradient occurs. (2) A temperature gradient as in a layer of sea water, in which the temperature decrease with depth is greater than that of the overlying and underlying water. Also known as metalimnion.

thermodynamic property. A quantity which is either an attribute of an entire system or is a function of position which is continuous and does not vary rapidly over microscopic distances, except possibly for abrupt changes at boundaries between phases of the system, *e.g.*, concentration, pressure, temperature, volume, surface tension, and viscosity. Also known as macroscopic property.

thermokarst. The formation in periglacial environments of a highly irregular ground surface, as a result of the thawing of masses of ground ice. Hummoocks , pits and larger enclosed depressions may become so numerous that there is a crude resemblace to the karstic features of limestone- though the processes of formation are totally different. Among the main features of thermokarst are alases (and alas valleys), pinggos ansd thaw lakes. The development of thermokarst landscapes results either form a warming of the climate or form a removal of the vegetation cover which insulates and protects ground ice bodies.

thermoluminescent. A term applied is a substance which emits characteristic light when warmed. It is not a true form of phosphorescence.

thermonatrite. 4[$Na_2CO_3.H_2O$]. An orthorhombic mineral composed of hydrous sodium carbonate; shows poor cleavage. Its hardness on Mohs scale is 1.0-1.5.

thick bedded. Pertaining to a sedimentary bed that ranges in thickness from 60 to 120 centimetres.

thin bands. A field term that, in accordance with a scale etablished for use in describing banded coal, denotes vitrain bands with a range of thickness from 0.5 to 20 mm.

thin bedded. A relative terms applied to sedimentary beds with a thickness in the range of 5-6 cm; a bed less than 5 cm but; more than 1 cm thick being very 'thin bedded'.

thinolite. (1) A yellow or brown variety of calcite, often terminated at both ends by pyramids. (2) A tufa deposit, consisting in part of layers of delicate prismatic skeletal crystals of the mineral thinolite upto 20 cm long and 1 cm thick; it occurs in domelike masses along the shores of extinct Lake Lahontan, northwest Nevada.

thin out. Gradual thinning of a stratum, vein or other body of rock until the upper and lower surfaces meet and the rock disappears.

thixotropy. Ability of a fluid to develop gel strength with time ; *i.e.*, property of a fluid causing it to build up a gel-strucuture if allowed to stand undisturbed.

tholeiite. An igneous rock of the gabbro clan. A medium-grained granular hypabyssal rock; it contains labradorite, augite, ilmenite and glass.

thomsonite. 4[$NaCa_2Al_5Si_5O_{20}.6H_2O$]. A snow-white zeolite mineral forming orthorhombic crystals and occurring in masses of radiating crystals. It shows uneven to subconchoidal fracture and perfect cleavage in one direction. Its hardness on Mohs scale is 5.0-5.5.

thorianite. 4[ThO_2]. A radioactive mineral that crystallizes in isometric system; occurs in worn cubic crystals. It shows uneven to subconchoidal fracture and poor cleavage. Its hardness on Mohs scale is 6.5.

thorite. $ThSiO_4$ A reddish-brown to black mineral consisting of silicate of thorium; occurring as irregular grains or masses. In thorite, some of the thorium may be replaced by uranium, and some of the silicate may be replaced by $(OH)_{44}$-. Nearly all the thorite is metamict, and is strongly radioactive. It is an accessory mineral in granites, syenites, and nepheline syenites, and in pegmatites associated with these rocks. It also occurs as a detrital mineral in sands.

throughflow. The sub-surface movement of water in a downslope direction, by contrast with the vertical movement of percolating rainwater towards the water table. Throughflow may be concentrated in the subsoil, especially where it lies above relatively impermeable bedrock, and may follow matural pipes in the soil (percolines). Throghflow is probabaly a far more important process than overland flow in the disposal of rainwater in hill slopes in humid temperate regions .

through valley. (1) A valley excavated by a through glacier. (2) A depression eroded across a divide by glacier ice or meltwater streams.

thrust fault. A low-angle (less than 45°) fault along which the hanging wall has moved up relative to the footwall. Also known as reverse fault ; reverse slip fault; thrust-slip fault.

thrust nappe. The body of rock that makes up the hanging wall of a thrust fault. Also known as thrust block; thrust plate; thrust sheet; thrust slice.

thrust plate. *See* thrust nappe.

thrust sheet. *See* thrust nappe.

thrust slice. *See* thrust nappe.

thrust-slip fault. *See* thrust fault.

thucholite. Mixture of hydrocarbons, ash, and uraninite, occurring as irregular nodules. It shows conchoidal fracture and does not show any cleavage. Its hardness on Mohs scale is 3.5-4.0. Also known as tucholite.

thulite. A pink or red variety of epidote that contains manganese ; used as an ornamental stone.

thuringite.2[(Fe(II),Fe(III),Mg,Al)$_6$(Si, Al)$_4$O$_{10}$(O,OH)$_8$]. A green, monoclinic mineral composed of hydrous iron magnesium aluminium aluminosilicate. It shows subcondchoidal fracture and good cleavage in one direction. Its hardness on Mohs scale is 2.5.

tidal channel. A major channel followed by tidal currents, extending from the oceaninto a tidal marsh or tidal flat.

tidal current. The alternating horizontal movement of water associted with the rise and fall of the tide caused by the astronomical tide-producing forces.

tidal delta. A sandbar of shoal formed in the entrance of an inlet by the action of reversing tidal currents.

tidal flat. A marshy, sandy, or muddy nearly horizontal coastal flatland which is alternately covered and exposed as the tide rise and falls.

tidal inlet. A natural inlet maintained by tidal currents.

tidalite. Any sediment transported and deposited by tidal currents.

tidal scour. Sea-floor erosion caused by strong tidal currents, resulting in removal of inshore sediment and formation of deep holes and channels. Also known as scour.

tide. The twice-daily (approximately) rise and fall in the level of the sea, resulting form the gravitational attraction on the earth's oceans exerted by the sun and moon. Tidal currents are stream-like flow of sea water into and out of bays and esturaries with the rising and falling of the tide (producing respectively flood currents and ebb currents) .

tideland. Land which is under water at high tide and uncovered at low tide.

tidewater glacier. A glacier that descnds into the sea and usually has a terminal ice cliff. Also known as tidal glacier.

tiemannite. 4[HgSe]. A granular mass composed of mercury selenide; shows uneven to conchoidal fracture and does not show any cleavage. Its hardness on Mohs scale is 2.5.

tiff. Originally a colloquial term for the mineral ($BaSO_4$) barite. The term has also been used in the field of drilling.

tiger's eye. A yellowish brown crystalline variety of quartz; a tranaslucent, fibrous, broadly chatoyant gemstone that may be dyed.

till. Unsorted and unstratified drift consisting of a hethrogeneous mixture of clay, sand, gravel, and boulders which is deposited by and undernath a glacier. Also known as boulder clay; glacier till; ice-land drift.

till-below. An undulating mass glacial drift that is disposed in an irregular pattern with regard to the direction of movement of the ice.

tillite. 1. Aterm used for pre-Pleistocene tills that have been indurated or conolidated by normal post-depositional diagenetic processes. 2. The term has also been used for cemented till, recognised by an almost complete lack of sorting, bedding, and generally by the presence of stiae on atleast a few of the largest pebbles and boulders.

tiloid. (1) A nonglacial till-like deposit. (2) A rock of uncertain origin which resembles tillite.

till plain. An extensive, relatively flat area overlying a till.

timazite. An altered igneous rock of the diorite clan. A fine- grained porphyritic extrusive rock; it contains phenocrysts of hornblende and biotite largely altered to augite and magnetite in a groundmass of plagioclase and quartz.

tinguaite. An igneous rock of the syenite clan. A fine grained porphyritic hypabyssal rock; it contains phenocrysts of orthoclase, nepheline and aegirine in a groundmass of orthoclase, nepheline, aegirine, leucite and katophorite.

titanite. $CaTiSiO_5$. A brown, grey, or green mineral consisting of silicate of titanium and calcium, also the calcium may be partly replaced by sodium or rare-earth elements, while titanium may be replaced by Al, Fe, Zr, Nb, Ta,Sn, and Sb; and a part of Si replaced by Al. Titanite is widely distributed as an accessory mineral in acid and intermediate igneous rocks, in their associated pegmatites, and several metamorhic rocks.

tollite. An igneous rock of the granite clan. A fine-grained porphyritic hypobyssal rock; contains phenocrysts of hornblende, andesine, biotite, muscovite, quartz, orthoclase and garnet in groundmass of quartz and feldspar.

tombolo. A bar or barrier joining an island with a mainland or another island; occurs along shorelines of submergence, where islands are common.

tonalite. An igneous rock of the diorite clan; a medium to fine-grained granular plutonic rock; contains quartz, oligoclase, biotite, and hornblende.

tonsbergite. An igneous rock of the syenite clan. A red coarse-grained granular plutonic rock; contains orthoclase, andesine, pyroxene, and biotote.

topaz. $4[Al_2SiO_4(OH,F)_2]$. A red, yellow, green, blue, or brown neosilicate mineral that crystallizes in the orthorhombic system and commonly occurs in prismatic crystals with pyramidal terminations; used as a gemstone. Its hardness on Mohs scale is 8.0.

topazite. Altered igneous rock of the granite clan. A medium grained granular hypabyssal rock; contains quartz and topaz with minor muscovite.

topocline. A cline related to a geographic zone and usually unrelated to any ecologic condition.

topographic adjustment. The condition existing where the gradient of a tributary is harmonious with that of the main stream.

topographic texture. Disposition, grouping, or average size of the topographic units composing a given topography; usually restricted to a description of the relative spacing of drainage lines in stream-dissected regions.

topography. The general configurtion of a land surface including its relief and the position of its natural and manmade features.

topology. (1) The analytical, detailed study of minor landforms, requiring fairly large scales of mapping. (2) The topographic study of a particular place; the history of region as indicated by its topography.

topset bed. One of the nearly horizontal sedimentary layers deposited on the surface of an advancing delta.

topsoil. (1) Surface soil usually corresponding with the *A* horizon, as distinguished from subsoil. (2) Soil presumed to be fertile and used to cover areas of special planting.

tor. A rocky mass, varying in scale but usually less than 30 metre in height , in which the exposed vertical and horizontal joints often produce a 'rude' architectural aspect. Tors commonly occur on hill-tops (summit or skyline tors) but also occupy valley sides (sub skyline tors) and particularly in some tropical landscapes rise above near-level plains of erosion.

torbanite. A mineral substance intermediate between oil shale and coal. Often referred to as a type of boghead coal, or as a synonym for boghead coal.

torbernite. $Cu(UO_2)_2(PO_4)_2.8\text{-}10H_2O$. A secondary mineral of copper, often formed as a result of oxidtion of uraninite in veins containing copper sulphide. Sometimes torbernite is associated with autunite.

tordrillite. An igneous rock of the granite clan. A fine-grained granular extrusive rock ; it contains quartz, sodaclase, albite, and minor ferromagnesian minerals.

torrent. A stream of water flowing at great velocity or turblence, as after a heavy rainfall or down a steep incline ; a cascade. Also any similar stream; as of lava.

toscanite. An igneous rock of the granite clan. A fine-grained porphyritic extrusive rock ; contains phenocrysts of quartz and labradorite in a groundmass of biotite, sanidine, hypersthene, labradorite, and quartz.

total porosity. The ratio of total void space in porous oil-reservoir rock to the bulk volume of the rock itself.

tourmaline. $MX_3Y_6(BO_3)_3(Si_6O_{18})$-$(OH,F)_4$, where M=Ca,Na, and/or K;

X=Al,Fe, Li, Mg, and/or Mn; and Y=Al, Cr, Fe, and/or V. Any of a group of cyclosilicate minerals with a complex chemical composition, shows vitreous to resinous lustre and variable colour; crystallizes in ditrigonal-pyramidal class of the hexagonal system. As suggested by the formula, tourmaline group minerals may differ greatly from each other. It is also true, however, that few tourmaline specimens have end-member compositions. It is commonly found as an accessory mineral in granitic pigmatites, and is widely distributed in acid igneous rocks and in metamorphic rocks. Schorl is the common black tourmaline most frequently seen in the rocks. When transparent and flawless, it may be cut into gems. The piezoelectric property of tourmaline is utilized in making pressure gauges to measure blast pressures both in air and underwater.

***T* phase.** A seismic phase with a period of 'see or less, which tavels through the ocean with the speed of sound in water. It is occassionally identified on the records of those earthquakes in which a large part of the path from epicentre to station is across the deep ocean.

***T* plane.** A term used in crystal plasticity to denote the crystal-lographic slip plane; known as translaion plane.

trace element. A nonessential element found in small quantities (usually less than 1 per cent in a mineral). Also known as accessory element ; guest element.

trace fossil. A fossilized track, trail, boring, or burrow formed by the movement of an animal in soft sediments.

trace slip. The component of net slip in a fault which is parallel to the trace of an index plane on a faul plane.

trace-slip fault. A fault whose net slip is trace slip.

trachyandesite. An extrusive rock with a composition between trachyte and andesite. It contains sodic plagioclase, alkali feldspar, and one or more mafic minerals (amphibole, biotite, pyroxene).

trachybasalt. An extrusive rock with a composition intermediate between trachyte and bsalt. Both calcic plagioclase and alkali feldspar are present, as well as augite and olivine. Leucite or analcime may be minor constituents.

trachyte. A fine-grained, extrusive alkaline rock, sometimes porphyritic, approximately silica-saturated and with a wide compositional range. It contains alkali feldspar, minor mafic minerals, and small amount of quartz. Trachytes range from slightly over-saturated types with less than 10% quartz to undersaturated types containing feldspathoids. With greater or less silica, they grade into rhyolites and phonolites.

trachytic. (1) A textural term applied to volcanic rocks in which feldspar microlites of the groundmass have a subparallel arrangement corresponding to the flow lines of the lava from which they are formed. (2) Pertaining to or composed of trachyte.

trachytoid. Physically or chemically similar to trachyte group of igneous rocks, *e.g.,* fine-grained with a structure of parallel rods of feldspars.

track. A fossil structure consisting of a mark left in soft material by the foot of an animal.

trail. (1) A line of rock fragments that were picked up by the glacial ice at a localized outcropping ang left scattered along a fairly well-defined tract during the movement of a glacier. (2) Crushed material along a fault surfaace that is used as an indication of direction of displacement. Such, material can be the source of mineral deposits. (3) The trace or mark left by a moving organism, *e.g.*, a worm tail.

train. (1) A series of oscillations on a seismograph record. (2) A narrow glacial deposit extending for a long distance, such as valley train or a boulder train.

transcurrent fault. A strike-slip fault characterized by a steeply inclined surface. Also known as transverse thrust.

transection glacier. A glacier that fills entire valley system, concealing the divides between the valleys.

transfer. A single process occurring continuosly in space-time in which erosion is followed by transportation and deposition of sediment.

transformation. A change from one crystal polymorph to another. Also known as inversion.

transformational breccia. A breccia occurring in a vertical body cutting through a stratigraphic section; believed to have been produced by collapse, as above a dissolved salt-bed.

transformation twin. A crystal twin developed by a growth transformation form a higher to a lower symmetry.

transform fault. (1) A plate boundary that ideally shows pure strike-ship displacement. (2) A special type of strikeslip fault, along which the displacement suddenly stops or changes form. Many transform faults are associated with mid-oceanic ridges, where the actual slip is opposite from the apparent displacement across the fault.

transformism. A theory that explains the origin of granite as a result of granitization, opposed to magmatism.

transgressive deposit. Sediment deposited during transgression of the sea or during subsidence of the land.

translation. A shift in position without rotation. When applied to plastic deformation it refers to the movement of one block of atoms past another.

translational fault. A fault in which there has been uniform movement in one direction and no rotational component of movement. Also known as translatory fault.

translational movement. Movement as of fault blocks, that is uniform, without rotation, so that parallel features maintain their orientation.

translation group. The collection of all translation operations which carry a crystal lattice into itself.

translation pane. *See T* plane.

translucent attritus. Attritus composed of mainly transparent humic degradation matter. Also known as humodurite.

transportation. A phase of sedimentation concerned with movement by natural agents of sediment or any loose or weathered material from one place to another.

transversal jade. A mineral that is not a true jade but a green grossularite garnet.

transverse bar. A slightly submerged sand bar extending perpendicular to the shoreline.

transverse dune. A sand dune having a gentle windward slope and a steep leeward slope, and elongated at right angles to the direction of prevailing winds.

transverse fault. A fault, whose strike is more or less perpendicular to the general structure trend of the region. *See also* fault.

transverse gallery. An auxiliary crosscut made in thick deposits across the ore body in order to divide it into sections along the strike.

transverse ripple mark. A ripple markformed nearly perpendicular to the direction of the current.

transverse thrust. *See* transcurrent fault.

transverse valley. (1) A valley cutting perpendicularly across a ridge, range, or chain of mountains. Also known as cross valley. (2) A valley perpendicular to the general strike of the underlying strata.

transverse wave. *See S* wave

trap-door fault. A curved fault bounding a block that is hinged along one edge, it is an intrusion displacement structure in Little Rocky Mountains of Montana.

trapezohedron. (1) An isomrtric crystal form of 24 faces, each face of which is ideally a foursided figure having no two sides parallel. (2) A crystal form consisting of six, eight, or twelve faces, half of which above are offset from the other half below.

travertine. A finely crystalline, massive deposit of calcium carbonate formed by chemical precipitation from solution in surface and ground waters, as around the mouth of springs, especially hot springs. It also occurs in limestone caves, where it forms a stalactites and stalagmites. A spongy or less compact variety is tufa.

tread. The flat or gently sloping surface of one of the series of natural steplike landforms, such as those of a glacial stairway or of successive terraces; a bench level.

trellis drainage. A drainage pattern paralleling deeply eroded folded strata; pattern; resembles a gardear trellis; a structurally controlled drainage pattern that forms where the underlying bedrock comprises a series of strata of alternating high and low resistance to erosion.

tremolite. $Ca_2Mg_5Si_8O_{22}(OH)_2$. Magnesium-rich monclinic calcium amphibole that forms one end member of a group of solid-solution series with sodium, iron and aluminium; occurs in long blade- shape or short stout prismatic crystals and also in masses or compound aggregates. It shows subconchoidal to uneven fracture and perfect cleavage in one dirretion. Its hardness on Mohs scale is 5.0-6.0.

tremolite asbestos. Massive fibrous variety of tremolite.

tremor. A minor earthquake.

trench. (1) A long, narrow, deep depression of the sea-floor, with relatively steep sides. (2) A narrow stream eroded canyon, gulley, or depression with steep sides.

trend. The direction of an outcrop of a layer, vein, fold, or other geologic feature.

treptomorphism. *See* isochemical metamorphism.

Triassic. A period of geological time that began 225 million year ago and lasted for about 30 million years.; it is the earliest period of the Mesozoic Era.

trichite. A black, straight or curved hairlike crystallite.

triclinic crystal. A crystal whose unit cell has axes which are not at right angles, and are unequal. Also known as an anorthic crystal.

triclinic system. A crystal system of one-fold symmetry, with three unequal axes intersecting obliquely. The prominent feature of triclinic crystals is lack of symmetry due to inclination of all axes at angles other than 90° and crystals are bounded by pinacoids, and only pairs of faces opposite and parallel match each other. Pinacoids may intersect one or more axes; Miller indices and may therefore contain one or two zones.

trigonal lattice. *See* rhombohedral lattice.

tridymite. SiO_2. A white or colourless form of silica occurring in minute, thin, tabular, small octahedral crystals, many of which are twinned on [111] and aggregated in round forms; a high-temperature polymorph of quartz. It occurs chiefly as phenocrysts in and as a groundmass component of silicic volcanic rocks and filling vesicles, fractures, and other cavities in these rocks, especially in rhyolites, dacites, and trachytes. In many instances, it can be shown to have formed originally as high-(beta-) trydimyte. It is commonly associated with such diverse minerals as sanidine, cristobalite, fayalite, cassiterite, kaolinite, and alunite. It has also been reported as a constituent of silicified wood, apparently formed by the crystallization of opaline silica at ordinary temperatures.

trihedral. Any figure obtained from three non-coplaner lines intersecting at a common point.

trilling. A cyclic crystal twin consisting of three individual crystals.

trimline. A sharp boundary line delimiting the maximum upper level of the margins of a glacier that has receded from an area.

trimorphism. A type of polymorphism in which there are three crystal forms, known as trimorphs.

trioctahedral. Pertaining to a layered-mineral structure in which all possible octahedral positions are occupied.

triphylite. 4[Le(Fe(II),Mn)PO_4]. A bluish-grey mineral crystallizing in orthorhombic system; it is isomorphous with lithiophilite. It shows uneven to sub-conchoidal fracture and good cleavage in one direction. Its hardness on Mohs scale is 4.0-5.0.

triple junction. A point where three lithospheric plates meets.

triple point. A particular temperature and pressure at which three different phases of a substance can coexist in equilibrium.

tripoli. A light weight, porous, friable, siliceous sedimentary rock that may have a white, grey, yellow, pink, or red colour. It is a variety of diatomite containing radiolaria.

troctolite. An igneous rock of the gabbro clan. A coarse-grained granulas plutonic rock; contains labradorite and olivine. Also known as forellenstein.

troilite. FeS. A hexagonal meteorite mineral composed of iron sulphide ; a variety of pyrrhotite commonly found in almost all meteorites.

trolly. A basin-shaped depression in strata. Also known as lum.

trona. 4[$Na_2(CO_3).Na(HCO_3).2H_2O$. A greyish-white or yellowish-white mineral that crystallizes in the monoclinic system and occurs in fibrous or columnar layers or masses. It shows uneven to sub-conchoidal fracture and perfect cleavaage in one direction. Its hardness on Mohs scale is 2.5- 3.0. Also known as urao.

troostite. A manganese-bearing variety of willemite.

tropept. A suborder of the order Inceptisol, characterized by moderately dark *A* horizons with modest additions of organic matter, *B* horizons with brown or reddish colours, and slightly pale *C* horizons; restricted to tropical regions with moderate or high rainfall.

tropic tide. A tide occurring when the moon is near maximum declination, diurnal inequality is then at a maximum.

tropopause. The boundary between the

troposphere and the stratosphere. It is about 16 km at the eqator, and about 6 km at the poles.

troposphere. That portion of the atmosphere next to the earth's surface, in which temperature generally decreases rapidly with altitude, clouds form, and convection is active. In middle latitudes the trposphere includes the first 10-12 km above the earth's surface.

trough. (1) The line connecting the lowest points of a fold. (2) Any narrow elongate depression in the surface of the earth. (3) A small, straight depression formed just offshore on the bottom of a sea or lake or on the landward side of a long-shore bar. 4. An elongate depression on the sea floor that is wider and shallower than a trench.

trough end. The abrupt head of a glaciated mountain valley (galcial trough), sometimes associated with near-vertical rock faces and spectacular waterfalls. Trough ends are believed to result from the immense powers of 'overdeepening' possessed by a large valley glacier which has been formd by the joining of a number of smaller, much less powerful glaciers flowing out of high-level cirques above the valley head.

trough fault. One of a set of two faults bounding a graben.

trough plane. *See* trough surface.

trough surface. A surface or a plain connecting the troughs of the bed of a syncline. Also known as synclinal axis; through plane.

trowlesworthite. An igneous rock of the syenite clan. A red coarse-grained granular plutonic rock ; it contains orthoclase, tourmaline, and fluorite.

truncation. Replacement of one edge or corner of a crystal by a plane.

tsingtauite. An igneous rock of the granite clan. A very fine- grained porphyritic hypabyssal rock ; contains phenocrysts of microperthite and plagioclase in a groundmass of granitic composition.

tsunami. A very large seismic sea-wave, generated by an earthquake shock on the ocean floor, Tsunamis travel for considerable distances across the sea as long, low wave (the wave length may be over 100 kilometres, the wave height as little as 1 metre , and the forward velocity as much as 600 kilometres per hour). As the tsumami approaches the shore, the wave height increases markedly, and sometimes exceeds 15 metres; it is thus capable of causing immense destruction to coastal settlements.

tucholite. *See* thucholite.

tufa. A spongy porous limestone formed by precipitation from evaporating spring and river waters

tufaceous. Pertaining to or similar to tufa.

tuff. Pyroclastic extrusive igneous rock composed of fragments with a diameter less than 2 mm. Its varieties are vitric tuff, lithic tuff, and crystal tuff.

tuff ball. *See* mud ball.

tuffite. A consolidated mixture of pyroclastic and sedimentary detritus usually well-bedded and sorted according to grain size.

tumulus. A mound built at the side of an ancient burial ground. The mound may be 10 to 20 metres in height and 20 to 50 metres in length. Tumuli date from pre-Roman times.

tungstite. H_2WO_4. An orthorhombic mineral composed of tungstic acid ; occurs as aggregates of acicular crystals or earthy masses. It shows perfect basal cleavage. Its hardness on Mohs scale is 2.5.

turbidite. Any sediment or rock transported and deposited by a turbidity current, generally characterized by graded bedding large amounts of matrix, and commonly exhibiting a Bouma sequence.

turbidity current. A density current in water or air ; specifically a bottom-flowing current laden with suspended sediment, moving swiftly down a subaqueous slope and spreading horizontally on the floor of the body of water, having been set in motion by locally stirred-up sediment that gives the water a density greater than that of the surrounding clear water. Such currents are known to occur in lakes, and are believed to have produced the submarine canyons notching the continental slope. The term is applied to a current due to turbidity, not to one showing that property.

Turbidity currents are capable of eroding the floor of the basin or slope and are responsible for transferring large quantites of shallow-water sediment into deeper zones. The recognition of the work of turbidity currents has revolutionized ideas on rates of sedimentation and the depths water in which sediments were deposited.

turtle stone. A flattened oval septarium released from its matrix and so weathered that the vein filled system of the cracks may be seen. Its form has a rough resemblance to that of a turtle and its polished surface bears a fancied resemblance to a turtle's back.

tuxtlite. Pyroxene intermediate in composition and properties between diopside and jadeite.

twinned crystal. A composite crystal consisting of a number of usually equal and similar crystals united in reversed positions with respect to each other, either with symmetrical parts in contact (contact twin) or by the parts growing out of one another (penetration twin). Repeated twinning in one plane leads to the formation of polysynthetic twins.

twin gliding. Crystal gliding that results in the formation of crystal twins

twin law. A statement relating two or more individuals of a twin to one another in terms of their crystallography.

twinning. The formation of twin crystals by growth, translation, or gliding.

twinning.A rational intergrowth of two or more single crystals of the same material in a mathematically describable manner so that some lattice are parallel whereas others are in reversed position. The symmetry of the two parts may be reflected about a common plane.

twinning axis. The crystal axis about which one individual of a twin crystal may be rotated ; usually 180° to bring it into coincidence with the other individual. It cannot be coincident with the axes of the twofold, fourfold, or six-fold symmetry.

twinning displacemlent. Displacement in a crystal due to twin gliding.

twinnilng plane. The common plane across which individual components of a crystal twin are symmetrically arranged or reflected. It is parallel to a crystal face but not to a plane of symmetry of a single crystal. It is usually identical with the composition plane.

two-cycle valley. A valley that is the result of two cylcles of erosion, as shown by a narrow inner valley bordered by high level terraces.

type locality. The location in which a stratotype is situated and from which it usually takes its name. It is contained within the type area and contains the type section.

type section. The sequence of strata identified as the original sequence for a location or area; the standard against which other stratigraphy of parts of the area are compared. Also known as section.

type specimen. The single specimen on which the original description of a particular species is based, which serves as a permanent point of nomenclature reference for application of the name of that species. The type specimen may be a holotype, a neotype, or a lectotype.

tyuyamunite. $4[Ca(UO_2)_2(VO_4)_2.10H_2O]$. A yellow orthorhombic mineral occurring in incrustations as a secondary mineral; composed of hydrous calcium uranyl vanadate. It shows perfect basal cleavage. Its hardness on Mohs scale is 2.0.

ubac. The shady side of a valley, normally the north-facing side, and receives sunshine during the summer only, and is likely to be forested or covered with grassland.

udalf. suborder of the soil order Alfisol ; brown soil formed in a udic moisture regime and in a mesic or warmer temperature regime.

udden grade scale. A logarithmic grade scale that uses 1 mm as the reference point and progresses by the fixed ratio of 1/2 in the direction of decreasing size and of 2 in the direction of increasing size, such as 0.25, 0.5, 1, 2, 4.

udert. A suborder of the soil order Vertilsoil ; formed in a humid region so that the surface cracks remain open only for 2-3 months.

udoll. A suborder of the Mollisol order ; found in humid, temperate, and warm regions where maximum rainfall occurrs during growing season ; has thick, very dark *A* horizons, brown *B* horizons, and paler *C* horizons.

udult. A suborder of the soil order Ultisol; organic-carbon content is low, argillic horizons are reddish or yellowish; formed in a udic moisture region.

ugrandite. Group name for uvarovite, grossular, andradite.

uintahite. Gilsonite that occurs primarily in veins in the Uinta Basin,Utah. It has a dark brown streak and shows conchoidal fraxcture. Its hardness on Mohs scale is 2.0-2.5.

ulexite. $2[NaCaB_5O_9.8H_2O]$. A white mineral that crystallizes in triclinic system and forms rounded reniform masses of extremely fine acicular crystals. It shows uneven fracture and perfact cleavage in one direction. Its hardness on Mohs scale is 2.5.

ullmannite. $4[NiSbS]$. A steel-grey to silver-white mineral that crystallizes in cubic system; shows uneven fracture and perfect basal cleavge. Its hardness on Mohs scale is 5.0-5.5.

ulrichite. An igneous rock of the syenite clan. A fine-grained porphyritic hypabyssal rock ; contains phenocrysts of nepheline, sanidine, barkevikite, and olivine in a groundmass of sanidine hornblende, and aegirine.

ulsterian. A stratigraphic stage name for the North American Lower Devonian.

ultimate landform. The theoretical or ideal landform that coincides with the end of a cycle of erosion.

ultisol. A soil order characterized by an argillic horizon that has a base saturation of less than 35 per cent at *p*H of 8.2.

ultrabasic. Said of igneous rock, having a low silica content, as opposed to the higher silica contents of acidic, basic, and intermediate rocks . The percent silica allowed in these rocks is not universally agreed upon but is generally, considered to be less than 45%.

ultramafic. Said of an igneous rock composed chiefly of mafic minerals, *e.g.*, monomineralic rocks composed of hypersthene augite, or olivine.

ultramarine. $(Na,Ca)_8(Al,Si)O_{24}(S, SO_4)$. A blue or voilet-blue feldspathoid mineral crystallizing in the isometric system ; the chief mineral constituent of lapis lazuli.

ultrasima. The supposedly ultrabasic layer of the earth below the sima, immediately beneath the Mohorovicic discontunity.

ultravulcanian. A type of volcanic eruption characterized by periodic violent gaseous explosions of lithic dust and solid blocks, with little if any, fiery scoria,

umber. A brown earth consisting of manganese oxides, hydrated ferric oxide, alumine, silica, and lime.

umptekite. An igneous rock of the syenite clan. A fine to coarse-grained granular plutonic rock; contains orthoclase, microcline microperthite, sodalite, arfvedsonite, aegirine, biotite, and nepheline.

unaka. A large residual form rising above a peneplain, and occasionally showing in its summit or surface the remnants of an even older peneplain. It is greater in height and size than a monadnock.

uncompahgrite. Feldspthoid igneous rock. A bluish-grey, coarse-rained granular plutonic rock; contains melilite. augite, and radite, and magnetite.

unconformable. Said of strata that do not suceed the underlying rocks in immeadiate order of age or in parallel position; especially younger strata that do not have the same dip and strike as the under-lying rocks. Also said of contact between unconformable roks.

unconformity. (1) A break or gap in the geologic record, such as interruption in the normal sequence of deposition of sedimentary rocks, or a break between eroded metamorphic rocks and younger sedimentary strata. An unconformity is of longer duration than a diastem. (2) The structural relationship between two groups of rock that are not in normal succession; also their surface of contact. Three types of unconformity are recognised; disconformity, angular conformity , and nonconformity. A disconformity is an unconformity in a sequence of essentially prallel beds; beds which should be present representing the time interval of their deposition are lacking either because of nondeposition or becuse they were eroded before sueeccding bads were deposited . An angular conformity is an erosion surface truncating folded or tilled strata and upon which a sequence of strata were deposited parallel to the surface. A nonconformity is created when sedimentary strata are deposited on a surface eroded on other plutionic rocks.

unconsolidated material. (1) A sediment that is loosely arranged or stratified, or whose particles are not cemented together, occurring either at the surface or at depth. (2) Soil material that is in a loosely agregated form.

unctuous. Greasy, oily, or scapy to touch.

unda. Said of the environment of sedimentation that lies in the zone of wave action. It may be used alone or as a combining form.

undathem. Rock units formed in the unda environment of deposition.

undation theory. A theory regarding building of mountains assuming that long and broad anticlines of basement rock generated by deep-sea magma, rose like huge wave in the crust. The sedimentary cover and sometimes the basement itself slide off to form the folds and faults observed in orogenic belts.

underclay. A layer of clay or other fine-grained detrital material underlying a coal bed or comprising the floor of a coal seam. Also known as coal clay; seat clay; seat earth; underearth; warrent.

underclay limestone. A thin, fresh-water limestone that is relatively free of fossils and is dense and nodular; found in underlying coal deposits.

undercliff. A subordinate cliff or terrace formed by material that has fallen or slide from above.
underconsolidation. Less than normal consolidation of sedimentary material for the existing overburden.
underearth. *See* underclay.
underfit stream. A misfit stream that appears to be too small to have eroded the valley in which it flows.
underground stream. A subsurface body of water flowing in a definite current in a distinct channel.
underlie. To lie or be situated under; to occupy a lower position, or to pass beneath.
undermeltilng. The melting from below of any floating ice.
undersaturated (1) Said of an igneous rock composed of silica-unsaturated minerals such as olivine and feldspathoids. (2) Said of a rock whose norm contains olivine and feldspathoids.
underthrust. A thrust fault in which the lower, actice rock mass has been moved under the upper, passive rock mass. *See also* faulting.
undertow. A subsurface seaward movement by gravity flow of water carried up on a sloping beach by waves or breakers.
unfreezing. The upward movement of stones to the surface as a result of repeated freezing and thawing of the containing soil.
uni-. A prefix denoting one.
uniaxial crystal. A doubly refracting crystal which has a single axis along which light can propagate without exhibiting double refaction.
uniclinal shifting. The lateral migration of a stream following a weak outcrop (of sand or clay) in a uniclinal structure. It is easier for the streams to shift laterally in the direction of dip, thus maintaining contact with a line of geological weakness, than to incise vertically into an underlying hard stratum.
uniclinal structure. A geological structure comprising a sequence of sedimentary strata dipping more or less uniformly in one direction. Uniclinal structures comprising alternating resistant and unresistant beds provide ideal condition for the development of scrap-and vale landscapes.
uniformity coefficient. An expression of variety in size of grains that constitute a granular material.
uniserial . Consisting of a single row or series, *e.g.*, the plates of a primitive crinoid arm.
unitary symmetry. A method (SU_3) of classifying elementary particles according to their properties in a similar manner to the classification of atomic properties in the periodic table. SU_3 has successfully predicted the existence of particles that have subsequently been detected experimentally.
unit cell. The unit of which a crystal lattice is constructed, for example, the body-centred and face centred lattices are forms of a cubic unit cell.
unit form. A crystal form in a system other than the cubic, having intercepts on the chosen crystal axis that define the axial ratio. Unit forms have Miller indices [111], [110], [011], and [101].
unpaired terrace. A stream terrace with no corresponding terrace on the opposite side of the valley, usually produced by a meandering stream swinging back and forth across a valley.
universal stage. A piece of apparatus which permits the rotation of a thin section or crystal grain about two (or more) horizontal axes, in addition to the nor-

mal vertical axis, of a petrologial microscope. It can be used in two main ways:(*i*) to determine the optic characters of a mineral accurately; and (*ii*) to determine the orientation of the crystal grains in an orientated section of a deformed rock.

unsaturated. Said of a mineral that does not form in the persence of free silica; *e.g.*, nepheline, leucite, feldspathoids.

unstratified. Not formed or deposited in strata; specifically said of massive rocks or sediments with an absence of layering, such as granite or glacial till.

updip block. The rocks on the upthrown side of a fault.

uphole shooting. In seismic exploration, the setting off of successive shots in a shot hole at varying depths in order to determine velocities and velocity variation of the materials forming the walls of the hole.

uplift. A structurally high area in the crust, produced by movements that raise the rocks, as in a broad dome or arch.

upper. Pertaining to rocks or strata that normally overlie those of earlier formations of the same subdivision of rocks.

upper Cambrian. The latest epoch of the Cambrian period of geologic time, beginning apperoximataly 510 million years ago.

upper Carboniferous. The European epoch of geologic time equivalent to the Pennsylvanian of North America.

upper Cretaceous. The latest epoch of the Cretaceous period of geologic time, beginning about 90 million years ago.

upper Devonian. The latest epoch of the Devonian period of geologic time, beginning about 365 million years ago.

upper Jurassic. The latest epoch of the Jurassic period of geologic time, beginning about 155 million years ago.

upper Mississippian. The latest epoch of the Mississippian period of geologic time.

upperOrdovician. The latest epoch of Ordovician period of geologic time, beginning about 440 million years ago.

upper Pennsylvanian. The latest epoch of the Pennsylvanian period of geologic time.

upper Permian. The latest epoch of the Permian period of geologic time, beginning about 245 million years ago.

upper Silurian. The latest epoch of the Silurian period of geologic time.

upper Triassic. The latest epoch of the Triassic period of geologic time, beginning about 200 million years ago.

upthrow. (1) The fault side has been thrown upward. (2) The amount of vartical fault displacement.

uralin emerald. Gem variety of andradite.

uralite. Actinolite pseudomorphed after pyroxene.

uralitization. The alteraation of an original pyroxene in an igneous rock to a mass of (usually) fibrous amphibole, commonly a type of hornblende, which was originally thought to be a distinct mineral species, 'uralite'. It is commonly forms during the lowgrade metamorphism of basic igneous rocks, or late stage hydrothermal alteration.

uraninite. UO_2. An isometric uranium mineral, that is highly radioactive and is the primary ore of uranium. It usually occurs in granular masses or aggregates Uraninite is found in pegmatites and high-temperature hydrothermaal veins in sedimentary clastic deposits with mineral of lead, tin, and copper. Its varieties are broeggerite, cleveite, pitchblende and nasturan.

uraanophane. $Ca(UO_2)_2Si_2O_7.6H_2O$. An orthorhombic, yellow mineral composed otf hydrous calcium uranyl silicate; occurs as acicular crystals. Its hardness on Mohs scale is 2.0-3.0. Also known as uranotile.

uranothorianite. Uranium rich variety of thorite.

uranotile. *See* uranophane.

uriconian. A stratigraphic name for the volcanic Precambrian underlying the Longmyndian in Shropshire, England.

urtite. Feldspathoid igneous rock. A medium-grained granular plutonic rock; contains nepheline, aegirine, and apatite.

ussingite. $Na_2AlSi_3O_8OH$. A triclinic mineral composed of basic sodium aluminium silicate; shows perfect cleavage in one direction.

utahlite. Compact nodular variety of variscite.

uvarorite. $8[Ca_3Cr_2Si_3O_{12}]$. The emerald-green, calcium chromicum end memmber of the garnet group; occurs as rhombohedral crystals. Its hardness on Mohs scale is 7.5.

V

vadose water. A term for water percolating downwards through the rock to the zone of saturation.

vadose-water discharge. The release, by evaporation, of water not origiating in the zone of saturation.

vaesite. NiS_2. An isometric mineral with pyrite structure; composed of nickel sulphide; occurs as octahedral crystals.

vagrant benthos. Bottom-dwelling organisms that are capable of movement on, in, or above the substratum.

valbellite. An igneous rock of the ultrabasic clan. A fine-grained granular hypabyssal rock; contains bronzite, olivine, hornblende, and magnetite.

valencianite. A variety of potassium feldspar from Mexico; a synonym of adularia.

valentinite. $4[Sb_2O_3]$. An orthorhombic mineral composed of antimony trioxide; occurs as prismatic or tabular crystal or in massive form . It shows white streak and perfect cleavage in one direction. Its hardness on Mohs scale is 2.5-3.0.

vallerite. $CuFeS_2$. A sulphide mineral composed of copper iron sulphide; also found in meteorites.

valley. Any hollow or low-lying land bounded by higher ground; usually traversd by a stream or river which receives the drainage of the surrounding heights.

valley flat. (1) A bedrock surface produced by lateral erosion, commonly veneered with the alluvium of a flood plain. (2) A low flat land lying between valley channal.

valley floor. The broad, flat bottom of a valley.

valley glacier. A glacier that flows down the walls of a mountain valley .

valley iceberg. An icebrg weathered in such a manner that a large *U*-shaped slot extends through the iceberg. Also known as drydock-iceberg.

valley plug. A local constriction in a stream channel, formed by any of several types of channel obstructions, which may cause rapid deposition.

valley train. A long, narrow body of outwash, deposited by melt-water streams for beyond the terminal moraine or the margin of an active glaciar and confined within in the walls of a valley below the glacier; it may or may not emerge from the mouth of the valley to join an outwash plain. Also known as outwash train.

vanadinite. $2[Pb_5(VO_4)_3Cl]$. A red yellow or brown mineral often occurring as globular masses encrusting other minerals in lead mines. It shows white or yellow streak and uneven to conchoidal fracture. Its hardness on Mohs scale is 2.7-3.0.

vanishing tide. When a high-water and low-water 'meet' together into a period for several hours with a nearly constant water level.

variation diagram. A graph or other similar diagram in which chemical features of a series of rocks are plotted against each other, or occasionally against some other property, *e.g.*, colour index, specific gravity. The commonest type is the Harker diagram, in which parcentages of elements (as oxides) are plotted against silica percentages. The diagrams are designed to reveal genetic and other relationships between rocks of a series.

variolite. An igneous rock of the diorite clan. A rock, originally glassy, which has developed variolitic structure throughout, contains varioles of plagioclase, augite and ilmenite in a groundmass of augite and devitrified glass.

variolitic . A term used to describe a texture or structure that occurs only in basaltic rocks ; equivalent to spherulitic texture in rhylolite. Variolitic structure is a common feature of the margins or certain pillow lavas. The varioles, which sometimes stand out as small knobs, are commonly composed of fan-like sprays of feldspar fibres.

variscite. 8[$AlPO_4.2H_2O$]. An orthorhombic mineral composed of hydrous aluminium phosphate; normally occurs in massive form. It shows uneven to splintery fracture and good cleavage in one direction. Its hardness on Mohs scale is 3.5-4.5.

varves. Regular layers of alterations of materials in sedimentary deposits caused by annual seasonal influenes. .Each varve is usually interpreted to represent the deposition during a year, and consists ordinarily of a lower part deposited in summer and an upper, fine-grained part deposited in winter . Varves of silt and clay-like minerals occur abundantly in lake sediments associated with glaciers and in certain marine shales and slates.

varve clay. A distinctly laminated lacustrine sediment consisting of clay-rich varves; also the upper fine-grained, 'winter' layer of a glacial varve.

varvite. Name given to a graded lamilnated clay which resembles varve clay but was produced by nonglacial processes.

vaterite. $CaCO_3$. A rare hexagonal mineral consisting of unstable calcium carbonate; it is trimorphous with calcite and aragonite.

vauclusian. A spring which emerges from underground in a limestone area. It takes its name form Fontaine de Vaucluse, a large and powerful spring in souhtern France.

vaughanite. A pure light-grey, fine-textured limestone, that breaks with a smooth conchoidal fracture; contains relatively few fossils, and has a white ,chalky appearence on weathered surfaces.

vaugnerite. An igneous rock of the diorite clan. A coarse-grained granular hypabyssal rock; contains andesine, biotite, hornblende, apatite, and quartz.

vein. A fissure or elongated opening in a rockmass filled with mineral matter formed later than the rock surrounding it; ore or other material filling a fissure in a rock.

vein dike. A pegmatitic intrusion that has the characteristics of both a vein and a dike.

veined gneiss. A composite gneiss with irregular layering. The term is generally used in the field and has no genetic implication.

veinite. A genetic type of veined gneiss in which the vein material was secreted from the rock itself.

vein quartz. A rock chiefly composed of quartz crystals of pegmatitic or hydrothermal origin and commonly of variable size.

velocity discontinuity. An abrupt change of the rate of propagation of seismic waves within the earth, as at a sesmic discontinuity.

velocity profile. A seismic arrangement used to record reflections over a large range of shot-to-geophone distances, which is used to determine seismic velocity from the time-distance relationship.

veneer. A thin layer of sediment covering a rock surface.

ventifact. A pebble, cobble, or boulder with its shape or surface modified by wind driven sand. Ventifacts are generally pitted, polished, and grooved.

Venus' hair. Golden-brown or yellow coloured, needle-like crystals of rutile, occurring embedded in quartz crystals.

vermiculite. A group of layer-lattice clay minerals having the general formula $2[(Mg,Fe(II),Fe(III),Al)_6(Si,Al)_8O_{20}(OH)_4.8H_2O]$, especially certain chlorites hydromicas, and clay minerals. It is derived from the alteration of biotite and phylogopite in the zone of weathering.

vertical accretion. Upward growth of a sedimentary deposit, *e.g.*, settling of sediment from suspension in a stream subject to overflow.

vertical corrasion. Corrasion on a rocky riverbed which works downwards to produce potholes. Gradually the potholes become enlarged until they merge with their neighbours. In this way the bed of the stream will be lowered.

vertical exaggeration. The extent to which the vertical scale is larger than the horizontal scale on a cross section or a stereo model.

vertical intensity. The magnitude of vertical component of any vector, e.g., of earth's magnetic or gravitational field at any point.

vertical separation. In a fault, the distance measured vertically between two parts of a displaced marker such as a bed.

vertical shift. In a fault, the vertical component of the shift .

vertical shift. In a fauting, the vertical component of the net slip, it is equal to the vertical component of dip slip.

vesicle. Small cavity in a fine-grained or glassy igneous rock, formed by a bubble of gas during the solidification of lava; openings in lava formed by expanding gas. Expansion of gases take place with reduction in pressure as lavas come up from depth to the earth's surface giving lava a spongy or vasicular appearance; if the bubbles are very small, producing a glassfroth, the structure is pumaceous; if the bubbles are larger, the structure is scoriaceous.

vesuvianite. $Ca_{10}Mg_2Al_4(SiO_4)_5(Si_2O_7)_2(OH)_4$. A brown, green or yellow mineral found in contact metamorphosed limestones; sometimes, contains iron and fluorine in idocrase.

vicinal face. A crystal faces that modifies a normal crystal face, which it closely approximates in angle.

vintlite. An igneous rock of the diorite clan. A greenish-black, fine-grained, porphyritic hypabyssal rock; contains phenocrysts of hornblende, epidote, oligoclase, and quartz in a groundmass of plagioclase, hornblende, diopside, bronzite and biotite.

violane. A dark violet-blue, fibrous variety of diopside.

violarite. $8[FeNi_2S_4]$. A cubic mineral mainly composed of iron and nickel sulphide. It shows uneven to subconchoidal fracture and poor cleavage. Its hardness on Mohs scale is 4.5-5.6.

virgation. A fold pattern in which the axial surfaces diverge or fan out from a central bundle.

viseite. $2[NaCa_5Al_{10}Si_3P_5O_{30}(OH,F)_{18}.16H_2O]$. A hydrous aluminium calcium sodium fluosilicophosphate. Its hardness on Mohs scale is 3.0-4.0.

vitrain. An ingradient of banded coal, characterized by brilliant vitreous lustre, black colour, and cubic cleavage with conchoidal fracture. Vitrain bands or lentiles are amorphous and thick enough to be visible to the unaided eye.

vitreous. Having the lustre and appearance of glass.

vitric tuff. Tuff composed principally of volcano glass fragments.

vitrification. Formation of glassy or noncrystaline material.

vitrinite. Name given to uniform grey-to-black anisotropic macerals that exhibit moderate reflectance and contitute the matrix throughout which liptinites and inertinites and disseminated. Vitrinites are generally thought to have been derived largely from humic acid fraction of woody tissues, and are major component of vitrain. Vitrinite reflectance is a frequently used standard reference value because it varies directly with degree of organic metamorphism.

vitrinoid. Vitrinite occurring in bituminous coking coals; characterized by a reflectance of 0.5-2.0 per cent.

vitro-. A prefix denoting glassy.

vitrodetrinite. Name given to clastic fragments of diverse vitrinites.

voidal concretion. Hollow limonitic concretion resulting from the weathering of clay ironstone.

vitroclastic. Pertaining to a pyroclastic rock structure charcterilzed by fragmented bits of glass; also said of a rock having such a structure.

vitrophyre. A porphyritic igneous rock having a glassy matrix.

vivianite. $2[Fe_2(PO_4)_2.8H_2O]$. A colourless monoclinic mineral composed of hydrous iron phosphate; occurs as prismatic crystals as well as earthy masses. It is amineral of secondary origin, occurs commonly in gossans of metallic ore deposits and in the surface zones of pegmatites that contain iron-manganese phosphates. It also occurs as concretions in clays and as crystal clusters in recent sedimentary deposits where it is associated with bone or other organic phosphatic remains. It shows vitreous or dull lustre in one direction. Its hardness on Mohs scale is 1.5-2.0.

vogesite. An igneous rock of the syenite clan. A fine-grained porphyritic hypabyssal rock; contains phenocrysts of horn-blende and diopside in a groundmass of orthoclase, hornblende, and diopside.

void ratio. The ratio of the volume of void space to that of the sedimentary rock.

volborthite. $Cu_3(VO_4)_2.3H_2O$. A green or yellow coloured mineral composed of hydrated copper vanadate; may also contain traces of calcium and barium.

volcam. The component of the earth's crust made up of volcanoes and various hypobyssal rocks.

volcanello. A small active cone within the centre of a volcano.

volcanic. Pertaining to the activities, structures, or rock types of a valcano.

volcanic arc. An arcuare string of volcanic islands associated with and parallel to a subduction zone along convergent plate boundaries, often found closely adjacent to a deep ocean trough.

volcanic ash. Fine pyroclastic material having particle diameter less than 4 mm.

volcanic bombs. Pyroclastic ejecta; the lava fragments, liquid or plastic at the time of ejection, acquire rounded forms, markings, or internal structure during flight or upon landing.

volcanic breccia. A pyroclastic rock that is composed of angular volcanic fragments having a diameter larger than 2 mm and that may not have a matrix.

volcanic chain. Linear arrangement of a number of volcanoes, apparently associated with a major geologic feature such as a fault or subduction zone.

volcanic chester. A group of volcanic vents without any apparent systematic arrangement.

volcanic conduit. The channel way that brings volcanic material up from depth.

volcanic cone. A conical hill of lava and/or pyroclastics that is built up around a volcanic vent.

volcanic conglomerate. A water-deposited conglomerate containing more that 50 per cent volcanic material.

volcanic dome. A steep-sided protrusion of viscous lava squeezed out from a volcano, forming a dome- or spine-shaped or bulbous mass above and around the vent. Older lavas may be lifted by the pressure of new lava rising from below. The structure generally develops inside a volcanic crater or on the flank of a large volcano, and is usually much fissured and brecciated. Most of the viscous lavas that form volcanic domes are very high in silica. They solidify as rhyolite or andesite if minerals crystallize, or as obsidian, if no minerals crystallize.

volcanic dust. Very fine particles of lava shot out of volcanoes; carried to great distances by wind and may remain suspended in air for a long time.

volcanic eruption. Forceful extrusion of molten lava on to the earth's surface with accompanying gases and pyroclastic materials. A volcanic cone usually develops around the point of eruption. The type of eruption and associated volcanic cone vary with the composition of the lava.

volcanic focus. The supposed centre of activity in a volcanic region or beneath a volcano.

volcanic gases. Volatile matter composed mainly of about 90 per cent watter vapour, hydrogen, nitrogen, carbon dioxide and carbon monoxide, relased during eruption of a volcano.

volcanic glass. Natural glass formed by the cooling of molten lava, or one of its liquid fractions, too rpidly to allow crystallization; constituent of extrusive igneous rocks.

volcaniclastic. Pertaining to a clastic rock containing volcanic material (in any proportion), without regard to its origin or environment.

volcanic mud. Sediment containing large quantities of ash from a volcanic eruption, mixed with water.

volcanic neck. A vertical pipe like intrusion that represents a former volcanic vent, especially, if standing as an erosional remment.

volcanic rent. A vey large volcanic depression bordered by fissures that are commonly concentric in form. It is a result of magmatic activity or the overburdening of cone material on a weaker substrarum.

volcanic rift zone. A zone comprising volcanic fissures with underlying dike assemblages.

volcanic rock. Finely crystalline or glassly igneous rock resulting from volcanic activity at or near the surface of the earth. Also known as extrusive rock.

volcanics. Igneous rock that solidified after reaching of nearing the earth's surface.

volcanic vent. Subcircular bodies composed of lava and fragmented volcanic rock; contacts are generally steep, and vents range in diameter from a few tens to thousands of metres.

volcanic water. Water within or derived from magma at the earth's surface or at a comparatively shollow depth. Also, juvenile water of volcanic origin.

volcanism. The movement of magma and its associated gases from the interior into the crust and to the surface of the earth. Also known as volcanicity.

volcano. A conical hill or mountain formed around an opening in the earth's surface through which hot rock fragments, gases materials deposit around the conduit, they build up a cone, increasing in size until a huge volcanic mountain is formed. A cone so constructed is known as volcano. The three major types of volcanoes (shield, cinder cone, and composite) are markedly distinct from one another in size, shape, and composition.

volcanogenic. Formed by processes directly connected with volcanism. Said of a mineral deposits considered to have been produced through volcanic agencies and demonstrable associated with volcanic phenomena.

volcanology. Study of the structure, origin, and petrology of volcanoes, and their role as contributors to earth's atmosphere and hydrosphere, and to the rock structure of earth's crust.

vorobyevite. A pinkish or purplish-red caesium and/or sodium containing variety of beryl; used as a gem. Also known as morganite; rosterite; vorobievite; worobieffite.

vredenburgite. Oriented intergrowth of jacobsite and hausmannite.

***V*-shaped valley.** A typical river valley, by contrast with a glaciated *U*-shaped valley. The precise angle of the '*V*' will depend on the relative rates of vertical erosion by the river and the valley shaped recession by weathering, rainwash, and mass transport. If river-downcutting is relatively rapid, the '*V*' will be narrow; if the slope recession is relatively rapid, the '*V*' will be more wide. Where lateral stream erosion is effective, the valley will be flat-floored; and a similar effect will be observed due to extensive deposition of alluvium on the valley floor.

vug. An opening in a rock usually 3-4 mm in diameter, a potential or actual site for the accumulation of petroleum or natural gas.

vulcanian eruption. *See* volcano.

vulpinite. A granular variety of anhydrite found at Vulpino in Lombardy, Italy.

vulsinite. An igneous rock of the syenite clan. A mottled-grey, fine-grained porphytitic extrusive rock, contains phenocrysts of sodaclase, anorthite, and augite in a groundmass of sodaclase, labradorite, and augite.

wacke. The sandstone composed of a mixture of angular unsorted or poorly sorted fragments of minerals and rocks and an abundant matrix of clay and fine silt. The term is properly used only in terms of greywacke and its use in any other sense is undesirable.

wad. A black or dark brown impure mixture of manganese and other oxides. Also known as bog manganese.

wagernite. $Mg_2(PO_4)F$. A yellow, red or brownish mineral composed of phosphate and flnoride of magnesium; sometimes also containing traces of iron and calciam.

wairabite. $CaAl_2Si_4O_{12}.2H_2O$. A mineral of zeolite family, isostructural with analchime.

wake dune. A sand dune occurring on the leeward side of a larger dune and moving away in the direction of the wind.

wall. (1) The side of a cave passage. (2) The side of a lode, or mine workings. (3) A series of columns along a joint crack that have fused togather into aa solid mass.

walled lake. A lake bordered along its shore by lake ramparts.

wallow. A depression or area, generallay filled with water and mud, that suggests a place where animals have wallowed.

wall reef. An elongate steep-sided coral.

wandering dune. A sand dune that has moved as a unit in leeward direction of the prevailing winds, and that is characterized by the lack of vegetation to anchor it. Also known as travelling dune.

want. *See* nip.

warp. (1) A layer of sediment deposited by water. (2) An upward or downward flexure of the earth's crust.

warrent. *See* underclay.

water-bearing strata. Ground layers below the standing water level.

waterborne. Floating on or transported by water.

water content. Water contained in porous sediment or sedimentary rock, generally expressed as a ratio of water-weight to dry sediment-weight.

water drive. Energy within an oil or gas pool, resulting from hydrostatic or hydrodynamic pressure transmitted from the surrounding aquifer.

waterfall. A steep to vertical descent of a stream course where it drops over the edge of a plateau or where erosion has worn away overhanging softer rock .Because of the force of the water, all waterfalls must necessarily be relatilvely ephemeral features.

water gaps. A deep and narrow pass that into to the base of a mountain ridge, and through which a stream flows.

waterline. (1) *See* shoreline. (2) *See* water table.

watermass. A body of a water identified by its temperature-salinity curve on chemical composition, and normally consisting of a mixture of two or more water types.

water opening. A break in sea-ice, leaving the sea surface.

water table. The position in the earth's crust below which cracks and other openings are filled with water. Above the water table (zone of aeration), the interstices in earth materials are partly filled with air . In the zone of saturation the interstices are completely filled with water. This zone extends from the water table downward. The position of the water table below the ground surface is a function of the topography of an area and of local climatic conditions.

water-table well. A well that taps into unconfined groundwater. Its water level does not nesessarily lie at the water-table level.

water type. Ocean water of a specified temperature and salinity.

water vein. 1. Ground water in a crevice or fissure in dense rock. 2. A term generally applied to any body of ground water, in part because dowsers commonly describe water occurring in veins. The term is not in common use in hydrology.

wave base. The depth at which sediments are not stirred by wave action, usually about 10 metres. Also known as wave depth.

wave basin. A basin close to the inner entrance of a harbour in which the waves from the outer absorbed, thus reducing the size of the waves.

wave-built plateform. A gently sloping coastal surface built by sediment and loose material at the seaward or lakeward edge of a wave-cut plateform. Also known as wave-built terrace.

wave-built terrace. *See* wave-built plateform.

wave-cut bench. A level to gently sloping narrow surface produced by wave erosion, extending outward from the wave -cut cliff; it is developed mainly above water level by storm waves , weathering, and rainwash. The bench may be bare rock or it may be temporarily covered by a beach.

wave-cut cliff. A cliff formed by the erosive action of waves on rock .

wave-cut plateform. A gently sloping surface produced by wave erosion, extending far into the sea or lake from the base of the wave-cut cliff. Also known as wave-cut terrace; wave plateform.

wave-cut terrace. *See* wave-cut plateform.

wave-depth. *See* wave base.

wavellite. $2[Al_6(PO_4)_4(OH)_6.9H_2O]$. A white, yellow, green or black mineral crystallizing in the orthorhombic system and occurring in small hemispherical aggregates. It shows perfect cleavage in one direction. Its hardness on Mohs scale is 3.5-4.0.

wave-up criteria. Symmetrical ripple marks have cusps which point upwards. Rain prints, trails, and foot-prints of organisms and falling pebbles all form depressions in the upper surface of the beds. Desiccation cracks taper downwards and are filled with sediment from the overlying beds. Washouts and channels cut intc the underlyling beds. Erosion on a sma' scale may also remove the tops of other features such as ripple marks, slump structures, etc. In volcanic rocks, tongues of lava may penetrate underlying sediments but care must be taken to distinguish this from deposition of sediment on an irregular surface of the lava, particularly when observing vertical sections. Vesicles in lava flows are generally concentrated at the top of a flow. A weathered or eroded upper surface may sometimes be recognized. Pipe amygdales, however, are more abundant in the lower parts of the flows, and if they are branched they bifurcate downwards. Pillow lavas have a rather rounded bulbous top and a more pointed, tapering, lower extremity. However, if the rocks have been extensively deformed these criteria may be difficult to establish.

***W*-chert.** Chert modules formed by weathering.

W-dolostone. Dolostone produced by weathering.

weathered layer. In seismology, the zone of the earth that is immediately below the surface, characterized by low seismic-wave velocities.

weathering. The initial stage of denudation, wearing down of rocks on the surface without their being moved by temperature effects, wind, precipitation, living agents, and chemical action. An essential feature of the process is that it affects rocks in situ; and no transportation is involved. This is the factor which distinguishes it clearly from erosion. The main types of weathering are : *(i)* frost wedging : resulting from the freezing and expanding of water in the joints breaking the rockmass into individual blocks or fragments; *(ii)* root wedging, resulting from the growth of tin roots which wedge joints apart from the pressure of their expansion; and *(iii)* spheroidal weathering, resulting from the expansion of the rockmass as it chemically weathers. The temperature variations related to forest and brush fires may physically weather rock by thermal expansion and dehydration.

websterite. An igneous rock of the ultrabasic clan. A coarse-grained granlular plutinilc rock; contains hylpersthlene and diopside.

wedge out. Disappearance of a thinning, tensing or truncated layer as its thickness reduces to zero.

weeping core. Applied to a core cut from a formation brought ot the surface covered with drops of fluid that have the general appearance of tears, such a formation is usually low in oil-content.

wehrlite. An igneous rock of the ultrabasic clan. A coarse-grained granular plutonic rock; contains olivine, dialage, and hornblende.

weinschenkite. (1) $YPO_4.2H_2O$. A white mineral consisting of hydrous yttriulm phosphate. Also known as churchite. (2) Variety of hornblende, that is enriched in iron and aluminium.

weiselbergite. An igneous rock of the diorite clan. A fine-grained porphyritic extrusive rock; contains phenocrysts of plagioclase and augite in a groundmass of augite, plagioclase, and glass.

Weissemberg method. A method of studing crystal struture by X-ray diffraction in which the crystal is rotates a beam of X-rays, and a photographic film is moved parallel to the axis of rotation; the crystal is surrounded by a sleeve which has a slot that passes only diffraction spots from a single layer of the reciprocal lattice. Permitting positive identification of each spot in the pattern.

welded tuff. A pyroclasitic deposit hardened by the action of heat, pressure from overlying material, and hot gasses. Also known as tuff lava.

welding. Consolidation of sediments by pressure; water is squeezed out and cohering partilcles are brought within the limits of mutual molecular attraction.

well. A hole or bore sunk into the ground with the objective of *(a)* bringing petroleum and natural gas to the surface , *(b)* injecting water into the ground to increase the petroleum production, and *(c)* detonating explosives during seismic prospecting.

well log. A record of variations with depth in a well of various geological features and parameters; including resistivity, spontaneous potential, radioactivity, acoustic velocity, gravity and lithology.

well shooting. A technique for measuring changes in acoustic velocity with depth in earth. This technique is prefered where it is desirable to avoid explosive damage to the well.

well ties. Wells existing or specially drilled yielding information to oil interpretation or test the accuracy of a seismic profile.

wellsite. $(Ba, Ca, K_2) Al_2Si_3O_{10}\text{-}3H_2O$. A colourless or white monoclinic mineral composed of hydrated barium potassium alumilnosilicate; occurs as prismatic crystals and does not show any type of cleavage. Its hardness on Mohs scale is 4.0-4.5.

wennebergite. An igneous rock of the syeniet clan. A fine-grained porphyritic hypabyssal rock; contains phenocrysts of orthoclase, biotite, and quartz in a groundmass of apatite, sphene, and glass.

Wentworth classification. A logarithmic grade for size classification of sediment particles starting at 1 mm and using the ratio of 1/2 in one direction and 2 in the other, providing diameter mits to the size classes of 1, 1/2, 1/4 etc. and 1,2, 4 etc.

Wentworth scale. A geometric grade scale for sedimentary particles ranging from clay particles (diameter less than 1/256 mm) to boulders (diameter greater than 256 mm), in which the size classe are related to one another by a constant ratio of 1/2 *i.e.*,(4,2,1,1/2,1/4, etc.).

wernerite. A mineral consisting of calcium sodium aluminosilicate, sulphate, carbonate and chloride (formula uncertain); occurs as pyramidal crystals and in massive form; shows subconchoidal fracture and distinct cleavage in two directions. Its hardness on Mohs scale is 5.0- 6.0.

westanite. A brick-red radiating crystalline variety of hydrous sillimanite.

white ice. Ice on or near the surface which still contains some air. At greater depths all the air is removed by compression, and the ice looks blue. As the ice chages form white to blue, there is melting followed by freezing, which is called regelation.

white mica. *See* muscovite.

white nickel. *See* rammelsberbergite.

white water. Frothy water as in whitecaps or breakers.

whitleyite. An achondritic meteorite consisting essentially of enstatite with fragements of black chondrite.

whitlockite. $Ca_9(Mg,Fe)H(PO_4)_7$. A rare mineral that exist in the form of hexagoral crystals.

whitneytie. Variety of native copper containing arsenic.

wiborgite. *See* rapakive.

wichtisite. *See* tachylite.

wilkeite. $2[Ca_5(Si,P,S)_3O_{12}(O,OH,F)]$. A pink or yellow hexagonal mineral composed of complex basic calcium silicate; occurs as prismatic crystals. It shows imperfect basal cleavage. Its hardness on Mohs scale is 5.0.

willemite. $6[Zn_2SiO_4]$. A white or greenish-yellow, or black hexagonal mineral composed of zinc silicate; occurs as prismatic crystals or in massive form. It shows vitreous to greasy lustre, conchoidal to uneven fracture and cleavage visible in one direction. It occurs with franklinite and zincite, constituting a green, red, or black mixture, also as large crystals in the marble host rock. It is not uncommon in the oxidised zone of zinc-bearing deposits, especially in arid regions, but it is generally inconspicuous and easily overlooked. Its hardness on

Mohs scale is 5.5.

williamsite. Impure apple-green lamellar variety of antigorite.

willyamite. Cobalt rich variety of ullmannite.

wilsonite. Magnesium potassilum alulminosllicate of uncertain formula. Its colour varies from pinkish-white to red; an altered scapolite.

wiluite. A synonoym of grossular and idocrase .

winchite. Blue variety of tremolite.

wind crust. A type of snow crust, formed by the packing action of wind on perviously deposited snow; wind crust may break locally but, unlike wind slab, does not constitute an avalanche hazard.

wind gap. A shallow, relatively high-level notch in the upper part of a mountain ridge usually an abandoned water gap. Most wind gaps are produced by glacial breaching; river capture; climatic changes, accidental diversion.

window. (1) A break caused by erosion of a thrust sheet or a large recumbent anticline that exposes the rocks beneath the thrust sheet. (2) The unfrozen part of a river surrounded by river ice during winter season.

window ice. A thin deposit of ice which forms by the freezing of many tiny drops of water that have condensed on the indoors side of a cold window surface.

windrow. Any accumulation of material formed by wind or tide action.

wind slab. A type of snow crust; a patch of hardpacked snow, which is packed as it is deposited in favoured spots by the wind, in contrast to wind crust.

windsorite. An igneous rock of the granite clan. A fine-grained granular aplite; contains milcro-perthie, orthoclase, oligoclase, quartz, and biotite.

winterbourne. A stream which flows in winter but dries up in summer when the water-table is lower . Such streams are found mainly in chalk areas. The name winterbourne is used only in sothern England.

Wisconsin. Pertaining to the fourth and last, glacial stage of the Pleistocene epoch in North American; followed by the Sangamon interglacial, beginning about 85000+15000 years ago and ending 7000 years ago.

winthamite. Red or yellow variety of epidote from Scotland.

witherite. $BaCO_3$. A barium mineral, found in hydrothermal veins associated with barytes and galena. Much of it occurs in low-temperature hydrothermal veins with fluorite, barite, and/or galena. It has also been found with sanbornite, gillespite, and other barium silicates in several deposits. It shows uneven fracture and good cleavage in one direction. Its hardness on Mohs scale is 3.0-3.5.

wolframite. $(Fe(II),Mn)WO_4$. An ore of tungsten found in hydrothermal and pneumatolite veins. It shows uneven fracture and perfect cleavage in one direction. The wolframite minerals form a complete series between ferberite ($FeWO_4$) and huebnerite ($MnWO_4$); nearly all specimens show intermediate composition, and wolframite with 20-80 atomic per cent iron (and 80-20 per cent manganese) is the mineral that occurs most widely. Most wolframite occurs in greisens, quartz-rich veins, or complex pegmatitic masses associated with granitic intrusive rocks. It also occurs in high-temperature hydrothermal veins associated with pyrrhotite, pyrite, chal copyrite and bismuthinite. The wolframite group minerals also occur in some moderate- and low-temperature veins along with sulphides, cassiterite,scheelite, quartz, and/or siderite; in contact metamorphic deposits adjacent to granitic intrusives; and in alluvial deposits. It is a main ore of tungsten. Its hardness on Mohs scale is 4.0-4.5

wollastonite. $CaSiO_3$. A mineral of the pyroxenoid group, formed by the thermal metamorphism of impure limestone. It occurs as tabular crystals; shows uneven fracture, and perfect cleavage in one direction , Its hardness on Mohs scale is 4.5-5.0.

woodendite. An igneous rock of the syenite clan. A black, fine-grained granular extrusive rock; contains augite, enstatite, olivine, and magnetite, in a glass of feldspar composition.

wood opal. Yellow or brown variety of opal petrifying wood.

woodruffite. $Zn,Mn(II)_2(Mn(IV)_5O_{12}.4H_2O$. A black mineral composed of hydrated manganese zinc oxide; occurs as botryoidal masses; shows brownish - black streak and conchoidal fracture. Its hardness on Mohs scale is 4.5.

wood tin. Botryoidal or renilform variety of cassterite with a radiating fibrous habit.

wrench fault. A near-vertical fault along which strike separation has occurred .

wulfenite. $4[PbMoO_4]$. A yellowish-orange tetragonal mineral composed of lead molybdate; occurs as tabular crystals as well as granular masses. It is a secondory mineral formed in the oxidise zone of ore deposits that contain lead and molybdenum minerals. It is commonly associated with pyromorphite, vanadinite, cerussite, limonite, calcite, and other secondary minerals It shows subconchoidal to uneven fracture and good cleavage in one direction. Its hardness on Mohs scale is 3.0.

wulff net. A coordinate system used in crystallography to plot a polar stereographic projection with conservation of equal angles, such as for plotting angular relations obtained from universal-stage measurements.

Wurm. The fourth of the four classical glacial stages of the Pleistocene of Europe, above the Riss.

wurtzilite. A black massive infusible asphaltic pyrobitumen, insoluble in turpentine; derived from metamorphism of petroleulm.

wurtzite. [Zn,FeS]. A brownish-black hexagonal mineral composed of zinc sulphide; occurs as pyramidal or prismatic crystals; shows brown streak and cleavage visible in one direction. Most wurtzite, like sphalerite, contains iron, and may also contain cadmium and manganese in substitution for zinc; CdS occurs with the wurtzite structure as the mineral greenocekite. The wurtzite structure can be described as a stacking of tetrahedral sheets of ZnS in a sequence *AB*, *AB*,........, parallel to (0001). The Zn_4S tetrahedra all point in the same direction, with a triangular face parallel to (0001). The structure is therefore hemimorphic and distinct from sphalerite, in which the ZnS tetrahedral sheets are stacked in a sequence *ABC*, *ABC*,........

wyomingite. A feldspathoid igneous, pinkish-red, fine-grained porphyritic extrusive rock. It contains phenocrysts of phlogopote in a groundmass of leucite and diopside.

xanthite. A yellowish-brown variety of idocrase.

x-cut. A quartz-crystal cut made in such a manner that the *x*-axis is perpendicular to the faces of the resulting slab.

xeno-. A prefix meaning strange; foreign.

xenocryst. A crystal resembling a phenocryst in an igneous rock, but unrelated to the rock body in which it occurs.

xenolith. A foreign inclusion in an igneous rock. In some cases the assimilation of xenoliths is only physical, whereas in other instances these foreign rock bodies are melted or dissolved, and thus result in chemical contamination.

xenomorphic. Without external crystal appearannce yet orystalline internally.

xenothermal. Pertaining to a mineral deposit formed at high temperature but at shallow to moderate depth.

xenotime. YPO_4. A yellowish-brown to reddish-brown, transparent to opaque rare earth mineral consisting of yattrium phosphate; occurs as a minor accessory mineral in acidic and alkalic igneous rocks, and as larger crystals in the associated pegmatites. It is also a common detrital mineral and has been recorded as occurring in mica- and quartz-rich gneisses.

xenotlite. $CaSiO_3.2H_2O$. A white, bluish-grey, or pink monoclinic mineral composed of hydrated calcium silicate; shows splintery fracture. Its hardness on Mohs scale is about 6.5.

Xeret. A suborder of the soil order Vertsol; formed in a xeric moisture regimle.

Xeret. A suborder of the soil order Vertisol; formed in a Mediterranean clilmate; wide surface cracks open and close once a year.

xeric. (1) Of or pertaining to a habit having a low or inadequate supply of moisture. (2) Of or pertaining to an organism living in such an environment.

Xeroll. A suborder of the soil order Mollisol, formed in a xeric moisture regime; may have a calcic, petrocalcic, or gypsic horizon, or duripan.

xerothermic period. A postglacial interval of a warm and dry climate.

Xerult. A suborder of the soil order Ultisol, formed in a xeric moisture regime; brownish or reddish soil with a low to moderate organic carbon-content.

xyloid coal. A variety of brown coal having the fibrous structure of wood. Also known as board coal; bituminous wood; xyloid lignite.

xyloid lignite. *See* xyloid coal.

yamaskite. An igneous rock of the gabbro clan. A coarse-grained, granular plutonic rock; contains hornblende, titanaugite, and anorthite.

yardang. An elongated ridge carved mostly from relatively weak earth materials by wind abrasion is a desert region. The ridges are separated by steep-sided troughs, both the ridges and troughs lying parallel to the prevailing winds.

yarmouth interglacial. The second interglacial stage of Pleistocene epoch in North America, following the Kansan glascial stage and before the Illinoian.

yatalite. An igneous rock of the diorite clan. A coarse-grained granular pegmatite; contains uralite, albite, titanite, and quartz.

yazoo stream. A tributary that flows parallel to the main stream for a considerable distance before joining it at a deferred junction; especially such a stream forced to flow along the base of a natural levee formed by the main stream.

y-cut. A quartz crystal cut such that the *y*-axis is perpendicular, to the faces of the resluting slab.

yellow ground. A superficial oxidized deposit covering blue ground; found at the surface of diamond pipes, *e.g.*, South Africa.

yenite. A synonym of ilvaite.

young mountain. A recently formed mountain. The young mountains include the highest mountains in the world. Fold mountains fomed by alpine earth movements have undergone only 40 or 50 million year's erosion and, on geological time scale, this makes them a young landscape feature. Although the alpine folding created several large mountain ranges, there a few smaller hills formed around the edges of the main ranges.

young valley. A valley in its early stages of development, it is characterized by a relatively straight orientation, *V*-shape cross- profile, and high gradient.

yoderite. $4[(Al,Mg,Fe)_2\,Si_2(O,OH)_5]$. A monoclinic mineral composed of basic aluminium magnesium iron silicate. Its hardness on Mohs scale is about 6.0.

young ice. Newly formed ice in the transitional stage of development from ice crust to winter ice.

young valley. A valley in its eary stages, when it is relatively straight and has a high gradient; a *V*-shaped cross-section, and short tributaries.

yttrocerite. A variety of fluorite with cerium replacing calcium, along with the presence of minor amounts of yttrium.

yttrofluorite. A granular massive variety of fluorite with yttrium replacing calcium.

yugawaralite. $Ca_4Al_7Si_{20}O_{54}..14H_2O$. A colourless or white monoclinic mineral composed of hydrated calcium aluminium silicate; shows imperfect cleavage. Its hardness on Mohs scale is 4.5.

yuksporite. A red mineral composed of calcium strontium barium and alkali silicate and fluoride of uncertain composition and crystal system; occurs as lamellar masses or fibres.

Z

zaratite. $Ni_3(CO_3))(OH)_4.4H_2O$. An emerald-green mineral consisting of a hydrated basic nickel carbonate; occurs in incrustations or compact masses. It shows green streak and conchoidal fracture. Its hardness on Mohs scale is about 3.5.

zeolite. A large group of hydrous aluminosilicates of alkali metals (especially sodium, potassium and calcium), characterized by their reversible loss of water of hydration, and by their fusion with a bubbling action when strongly heated. They form a well-defined group of species that are closely related to one another in composition, in conditions of formation, and hence the mode of occurrence. Their general formula is: $A_mB_rO_{2r}.xH2O$, where, *A* is mainly Na and Ca (K, Ba and Sr to a lesser extent), *B* is Si+Al where, Si:Al is 1:1, and the (Al+Si):O ratio is always 1:2. The compositional ranges of individual eolite species, however, are narrow, which is indicative of a narrow range of stability for each of the species and also may serve as an explanation for the existence of a larger number of species within the group. Zeolites may be deposited from the late magmatic solutions, by hydrothermal solutions, or they may be formed during diagenesis. Many are products of alteration of feldspars, feldspathoids, or volcanic glass. They are especially common as amygdules, fracture fillings, and fillings in other cavities in many kinds of rocks, especially in basalts. They are characteristic of a low-temperature, low-pressure metamorphic facies, known as **zeolite facies.** They also occur in some metalliferous veins, especially where the gangue is calcite, and in the deposits of hot springs, where they have in some cases been formed within the historic time and are still being formed. The zeolites are commonly associated with calcite, prehnite, pectolite, apophyllite, and datolite. Zeolites find numerous applications as ion-exchangers, as soil conditioners, and as acid-resistant adsorbers in gas drying.

zeolite facies. Metamorphic rocks formed in the transitional period from diagensis to metamorphism, at pressures of about 200-3000 bars and temperatures of 475-575K.

zeolitization. Introduction of, or replacement by a zeolite mineral.

zero curtain. The layer of ground between the active layer and permafrost where the temperature remains nearly constant at about 273 K.

zeugen. Tabular erosional remnants formed in desert regions by the erosional effect of sand-laden winds; masses of resistant rock resting on supports of softer rock, becoming separated by erosion from the once continuous horizontal stratum.

zinc blende. *See* sphalerite.

zincite. 2[Zn,MnO]. An orangish-yellow to red hexagonal mineral composed of zinc oxide; shows conchoidal fracture and perfect cleavage in one direction. Its hardness on Mohs scale is about 4.0. Zincite is a rare mineral, and occurs with willemite nad franklinite in calcite, largely in granular form.

zinckenite. $12[Pb_2Sb_4S_{27}]$. A steel-grey hexagonal mineral composed of lead antimony sulphide; occurs as prismatic crystals. It shows uneven fracture and indistinct cleavage. Its hardness on Mohs scale is 3.0-3.5.

zinnwaldite. $2[K_2(Li,Fe,Al)_6(Si,Al)_8O_{20}(F,OH)_4]$. A brown, yellow,or pale-violet monoclinic mineral composed of basic potassium, lithium, iron and aluminium fluoroaluminosilicate; shows perfect basal cleavage. Its hardness on Mohs scale is 2.5-3.0.

zippeite. $2[(UO_2)_3(SO_4)_2(OH)_2.8H_2O]$. An organish-yellow orthohombic mineral composed of hydrated basic uranyl sulphate; crystals are rare, normally occurs as crusty or earthy masses; shows perfect cleavage in one direction.

zircon. $ZrSiO_4$. A brown, green, pale-blue, red, orange, golden-yellow or colourless neosilicate mineral occurring as tetragonal prisms; shows conchoidal fracture and cleavage visible in one direction. Its hardness on Mohs scale is 7.0. It occurs in granites, syenites, schists, and crystalline limestones; its varieties are cyrtolite, hyacinth, jargon, and zirconite. Zircon is occasionally radioactive because of the presence of thorium and uranium, replacing zirconium in the structure. The presence of thorium and uranium makes it a useful mineral for the age-determination of such rocks. It is a principal source of zirconium and hafnium for industry, and is extracted from sands, the most important sources being beach deposits.

zirconite. A grey or brown variety of zircon.

zobtenite. A variety of flaser gabbaro from Germany.

zoisite. $4[Ca_2Al_3Si_3O_{12}OH]$. A white, brown, green, grey, or rose-red orthorhombic mineral of the epidote group; consisting of a basic calcium aluminium silicate and occurring massive or in prismatic crystals; shows uneven to subconchoidal fracture and perfect cleavage in one direction. Its variety tanzanite, found in Tanzania, is a prized gem, which is an essentially iron-free zoisite, the colour of which is attributed to the presence of vanadium.

zonal soil. Accoarding to early classification systems in the United States, a soil order including soils with well-developed characteristsics that reflect the influence of agents of soil genesis. Also known as mature soil.

zonation. The condition of being arranged in zones.

zone. A band, belt, layer, or strip of earth material such as rock or soil.

zone axis. The line of crystallographic direction through the centre of a crystal which is parallel to the intersection edges of the crystal faces defining the crystal zone.

zone of aeration. A subsurface zone containing water below atmospheric pressure and air or gases at atmosheric pressure.

zone of cemenataton. The layer of the earth's crust in which unconsolidated deposits are cemented by percolating water containing dissolved minerals from the overlying zone of weathering.

zone of fracture. (1) The outer rigid part of a glacier, in which the ice is much fractured. (2) The upper, brittle part of the earth's crust, in which deformation is by fracture.

zone of leaching. The top layer of the soil, characterized by the downward movement of water. Part of the rain falling on the ground percolates downward through the soil, which tends to leach or wash down, some of the soil materials movement of water. Part of the rain falling on the ground percolates downward through the soil, which tends to leach or wash down, some of the soil materials to the lower levels. In a wet and humid climate, clay minerals and iron oxides, and dissolved calcite, are leached downward.

zone of plastic flow. The part of the earth's crust that is under sufficient pressure to prevent fracturing. *i.e.*, ductile, so that deformation is by flow.

zone of saturation. A subsurface zone in which water fills the interstices and is under pressure greater that atmospheric pressure. If a well were filled in this zone, ground water would fill the lower part of the well. The water level inside the well marks the upper surface of the zone of saturation. Above the zone of saturation, the rock openings are filled partly with water and partly with air.

zone of sphere. A portion of the surface of a sphere cut off by two parallel planes. Its area is given by $2\pi rd$, where r = radius of the sphere, and d = distance between the two planes.

zone of weathering. The layer of the earth's crust above the water table that is subject to destructive atmospheric processes, and in which soils develop.

zorsite. $Ca_2Al_3Si_3O_{12}(OH)$. A white, grey, brown, green or red, orthorhombic mineral of the epidote group; an essential constituent of saussurite.

Z-phenomenon. A possible time lag (a few seconds or less) between the issuanace of *P* and *S* waves from an earthquake focus.

zunyite. $4[Al_{13}Si_5O_{20}(OH,F)_{18}Cl]$. A cubic mineral composed of basic aluminium fluosilicate and chloride; occurs as tetrahedral crystals. Its hardness on Mohs scale is about 7.0.

zussmanite. $K(Fe(III),Mg,Mn,Al)_{13}(Si,Al)_{18}O_{42}(OH)_{14}$.. A basic aluminosilicate of potassium and iron; shows perfect cleavage in one direction; occurs in metamorphosed shales, ironstones, and limestones.

zweikanter. A ventifact having two faces intersecting in two sharp edges.

Determinative Tables

In the identification of minerals, a set of determinative tables is of great importance. The whole detective process is stimulating, and determining the identity of an unknown mineral is a most satisfying triumph.

No set of tables does away with the necessity of individual enterprise. Although the user who follows the general precepts and special advice of determinative tables will certainly be helped to avoid the waste of time, there is no foolproof system in the domain of mineral identification. The best way of mineral identification is to "be smart" and "remember everything"!

The primary basis of discrimination used in the following tables is lustre—that is, whether the mineral appears metallic or nonmetallic. A few opaque or nearly opaque minerals with a dark brown to black colour are sometimes described as having submetallic lustre; they are grouped with the minerals with metallic lustre, but some of them are also tabulated with the nonmetallic group.

The secondary basis of discrimination is hardness. Relative hardness is easily determined with the simplest of tools, and it is one of the few macroscopically determined properties capable of being expressed in numerical terms. The tables are divided into three groups: minerals with the hardness less than 2.5– (can be scratched with the fingernail); minerals with hardness between 2.5 and 5.5 (will scratch the fingernail but can be scratched with a knife blade); and minerals with hardness greater than 5.5—(will scratch the knife blade). With a little practice, a student should be able to place figure on the hardness of a mineral in the 2.5 — 5.5 group on the basis of its relative ease of scratching. The hardness given for each mineral is that of clean smooth surface; minerals in earthy varieties, such as some forms of hematite, appear to be softer than the same minerals in good crystals.

Minerals with nonmetallic lustre far exceed in numbers those with metallic lustre. Therefore, it is convenient to find a further criterion for subdeciding the nonmetallic minerals. The following tables for the nonmetallic minerals are subdivided into two sections based on colour; one section includes those minerals that are colourless or white, unless tinted by impurities. The latter also have coloured streaks whereas the former, even if coloured, tend to have white streaks.

By using lustre and hardness, and dividing the nometallic minerals into two sections on the basis of colour, we thus obtain a set of tables comprising nine groups, as follows:

Lustre	Hardness	Number of Minerals
	H < 2.5	4
Metallic	H = 2.5-5.5	29
	H > 5.5	19
	H <2.5	8
Nonmetallic	H = 2.5-5.5	36
Colorless or White	H > 5.5	20
	H < 2.5	11
	H = 2.5-5.5	32
Colored	H > 5.5	46

Within each group, the minerals are described in order of least to greatest specific gravity. Specific gravity is, of course, not so easily estimated as hardness is; however, practice in handling minerals of known density generally develops one's ability to distinguish specific gravities differing by one unit; thus, the user should eventually be able to say whether a specimen has a specific gravity near 3, or 4, or 5, or greater than 5. With a small bottle of tetrabromoethane (acetylene tetrabromide), which has a density of 2.96, it takes only a moment to determine whether a fragment of an unknown mineral sinks or floats—that is, whether it is a heavy or a light mineral.

Having determined lustre and hardness and made a reasonable estimate of the specific gravity, the choice is narrowed to comparatively few species; thence, the other physical properties—particularly crystal form, cleavage, and colour—will often lead to the correct identity.

Lustre metallic; H < 2.5 [SG ≡ Specific gravity ; H ≡ Hardness]

Name Formula	Colour	Streak	SG	H	System and habit	Cleavage	Observations
Graphite C	Black	Black; black on glazed porcelain	2.09 –2.23	1–2	Hexagonal; hexagonal plates or massive, foliated or earthy	{0001} perfect	Greasy feel; plates flexible, inelastic; marks paper; commonly occurs with calcite in metamorphic rocks
Stibnite Sb_2S_3	Black	Grey	4.6	2	Orthorhombic; prismatic crystals enlogated ‖ *c*; Bladed masses	{010} perfect	Brittle; occurs in quartz veins
Molybdenite MOS_2	Blue-grey	Blue-grey; greenish black on glazed porcelain	4.62 –4.73	1–1.5	Hexagonal; hexagonal plates or massive, foliated	{0001} perfect	Like graphite, but heavier and more metallic; marks paper; generally associated with quartz or feldspar
Sylvanite $(Ag,Au)Te_2$	White to grey	Grey	8.2	1.5–2	Monoclinic; prismatic and tabular crystals, also massive bladed	{010} perfect	Occurs in quartz veins with gold on telluride minerals

Lustre metallic; H = 2 .5 – 5.5

Nams Formula	Colour	Streak	SG	H	System and Habit	Cleavage	Observations
Goethite FeO(OH)	Brown, black	Yellow to brown 4.3	3.3–	5–5.25	Orthorhombic; typically massive, botryoidal or stalacitic	{010} perfect; {100} good	Earthy varieties known as limonite; cement and colouring matter in sedimentary rocks; forms pseudomorphs after pyrite
Chalcorpyrite $CuFeS_2$	Brass-yellow (irides-cent tarnish)	Greenish, black	4.1–4.3	3 .5–4	Tetragonal; sphenoidal crystals; and typically m assive, compact	{011} indistinet	Commonly tar-nished blue or grey; occurs with pyrite, galena, sphalerite; yellower and softer than pyrite
Manganite MnO(OH)	Black	Brown	4.3	4	Monclinic (pseudoorthorhomic) prismatic crystals; commonly elongated ǁc	{010} perfect; {110}, {001} good	Occurs in hydrothermal depsoits; alters to pyrolusite, the streak then becoming black
Enargite Cu_3AsS_4	Black	Black	4.4	3	Orthorhombic; tabular or prismatic crystals; also massive, compact or granular	{110} perfect; {100}, {010} distinct	Widely associated with chalcocite and bornite

Continued

Name Formula	Colour	Streak	SG	H	System and habit	Cleavage	Observations
Pyrrhotite Fe1—*S	Bronze-yellow-	Grey-black	4.58–4.65	3-.5–4.5	Hexagonal; typically massive, granular or compact	None	Magnetic; surface generally tarnished brown
Psilomelane Romanechite $BaMn_9O_{16}$ $(OH)_4$	Black	Black	4.7	5–6	Monoclinic; massive, commonly botryoidal	None visible	Secondary mineral, weathering product of other manganese minerals
Covellite CuS	Indigo-blue to black	Black-grey	4.6–4.76	1.5–2	Hexagonal; hexagonal plates; also massive, foliated or granular	{0001} perfect	Uncommon copper mineral; generally associated with chalcocite or enargite
Pentlandite $(Fe,Ni)_9S_8$	bronze-yellow	Brown	4.6–5.0	3.5–4	Isomatric; massive	None; parting on {111}	Occurs intermixed with pyrrhotite
Tetrahedrite $Cu_{12}Sb_4S_{13}$ Tennantite $Cu_{12}As_4S_{13}$	Black	Brown to black	4.6–5.1	3–4.5	Isometric; tetrahedral crystals; also massive, granular or compact	None	Widely distributed in hydrothermal Cu deposits; commonly contains silver
Bornite Cu_5FeS_4	Bronze brown, purple tarnish	Grey-black	5.07	3	Isometric; typically massive, granular or compact	{111} in traces	Widely associated with chalcopyrite or chalcocite; purple tarnish characteristic

Continued

Name Formula	Colour	Streak	SG	H	System and habit	Cleavage	Observations
Pyrolusite MnO_2	Grey to Black	Black	5.0–5.2	5 .5	Tetragonal; massive, columnar, fibrous, or powdery	{110} perfect	Secondary after other Mn minerals; typically soft and earthy
Millerite NiS	Pale brass-yellow	Greenish black	5.5	3– 3.5	Trigonal; typically as hairlike crystals, generally in radiating groups	{10$\bar{1}$1}, {01112} perfect	Commonly occurs in cavities associated with calcite or siderite
Chalcocite Cu_2S	Grey to black	Grey to black	5.5–5.8	2.5–3	Orthorhombic; typically massive, compact or granular	{110} indistinct	Sectile; typically has sooty-appearing coating
Arsenic As	Grey-White tarning to dark grey	Grey	5.7	3.5	Hexagonal; typically massive, often botyroidal	{0001} perfect	Volatilizes without melting
Pyragryrite Ag_3SbS_3	Deep red to black	Red	5.6–5.9	2–2.5	Trigonal; prismatic ‘‘crystals; also massive, granular	{10$\bar{1}$1} good	Widely associated with galena in Pb and Ag ores
Bournonite $PbCuSbS_3$	Grey to black	Grey-black	5.83	2.5–3	Orthorhombic; short prismatic or tabular crystals; also massive, granular or compact	{010} indistinct	Occurs in veins with other sulphides, commonly with galena and tetrahedrite
Boulangerite $Pb_5Sb_4S_{11}$	Bluish lead-grey	Brownish grey	6.23	2 .5–3	Monoclinic; long prismatic crystals, also fibrous and hairlike	{100} good	Occurs in vein deposites associated with other Pb minerals

Continued

Name Formula	Colour	Streak	SG	H	System and habit	Cleavage	Observations
Antimony Sb	White to grey	Grey	6.7	3–3.5	Hexagonal; massive, lamellar	{0001} perfect	Commonly coated with white antimony oxide
Argentite Ag_2S	Black	Dark grey-black	7.2	2–2.5	Isometric; cubes or octahedra; also massive	{001} {011} indistinct	Sectile; occurs with other silver minerals
Wolframite $(Fe,Mn)WO_4$	Brownish black	Brown-black	7.1–7.5	4–4.5	Monoclinic; crystals commonly tabular ∥ {100}; also massive, bladed or compact	{010} perfect	Resembles columbite-tantalite, but distringuished by perfect cleavage; occurs inuartz veins
Galena PbS	Lead-grey	Lead-grey	7.6	2.5	Isometric; cubes or cubo-octahedra; generally massive, granular	{001} perfect	The commonest lead mineral; commonly occurs with sphalerite
Iron Fe	Grey	Grey	7.8	4	Isometric; massive	{100} not prominent	Generally as meteorites, which al- ways contains Ni; magnetic
Nickeline NiAs	Copper-red	Brown-black	7.8	5–5.5	Hexagonal; typically massive, compact or granular	None	Generally coated with green annabergite
Copper Cu	Copper-red, typically tarnished	Red; metallic	8.9	2.5-3	Isometric; massive and in dendritic or wirelike forms	None	Commonly in cavities in basic igneous rocks; malleable

Continued

Name Formula	Colour	Streak	SG	H	System and habit	Cleavage	Observations
Calaverite $AuTe_2$	Brass-yellow	Green-grey	9.2	2.5–3	Monoclinic; bladed crystals or massive	None	Melts easily giving globules of Au
Bismuth Bi	Silver-White, creamy, reddish	Silver-white	9.7–9.8	2–2.5	Hexagonal indistinct crystals, commonly hopper-shaped; typically in platy masses	{0001} perfect	Occurs in quartz viens and some pegmatites; generally possess brassy tarnish
Silver Ag	Silver-white, commonly tarnished black	Silver	10.1–11.1	2.5–3	Isometric commonly in dendritic or wirelike forms	None	Occurs widely with barite and calcite in veins; tarnishes grey or black; malleable
Platinum Pt	Steel-grey	Steel-grey	14–19	4–4.5	Isometric; typically as nuggets or grains	None	Typically as grains in alluvial deposits may be magnetic because it contains iron
Gold Au	Yellow	Yellow	15–19	2.5–3	Isometric; typically in dendritic forms, or as alluvial grains or nuggets	None	Easiy recognized by its colour, softness, and malleability

Lustre metallic; H>5.5

Name Formula	Colour	Streak	SG	H	System and Habit	Cleavage	Observations
Anatase TiO_2	Commonly brown	White	3.9	5.5–6	Tetragonal; typically in acute pyramidal crystals	{001}, {101} perfect	Occurs in small crystals in joint planes and veins in schists and gneisses
Brookite TiO_2	Brown to black	White to grey	4.1	5.5–6	Orthorhombic; tabular and prismatic crystals	{120} indistinct	Occurs in same manner as anatase
Rutile TiO_2	Red-brown	Yellow to grey	4.2–4.3	6–6.5	Tetragonal; typically in long prismatic crystals; also massive, granular	{110} distinct	Common accessory mineral of igneous and metamorphic rocks
Chromite $FeCr_2O_4$	Black	Brown	4.5–4.8	5.5–6	Isometric; octahedral crystals rare; typically massive, granular or compact	None; {111} parting	Occurs in ultrabasic rocks and serpentinites
Ilmenite $FeTiO_3$	Black	Black	4.6–4.7	6	Trigonal; tabular crystals; typically massive, compact or granular	None; parting on {0001}, {011T2}	Occurs associated with basic igneous rocks and in sands derived therefrom
Branuite $3Mn_2O_3$. $MnSiO_3$	Grey to black	Grey to Black	4.7–4.8	6– 5.5	Tetragonal; pyramidal crystals or massive granular	{112} perfect	Ore mineral of manganese; occurs in veins and in metamorphic deposits
Hausmannite Mn_3O_4	Brownish black	Brown	4.84	5.5–6	tetragonal; pyramidal crystals or massive granular	{001} perfect	Occurs in same way as braunite; disitinguished from that mineral by brown streak

Continued

Name Formula	Colour	Streak	SG	H	System and Habit	Cleavage	Observations
Marcasite FeS_2	Light brass-yellow	Grey brown	4.89	6–6.5	Orthorombic; complex twinned crystals; also massive, stalactic or nodular	{101} distinct	Occurs in low-temperature surface and near-surface deposits
Pyrite FeS_2	Light brass-Yellow	Brown-black	5.0	6.5	Ismoetric; pyritohedra, cubes, or octahedra; also massive, compact or granular	{001} indistinct	The commonest sulfide mineral; alters to limonite
Franklinite $ZnFe_2O_4$	Black	Brown	5.1–5.3	5.5 – 6.5	Isometric; pyritohedra, cubes, or octahedra or massive granular	None {111} parting	Occurs in metamorphosed limestone; weakly magnetic
Magnetite Fe_3O_4	Black	Black	5.2	5.5 – 6.5	Isometric; octahedra or massive granular	None; parting on {111}	Accessory mineral of igneous and metamorphic rock, and in sands; strongly magnetic
Hematite Fe_2O_3	Red to black	Red-brown	5.26	6	Trigonal; tabular crystals; massive, earthy	None; parting on {0001}, {011$\bar{1}$2}	Common iron mineral of sediments and metamorphosed sediments
Arsenopyrite FeAsS	Silver-white	Grey balck	6.1	6	Monoclinic; prismatic crystals; typically massive, columnar or granular	{101} distinct	Generally occurs in metalliferous veins; gives garlic-like odour when scratched or pulverized
Cobaltite CoAsS	Silver White	Grey black	6.3	5.5–6	Isometric; pyritohedra,cubes, or massive, granular or compact	{001} perfect	Occurs is vein deposits; in some cases coated with pink erythrite

Continued

Name Formula	Colour	Streak	SG	H	System and Habit	Cleavage	Observations
Skutterudite $(Co,Ni)As_{2\text{-}3}$	Silver-white	Black	6.5	5.5	isometric; cubes and octahedra, also massive, granular or compact	{100}, {111} distinct	Occurs in vein deposits; in some cases coated with pink erythrite
Columbite-Tantalite $(Fe,Mn)(Nb,Ta)_2O_6$	Red-brown to black	Brown to black	5.2–8.0	6.5	Orthorhombic; taular or prismatic crystals, also massive	{010} distinct; {100} less distinct	Occurs in granite pegmatites
Cassiterite SnO_2	Brown to black	White	7.0	6.5	Tetragonal; prismatic crystals and massive, botyryoidal or concretionary	{100} distinct; {110} indistinct	Occurs in quartz viens and granite pegmatites; also detrital
Uraninite (pitchblende) UO_2	Black	Brown-black	8–10	6.5	Isometric; cubes; also massive, in some cases botryoidal	None	Occurs in granite pegmatites, hydrothermal veins, and disseminated in sandstones and limestones; yellow alteration products commonly present
Sperrylite $PtAs_2$	Tin-white	Black	10.6	6–7	Isometric; cubic crystals and rounded grains	{001} indistinct	Occurs in basic igneous rocks and in sands derived therefrom

Lustre nonmetallic; typically colourless or white unless tinted by impurities; streak white : H < 2.5

Name Formula	Colour	SG	H	System and Habit	Cleavage	Observations
Epsomite $MgSO_4.7H_2O$	White	1.67	2	orthorhombic; in acicular crystals, elongated ‖ c, or powdery crusts	{010} perfect; {101} distinct	Secondary mineral; soluble in H_2O; bitter and salty taste
Nitratite $NaNO_3$	Colourless, white	2.25	1.5–2	Trigonal; typically in granular masses	{10$\bar{1}$1} perfect	Soluble in H_2O; cooling and saline taste; occursin rain less deserts
Gypsum $CaSO_4$. $2H_2O$	Colourless, white; some is tinted pink, yellow, grey	2.32	2	Monoclinic; tabular Prismatic crystals. elongated ‖c, tabular ‖ {010}; swallow-tail twins; commonly granular, massive or fibrous	{010} perfect; {100, {011} distinct	Cleavage laminate flexible but not elastic
Monotmorillonite $Al_2Si_4O_{10}$-$(OH)_2$. xH_2O	White grey, greenish-grey	2.0–2.7	2	Monoclinic; clayey masses	{001}perfect, notmegascopically have soapy feel	Clay mineral; also orbs swater with swelling;
Brucite $Mg(OH)_2$	Colourless white, pale green	2.4	2	Trigonal; in hexagonal plates ro massive compact; in some cases fibrous	{0001} perfect	Occurs in association with dolomite or serpentine or in metamorphosed dolostone
Kaolinite $Al_4Si_4O_{10}$-$(OH)_8$	White, grey, brown	2.6	2	Triclinic; clayey masses	{001} perfect, not usually observable	Clay mineral; clayey odor when breathed upon; does not swell in water
Talc $Mg_3Si_4O_{10}(OH)_2$	Green	2.82	1	Monoclinic; in foliated or fine-grained	{001} perfect	Greasy feel; occurs in Al-rich metamorphic rocks

Lustre nonmetallic; typically colorless or white unless tinted by impurities; streak white; H = 2.5–5.5

Name formula	Colour	SG	H	System and Habit	Cleavage	Observations
Carnallite $KMgCl_3.6H_2O$	Colourless or white	1.60	2.5	Orthorhombic; typically in coarsely granular masses	None	Bitter taste, occurs in salt deposites;
Broax $Na_2B_4O_7.10H_2O$	White	1.71	2.5	Monoclinic; short Prismatic crystals	{100}, {110} perfect	Sweetish alkaline taste; deposited from salt lakes; effloresces in dry air
Sylvite KCl	Colourless, sometimes light orangish brown, or grey	1.99	2.5	Isometric; cubic crystals; typically massive, granular	{100} perfect	Salty taste like halite, but somewhat bitter, sectile
Opal $SiO_2.nH_2O$	Colourless, white	2.0–2.2	5.5	Amorphous; massive Compact, commonly mammillary	None	Secondary silica deposited at low. temperatures
Chabazite $CaAl_2Si_4O_{12}.6H_2O$	Colourless, white, pink	2.0–2.2	4	Trigonal; typically in cubslike rhombohedral crystals	{10111} indistinct	Occurs in cavities in basalts; crystal form distinguishes it from other zeolites
Halite NaCl	Colourless, white, red	2.16	2.5	Isometric; cubic crystals; typically massive, granular	{100} perfect	Characteristic salty taste
Heulandite $CaAl_2Si_7O_{18}.6H_2O$	White, light brown	2.2	3–4	Monoclinic; trapezoidal crystals tabular ǁ {010}	{010} perfect	Pearly lustre on cleavage; occurs in cavities in basalts; crystal form distinguishes it from other zeolites

Continued

Name formula	Colour	SG	H	System and Habit	Cleavage	Observations
Natrolite $Na_2Al_2Si_3O_{10}.2H_2O$	Colourless, white	2.25	5	Orthorhombic; long prismatic crystals and radiating nodular forms	{110} perfect	Occurs in cavities in basalts; crystal from distinguishes it from other zeolites
Analcime $NaAlSi_2O_6.H_2O$	Colourless white	2.26	5	Isometric; trapezohedral crystals; also massive, compact or granular	None	Occurs in cavities in basalts and in some igneous and sedimentary rocks
Laumontite $CaAl_2Si_4O_{12}.4H_2O$	White	2.25–2.30	4	Monoclinic; small prismatic crystals; also massive, commonly powdery	{010}, {110} perfect	Occurs in cavities in basalt; crystal form distinguishes it from other zeolites
Apophyllite $KCa_4(Si_4O_{10})_2(OH,F).8H_2O$	Colourless, White	2.35	5	Tetragonal; prismatic crystals with basal plane and pyramids	{001} perfect	Occurs in cavaties in basalts and in hydrothermal veins, generally with zeolites
Gibbsite $Ai(OH)_3$	White, red, brown	2.3–2.4	2.5–3.5	Monoclinic; typically massive, compact, pisolitic, or earthy	{001} perfect	A major component of bauxite
Colemanite $Ca_2B_6O_{11}.5H_2O$	Colourless, white	2.42	4.5	Monoclinic; short prismatic crystals, generally with complex terminations; also massive	{010} perfect {001}	Occurs as geodelike masses in sedimentary rocks
Calcite $CaCO_3$	Colourless, white	2.71	3	Trigonal; scalenohedral and rhombohedral crystals; also massive	{1011} perfect	May be coloured brown, grey, or black with impurities
Alunite $KAl_3(SO_4)_2–(OH)_6$	White, pink	2.6–2.9	3.5–4	Trigonal; massive, granular or compact	{0001} distinct	Generally occurs as hydrothermal alteration of volcanic rocks

Continued

Name formula	Colour	SG	H	System and Habit	Cleavage	Observations
Muscovite $KAl_2(AlSi_3O_{10})(OH)_2$	Colourless or pale tints	2.8–2.9	2.5–4	Monoclinic; typically in irregular platy crystals; also massive	{001} perfect	Large crsytals in granite pegmatites; small flakes in many metamorphic and some igneous rocks
Dolomite $CaMg(CO_3)_2$	White, yellow, grey, pink	2.85	3.5–4	Trigonal; small curved rhombohedral crystals and saddle-shaped shaped groups; also massive, granular	{10$\bar{1}$1} perfect	Effervesces in warm dilute HCl or in cold dilute HCl in cold freshly powdered
Pectolite $NaCa_2Si_3O_8(OH)$	White	2.86	5	Triclinic; typically in radiating aggreates of needlelike crystals	{001}, {100} perfect	Typically occurs associated with zeolites in cavities in basalt
Wollastonite $CaSiO_3$	White	2.9	5	Triclinic; typically in columnar or fibrous masses	{100} pefect {1102}, {001} good	Occurs in metamorphosed limestones
Aragonite $CaCO_3$	White	2.93	3.5–4	Orthorhombic; hexagonal twinned crystals; also massive, coralloid	{010} distinct	Effervesces in cold dilute HCl; much less common than calcite
Cryolite Na_3AlF_6	White	2.97	2.5	Monoclinic; massive, coarsely granular	None; parting on {001} and {110}	Wet-snow lustre; occurs in association with siderite
Anhydrite $CaSO_4$	White, grey, pale blue	2.9–3.0	3.5	Orthorhombic; typically massive, granular	{010} perfect {100}, {001} good	Forms extensive beds in sedimentary rocks; alters to gypsum
Magnesite $MgCO_3$	White	3.0	4	Trigonal; typically massive, compact	{10$\bar{1}$1} perfect	Cleavage not seen in compact forms, which tend to have conchoidal fracture

Continued

Name formula	Colour	SG	H	System and Habit	Cleavage	Observations
Boehmite $AlO(OH)$	White, yellow, brown	3.0–3.1	3	Orthornombic; massive, compact; commonly pisolitic	{010} not visible	Important constituent of bauxite
Hemimorphite $Zn_4Si_2O_7(OH)_2$–H_2O	White	3.4–3.5	5	Orthorhombic; typically massive, commonly in mammillary forms	{110} perfect	Alteration product of sphalerite; generally associated with smithsonite
Periclase MgO	White	3.58	5	Isometric; typically in small rounded grains	{100} perfect	Occurs as small grains commonly altered to brucite in metamorphosed dolomite
Strontianite $SrCO_3$	White	3.78	3.5	Orthorhombic; typically massive, columnar to fibrous	{110} good	Effervesces in cold dilute HCl; colours flamered
Celesite $SrSO_4$	White, pale blue	3.9	3	Orthorhombic; crystals tabular ‖ {001}, also massive, often fibrous	{001} perfect; {210} good	Red flame colouration
Witherite $BaCO_3$	White yellow,	4.3	3	Orthorhombic; hexagonal twinned crystals, also massive, columnar or granular	{010} distinct	Effervesces in cold dilute HCl; colours flame green
Smithsonite $ZnCO_3$	White, yellow, blue, green	4.4	4	Trigonal; typically massive compact, commonly concretionary	{10$\bar{1}$1} good	Efferversces in warm dilute HCl
Barite $BaSO_4$	Colourless, white, grey	4.5	3	Orthorhombic; crystals massive compact, also massive, granular	{001}, {210} perfect	Green flame colouration

Continued

Name formula	Colour	SG	H	System and Habit	Cleavage	Observations
Scheelite $CaWO_4$	White, yellow, brown	6.1	4.5 – 5	Tetragonal; typically massive, granular	{101} distinct	Fluoresces white or yellow in ultraviolet light
Anglesite $PbSO_4$	White grey	6.3	3	Orthorhombic; crystals prismatic or also massive, compact	{001} good {210} distinct	Commonly has a nucleus of galena
Cerussite $PbCO_3$	White, grey	6.5–6.6	3–3.5	Orthorhombic; in latticelike aggregates of twinned crystals; also massive, granular	{110}, {021} distinct	Efferversces weakly in cold dilute HCl, strongly in warm dilute HCl
Mimetite $Pb_5(AsO_4)_3Cl$	White, Yellow	7.2–7.3	3.5–4	Hexagonal; in small prismatic crystals; also massive, commonly globular and botryoidal	None	Occurs in oxidised zone of lead deposits

Lustre nonmetallic, typically colourless or white unless tinted by impurities; streak white; H > 5.5

Name Formula	Colour	SG	H	System and Habit	Cleavage	Observations
Tridymite SiO_2	Colourless, White	2.26	6.5–7	Hexagonal; small thin hexagonal plates, commonly twinned	None	Occurs in cavities in volcanic rocks
Cristobalite SiO_2	White	2.32	6.5	Isometric; minute octahedra or small rounded aggregates	None	Occurs in cavaities in volcanic rocks
Leucite $KAlSi_2O_6$	White grey	2.45	6	Isometric; trapezo-hedral crystals	{110} indistinct	Occurs as crystals in basic volcanic rocks
Nepheline $NaAlSiO_4$	White, brown grey	2.5–2.6	6	Monoclinic; prismatic also massive granular	{10$\bar{1}$0} indistinct	Occurs in many silica deficient igneous rocks gelantinizes in HCl
Orthoclase $KAlSi_3O_8$	White, flesh to salmon pink	2.56	6	Monoclinic; prismatic crystals, flattened ‖ {010} or elongated ‖ a; also massive, granular	{001} perfect; {010} good	Occurs in many common igneous and metamorphic rocks
Microcline $KAlSi_3O_8$	White, pink, green	2.56	6	Triclinic; like orthoclase	{001} perfect {010} good	Like orthoclase, but may show imperfect polysynthetic twinning striations or grid on {001}
Quartz SiO_2	Colourless, White, grey, pink, green	2.65	7	Trigonal; prismatic crystals terminated by rhombohedra; also massive, granular or compact (chalcedony)	None	Occurs in Si-rich igneous rocks, granite pegmatites, sedimentary and metamorphic rocks

Continued

Name Formula	Colour	SG	H	System and Habit	Cleavage	Observations
Scapolite $(Na,Ca)_4$–$[(Al,Si)_4O_8]_3$–(Cl,CO_3)	White, yellow, pink, grey	2.56–2.77	6	Tetragonal, prismatic crystals terminated by pyramids; also massive, granular or columnar	{100}, {110} distinct	Occurs in Ca-rich metamorphic rocks
Plagioclase (Na,Ca)–$(Al,Si)_4O_8$	White, grey, brown	2.62–2.76	6	Triclinic: prismatic crystals flattened {010}; also massive, granular	{001} perfect {010} good	Polysynthetic twinning, generally visible on {001} cleavage surfaces
Amblygonite $(Li,Na)Al(PO_4)(F,OH)$	White, tints of other colours	2.9–3.1	5.5–6	Triclinic, equant crystals; compact or cleavable masses	{100} perfect; {110} good; {0$\bar{1}$1} distinct	Occurs in pegmatites and greisens; commonly has twinning striae on cleavages
Tremolite $Ca_2Mg_5Si_8O_{22}$–$(OH)_2$	White	3.0	6	Monoclinic; typically as columnar or fibrous aggregates	(110) perfect	Occurs in calcareous and magnesian metamorphic rocks, associated with calcite, dolomite, talc
Datolite $CaBSiO_4(OH)$	Colourless, Pale yellow, pale green, pink, white	3.0	6.5	Monoclinic, short prismatic crystals; also massive, granular or compact	None	Occurs in cavities in basic igneous rocks, typically associated with zeolites
Anthophyllite $(Mg,Fe)_7Si_8O_{22}$–$(OH)_2$	White, grey pink	2.9–3.3	6	Orthorhombic; massive, bladed or fibrous aggregates	{210} perfect	Occurs in Mg-rich metamorphic rocks
Lawsonite $CaAl_2Si_2O_7$–$(OH)_2.H_2O$	White, pink grey	3.1	8	Orthorhombic: prismatic and tabular crystals; also massive granular	{100}, {001) perfect {010} indistinct	Occurs in schists and gneisses with glaucophane

Continued

Name Formula	Colour	SG	H	System and Habit	Cleavage	Observations
Spodumene $LiAl_2Si_2O_6$	Colourless, white pink	3.1–3.2	6.5	Monoclinic; long prismatic crystals, also platy masses	{110} perfect; parting on {100}	Occurs in granite pegmatites; will colour a flame red
Andalusite Al_2SiO_5	White, grey, pink, brown	3.15	7.5	Orthorhombic; coarse prismatic crystals, square cross section; also massive, columnar or granular	{110} good	Many grains contain inclusions symmetrically arranged typically occurs in hornfelses
Silimanite Al_2SiO_5	White, brown, grey	3.24	7	Orthorhombic; typically as fibrous aggregate	{010} perfect	Occurs in high-grade schists and gneisses
Diaspore $AlO(OH)$	White, grey, brown	3.3–3.4	6.5	Orthorhombic; massive, bladed, foliated, or compact	{010} perfect	Occurs in Al-rich metamorphic rocks
Diamond C	Colourless, yellow, Pink grey, yellow, black	3.5	10	Isometric; rounded octahedral crystals, commonly twinned	{111} perfect	Occurs in ultrabasic igneous rocks or sands derived from them
Topaz Al_2SiO_4-$(OH,F)_2$	colourless, white, yellow, blue	3.5–3.6	8	Orthorhombic; prismatic crystals terminated by pyramids; also massive, granular	{001} perfect	Occurs in veins and cavities in Si-rich igneous rocks, greisens, and pegmatites

Lustre nonmetallic; typically coloured; streak white unless otherwise stated; $H < 2.5$

Name formula	Colour	SG	H	System and Habit	Cleavage	Observations
Melanterite $FeSO_4 \cdot 7H_2O$	Pale green	1.90	2	Monoclinic; typically stalactitic or powdery crusts	{001} perfect, {110} distinct	Secondary mineral after iron sulphides; soluble in H_2O; tastes like ink
Vermiculite $\sim Mg_3Si_4O_{10}(OH)_2 \cdot 4H_2O$	yellow, brown, green	2.4	1.5	Monoclinic; pseudo-hexagonal plates, pseudomorphs after biotite	{001} perfect	Expands enormously when heated; occurs in Mg-rich rocks
Glauconit $\sim K(Fe,Mg,Al)_2Si_4O_{10}(OH)_2$	Dark green to black	2.5–2.8	2	Monoclinic; rounded grains	{001} perfect not megascopically observable	Streak pale green; occurs in marine sedimentary rocks as pure beds or as grains in limestone, sandstone, mudstone
Vivianite $Fe_3(PO_4)_2 \cdot 8H_2O$	Blue, blue-green	2.7	1–1.5	Monoclinic; in prismatic crystals, elongated ǁc, or earthy masses	{010} perfect	Colour darkens on exposure; streak blue; earthy masses in clay and fossil bones; crystals associated with iron sulphides in ore deposits
Chlorite $(Mg,Fe,Al)_{5-6}(Al,Si)_4O_{10}(OH)_8$	Green	2.6–3.4	2	Monoclinic; in pseudo hexagonal plates or micaceous masses	{001} perfect	Cleavage laminae flexible but not elastic; common in schists
Erythrite $Co_3(AsO_4)_2 \cdot 8H_2O$	Carmine-red	3.06	1.5–2.5	Monoclinic; prismatic crystals and powderly masses	{010} perfect	Streak pink; occurs as coatings or crusts on cobalt arsenides

Continued

Name Formula	Colour	SG	H	System and Habit	Cleavage	Observations
Autunite $Ca(UO_2)_2$-(PO_4)–$\cdot 10\text{-}12H_2O$	Lemon-yellow	3.1–3.2	2–2.5	Tetragonal; thin tabular on {001}; scaly aggregates	{001} perfect	Streak yellow; occurs as an oxidation product from veins or pegematites containing uraninite
Torbernite, Metatorbernite $Cu(UO_2)_2(PO_4)_2$ $12H_2O$	Emerald-green to apple-green	3.2–3.7	2–2.5	Tetragonal; tabular on {001}; foliated or micaceous	{001} perfect	Streak green; occurs as an oxidation product over veins containing uraninite and copper sulphides
Orpiment As_2S_3	Yellow	3.49	1.5–2	Orthorhombic; acicular crystals, elongated ‖c, typically in foliated or granular masses	{010} perfect	Streak yellow; commonly associated with realgar
Realgar AsS	Red	3.56	1.5–2	Monoclinic; short prismatic crystals; typically in granular masses	{010} good	Streak orange; commonly occurs with orpiment
Carnotite $K_2(UO_2)_2(VO_4)_2$-$3H_2O$	Yellow	4–5	2	Monoclinic; typically massive, powdery	{001} perfect, seldom visible	Streak yellow; occurs disseminated in sandstone; like tyuyamunite, but rare

Lustre nonmetallic; typically coloured; streak white unless otherwise stated; H = 2.5 – 5.5

Name Formula	colour	SG	H	System and Habit	Cleavage	Observations
Sulphur S	Yellow	2.07	2.5	Orthorhombic; pyramidal crystals; typically massive, granular or compact	{001}, {110}. indistinct	Burns, giving choking fumes of SO_2; deposited by fumaroles and occurs in limestones and rock gypsum
Chrysocolla $(Cu,Al)_2H_2$-$Si_2O_5(OH)_4 \cdot nH_2O$	Blue, green	2.0–2.4	2–4	Orthorhombic (?); massive, compact	None	Secondary mineral in oxidized copper ores
Chalcanthite $CoSO_4.5H_2O$	Blue	2.28	2.5	Triclinic; short prismatic crystals; also massive, granular	{1$\bar{1}$0} indistinct	Streak pale blue; metallic nauseous taste; occurs in oxidized zone of sulphide copper ores
Serpentine $Mg_3Si_4O_{10}$–$(OH)_8 \cdot 4H_2O$	Green, yellow, brown, greu	2.5–2.6	2.5	Monoclinic; massive, Compact or fibrous (asbestos)	None visible	Alteration product of olivine, typically in ultrabasic igneous rocks
Stilpnomelane $K(Fe,Al)_{10}$–$Si_{12}O_{30}(OH)_8$	Black, greenish black, yellowish	2.59–2.96	3–4	Monoclinic; foliated masses, thin scales	(001) perfect {010} indistinct	Cleavage less perfect than that of biotite, second cleavage perpendicular to first;
Lepidolite $KLi_2Al(Si_4O_{10})$–$(OH)_2$	Colourless, pink, mauve, violet, grey	2.8–2.9	2.5–3	Monoclinic; massive, coarse-to-fine granular	{001} perfect	Pale mauve colour often diagnostic; occurs in granite pegmatites

Continued

Name Formula	colour	SG	H	System and Habit	Cleavage	Observations
Phlogopite $KMg_3(AlSi_3O_{10})(OH)_2$	Pale brown	2.8–2.9	2.5–3	Monoclinic; typically in irregular platy grains	{001} perfect	Occurs in ultrabasic rocks and metamorphosed dolostones
Jarosite $KFe_3(SO_4)_4(OH)_6$	Brown	2.9–3.3	3	Trigonal; massive, compact	{0001} distinct	Resembles limonite, but contains sulphate; yellow streak
Biotite $K(Mg,Fe)_3AlSi_3O_{10}(OH)_2$	Green, brown, black	2.9–3.4	2.5	Monoclinic; typically in irregular platy crystals	{001} perfect	Occurs in granitic igneous rocks, pegmatites and metamorphic rocks
Apatite $Ca_5(PO_4)_3F$	Green blue, brown, yellow, white	3.1–3.2	5	Hexagonal; prismatic crystals, commonlywith "rounded" edges; also massive, granular or compact	{0001} indistinct	Occurs in pegmatities and metamorphosed limestones; also in sedimentary rocks (phosphorite)
Fluorite CaF_2	Purple, green, yellow, colourless	3.18	4	Isometric; cubic crystals, commonly twinned on (111); also massive, granular	{111} perfect	Common vein mineral, and replacing limestones and dolostones
Kyanite Al_2SiO_5	Blue, green, white	3.63	4–7	Triclinic; typically as bladed crystals	{100} perfect; {010} good	Occurs in Al-rich schists and gneisses
Rhodochrosite $MnCO_3$	Pink	3.7	3.5–4	Trigonal; massive, granular	{10$\bar{1}$1} perfect	Effervesces in warm dilute HCl; occurs
Azurite $Cu_3(CO_3)_2(OH)_2$	Blue	3.77	3.5–4	Monoclinic; crystals of various habit, also massive, spherical aggregates	{011} perfect; {100} distinct	Streak blue; effervesces in cold dilute HCl
Siderite $FeCO_3$	Yellow, brown	3.8–4.0	4	Monoclinic; crystals bohedral aggregates, also massive, granular to compact	{10$\bar{1}$1} perfect	Effervesces in warm dilute HCl; occurs in veins and as replacement of limestone

Continued

Name Formula	colour	SG	H	System and Habit	Cleavage	Observations
Antlerite $Cu_3(SO_4)(OH)_4$	Dark green	3.88	3.5	Orthorhombic; typically in granular aggreates or cross-fibre veinlets	{010} perfect	Streak green; soluble in HCl without efervescence
Malachite $Cu_3(CO_3)(OH)_2$	Green	3.6–4.1	3.5–4	Monoclinic; massive compact, commonly stalactitic or botryoidal	{$\bar{2}$01} perfect; {010} distinct	Streak green; effervesces in cold and dilute HCl
Goethite (Limonite) FeO(OH)	Yellow, brown	3.3–4.1	5–5.5	Orthorhombic; massive, compact or earthy, also stalactitic	{010} perfect; {100} good	Streak yellow; dissolves in dilute HCl giving a yellow solution
Brochantite $Cu_4(SO_4)(OH)_6$	Green	4.0	3.5 –4	Monoclinic; prismatic to acicular crystals; also massive granular	{100} perfect	Streak green; dissolves in cold dilute HCl without effervescence
Sphalerite (Zn,Fe)s	Yellow, brown, black	4.0–4.1	3.5 – 4	Isometric; typically massive, coarse-to fine granular	{110} perfect	Dissolves in warm HCl giving off H_2S
Wurtzite ZnS	Brownish black	4.09	3.5 – 4	Hexagonal; pyramidal or prismatic crystals; typically massive, fibrous, columnar, or as banded cursts	{11$\bar{2}$0} perfect	Streak brown; much less common than sphalerite
Xenotime YPO_4	Yellowish to reddish brown	4.4–5.1	4–5	Tetragonal prismatic with pyramidal faces resembling zircon	{100} perfect	Accessory mineral in acidic igneous rocks, pagmatites, and gneisses
Monazite $CePO_4$	Yellow, brown	4.6–5.4	5–5.5	Monoclinic; small crystals flattened ‖ {100}, also as detrital grains	{100} distinct; parting on {001}	Accessory mineral in granites and pegmatites; as concentrates in sands and sandstones

Continued

Name Formula	colour	SG	H	System and Habit	Cleavage	Observations
Pyrochlore-microlite $NaCa(Nb,Ta)_2$-O_6F_2	Yellow, brown, black	4.2–6.4	5–5.5	Isometric; octahedral crystals and rounded grains	{111} distinct	Occurs in carbonatites (pyrochlore) and granite pegmatites
Thorite $ThSiO_4$	Orange, brown, black	4– 6.7	5	Tetragonal; pyramidal crystals; also massive, compact	{100} distinct	Strongly radioactive; occurs in pegmatites and as detrital grains
Zincite ZnO	Red	5.7	4	Hexagonal; platy masses or rounded grains	{10$\bar{1}$0} perfect	Streak orange occurs with franklinite and willemite
Crocoite $PbCrO_4$	Orange, red	6.1	2.5 – 3	Monoclinic; platy as prismatic crystals	{110} distinct	Streak orange–yellow; rare secondary mineral in lead viens
Cuprite Cu_2O	Red, often tarnished	6.1	3.5 – 4	Isomatric; cubic crystals, also massive, granular or hairlike	{111} indistinct	Streak red-brown; easily reduced to copper on charcoal
Wulfenite $PbMoO_4$	Yellow, red, brown	6.5–7.0	3	Tetragonal; typically square, platy crystals, tabular ‖ {001}	{101} distinct	Commonly associated with vanadinite
Vanadinite $Pb_5(VO_4)_3Cl$	Yellow, red, brown	6.9	2.5 –3	Hexagonal; typically small hexagonal prisms	None	Streak yelow; occurs in the oxidation zone of lead deposites
Pyromorphite $Pb_5(PO_4)_3Cl$	Green, Yellow, brown	7.0–7.1	3.5 –4	Hexagonal; in small prismatic crystals; also massive, globular and botryoidal	None	Occurs in the oxidized zone of lead deposits
Cinnabar HgS	red	8.1	2–2.5	Hexagonal; massive, granular	{10$\bar{1}$0} perfect	Streak red; volatilizes on heating

Lustre nonmetallic; typically coloured; streak white unless otherwise stated; H > 5.5

Name Formula	Colour	SG	H	System and Habit	Cleavage	Observations
Sodalite $Na_8(AlSiO_4)_6Cl_2$	Blue, white,	2.3	6	Isometric; massive, granular	{110} indistinct	Blue colour characteristic; occurs in nepheline syenites
Cancrinite $Na_6Ca_2(AlSiO_4)_6(CO_3)_2$	Yellow, white, pink	2.4–2.5	6	Hexagonal; massive, granular	{10$\bar{1}$0} perfect	Effervesces in warm dilute HCl; yellow colour
Cordierite $Al_3(Mg,Fe)_2AlSi_5O_{18}$	Colourless, blue, grey, brown	2.55–2.75	7	Orthorhombic; massive, granular	{010} indistinct; parting on {001}	Occurs in aluminous metamorphic rocks
Turqousie $CuAl_6(PO_4)_4(OH)_8.4H_2O$	Blue, blue, White, yellow, green.	2.6–2.8	5–6	Triclinic; massive, crystals or massive granular	{001} distinct	Occurs in veinlets in pegmatites
Prehnite $Ca_2Al_2Si_3O_{10}(OH)_2$	White, pale to apple-green.	2.9	6.5	Orthorhombic; generally massive; encrusing with crystalline surface	{001} good	Occurs in cavities in basic igneous rocks associated with zeolites and in low-grade metamorphic rocks
Melilite $Ca_2Al(AlSiO_7)$-$Ca_2Mg(Si_2O_7)$	Nearly colourless, yellow or reddish brown, greenish, grey	2.94–3.05	5–6	Tetragonal; short square prisms	{001}, indistinct {110}	Conchoidal fracture; in metamorphosed carbonate rocks and Si-deficient igneous rocks

Continued

Name Formula	Colour	SG	H	System and Habit	Cleavage	Observations
Tourmaline Na,Mg,Fe,Al borosilicate	Black, brown, pink, green, blue	3.0–3.2	7.5	Trigonal; prismatic crystals with rounded trigonal cross section; also massive, columnar	{11$\bar{2}$0}, {1011} indistinct	Prism faces vertically striated; occurs in granite pegmatites, schists, and gneisses
Glaucophane $Na_2Mg_3Al_2Si_8$–$O_{22}(OH)_2$	Blue, lavender, blue-black	3.0–3.3	6	Monoclinic; commonly in columnar or acicular crystals, or as felt or silky needles	{110} perfect	Occurs in schists, and gneisses
Actinolite $Ca_2(Mg,Fe)_5$ $Si_8O_{22}(OH)_2$	Green	3.0–3.4	6	Monoclinic; commonly in columnar or fibrous aggregates	{110} perfect	Occurs in schists, gneisses, and amphibolites
Hornblende $NaCa_2(Mg,Fe,$–$Al)_5(Al,Si)_8O_{22}$–$(OH)_2$	Green, brown, black	3.0–3.4	6	Monoclinic; long prismatic crystals, also columnar, fibrous or granular	{110} perfect	A common mineral in both igneous and metamorphic rocks
Pumpellyite $Ca_2MgAl_2(SiO_4)$–$(Si_2O_7)(OH)_2$–$.H_2O$	Bluish–green to brown to nearly white	3.18–3.23	5.5—6	Monoclinic; minute fibres or narrow plates	{001} distinct; {100} less distinct	Occurs with prehnite and zeolites in amygdules and low-grade metamorphic rocks
Humite $(Mg, Fe)_7(SiO_4)_3$–$(F,OH)_2$	Yellow, honey-yellow, reddish yellow, brownish red	3.15–3.35	6–6.5	Monoclinic (clinohumite and chondrodite); orthorhombic (humite and norbergite); irregular masses	{100} indistinct	Brittle; subconchoedal fracture; occurs in contact metamorphosed dolostones

Continued

Name Formula	Colour	SG	H	System and Habit	Cleavage	Observations
Axinite Ca, Mn, Fe, Al borosilicate	Brown, Yellow	3.2–3.3	7	Triclinic; wedge-shaped crystals; also massive granular	{100} good	Occurs in contact zones of igneous intrusions
Diopside $Ca(Mg, Fe)Si_2O_6$	White, green	3.25–3.40	6	Monoclinic; prismatic crystals, subspherical grains; massive, granular	{110} good	Occurs in metamorphosed dolostones
Jadeite $NaAlSi_2O_6$	White, green	3.3	6–6.5	Monoclinic; massive, granular to compact	(110) good	Metamorphic mineral
Zoisite $Ca_2Al_3Si_3O_{12}$–(OH)	Grey, green, pink	3.3	7	Orthorhombic; massive; columnar or bladed	{001} perfect	Occurs in metamorphic rocks rich in lime and alumina
Allanite $(Ca,Ce)_2(Al,$–$Fe)_3Si_3O_{12}OH$	Black, dark brown	2.7–4.0	6.5	Monoclinic; platy or columnar crystals; also massive, granular	{001} seldom observable	Streak grey or green-grey; typically somewhat radioactive
Titanite $CaTiSiO_5$	Brown, yellow, green	3.5	6	Monoclinic; typically in diamond-wedge-shaped crystals	{110} or distinct	An accessory mineral of igneous and metamorphic rocks
Rhodonite $MnSiO_3$	Pink	3.5–3.7	6	Triclinic; tabular crystals; massive, granular to compact	{100}, {010} perfect	Commonly veined with black manganese oxides
Hypersthene $(Mg,Fe)SiO_3$	Brown, black	3.4–4.0	6	Orthorhombic; massive, rarely platy	{210} good	Occurs in basic igneous rocks
Cummingtonite $(Fe,Mg)_7Si_8O_{22}(OH)_2$	Brown	3.2–3.6	6	Monoclinic; typically massive or as fibrous	{110} perfect	Occurs in Fe-rich schists and gneisses
Augite $(Ca,Na)(Mg,Fe,Al,Ti)(Si,Al)_2O_6$	Black, dark green	3.25–3.55	6	Monoclinic; short prismatic crystals; also massive, granular	{110} good	The common pyroxene of igneous rocks

Continued

Name Formula	Colour	SG	H	System and Habit	Cleavage	Observations
Vesuvianite (Ca,Mg,Al) silicate	Brown, green, yellow, blue	3.3–3.5	7	Tetragonal; short prismatic or low pyramidal crystals; also massive granular	{110} indistinct	Occurs in metamorphosed limestones
Riebeckite $Na_2(Mg,Fe)_5Si_8O_{22}(OH)_2$	Dark blue, black	3.3–3.6	6	Monoclinic; bladed crystals, also massive granular; in some	{110} perfect	Blue-grey streak; occurs in Na-rich igneous rocks
Olivine $(Mg,Fe)_2SiO_4$	Olive-green	3.3–3.6	6.5	Orthorhombic; granular, massive	{010} indistinct	Occurs in basic and ultrabasic igneous rocks
Clinozoisite, Epidote $Ca_2(Al,Fe)_3$–$Si_3O_{12}(OH)$	Bilious green to black	3.3–3.6	7	Monoclinic; prismatic crystals elongated; also finely massive, columnar or granular	{001} perfect	Yellow-green colour typical; occurs in metamorphic rocks and veins
Hedenbergite $Ca(Fe,Mg)Si_2O_6$	Green to black	3.40–3.55	6	Monocllinic; massive, Isometric; or fibrous	{110} good	Occurs in skarn associated with ore
Spinel $MgAl_2O_4$	Black, brown, green, red, blue	3.5–3.6	7.5–	Isometric; octahedral crystals, commonly twinned on {111}	{111} None; parting	Occurs in metamorphic rocks and ultrabasic igneous rocks and in sands derived from them
Chloritoid–$(Fe,Mg,Mn)_2$–$(Al,Fe)Al_3$–$O_2(SiO_4)_2$–$(OH)_4$	Green (commonly grassy green)	3.58–3.61	6.5	Monoclinic; commonly foliated masses	{001} perfect; {110} distinct	Cleavage flakes are brittle; occurs in low-to-medium-grade metamorphic Fe-and Al-rich sedimentary rocks

Continued

Name Formula	Colour	SG	H	System and Habit	Cleavage	Observations
Aegirine $NaFeSi_2O_6$	Dark green, brown, black	3.6	6	Monoclinic; typically in prismatic crystals, often elongated ‖ c	{110} perfect	Occurs in Na-rich igneous rocks
Grossular $Ca_3Al_2Si_3O_{12}$	White, yellow, green, pink, brown	3.0	7–7.5	Isometric; dodecahedra and trapezohedra, and combinations; also massive, granular or compact	None	Occurs in metamorphosed limestones
Pyrope $Mg_3Al_2Si_3O_{12}$	Red to nearly black	3.6	7–7.5	Isometric typically in anhedral grains	None	Occurs in high-grade metamorphic rocks
Chrysoberyl $BeAl_2O_4$	Yellow, green, brown,	3.7–3.8	8.5	Orthorhombic; crystals tabular ‖ {001}, and pseudohexagonal twins	{110} distinct	Occurs in granite pegmatites and in alluvial deposits
Staurolite $(Fe,Mg,Zn)_2Al_9$–$(Si,Al)_4O_{22}(OH)_2$	Brown	3.7–3.8	7	Orthorhombic; prismatic crystals	{010} distinct gneisses	Occurs in medium-grade schists and
Uvarovite $Ca_3Cr_2Si_3O_{12}$	Green	3.8–	7–7.5	Isometric; typically as dodecahedra	None	Occurs in metamorphosed with serpentine and chromite
Andradite $Ca_3Fe_2Si_3O_{12}$	Yellow green, brown, black	3.8–3.9	7–7.5	Isometric; dodecahedr and trapezohedra, and combinations; also massive	None	Occurs in metamorphosed limestones and associated with ore deposits
Anatase TiO_2	Brown to colourless	3.9	5.5–6	Tetragonal; typically as acute pyramidal crystals	{001}, {101} perfect	Occurs is small crystals in joint planes and veins in schists and gneisses

Continued

Name Formula	Colour	SG	H	System and Habit	Cleavage	Observations
Corundum Al_2O_3	Grey, blue, pink, brown	4.0	9	Trigonal; hexagonal prisms and pyramids; also massive, granular	None; parting on {0001}, {0112}	Occurs in metamorphic rocks, and in some Al-rich igneous rocks
Willemite Zn_2SiO_4	yellow green, brown, black	3.9–4.2	5.5–6	Trigonal; massive granular indistinct	{0001} good; {1120}	Much fluoresces green in ultraviolet radiations
Spessartine $Mn_3Al_2Si_3O_{12}$	Orange, red, brown	4.0–4.2	7–7.5	Isometric; trapezohedra, some in combination with dodecahedra; also massive granulor	None	Occurs in granite pegmatites and metamorphosed Mn-rich rocks
Brookite TiO_2	Brown to black	4.1	5.5–6	Orthorhombic; tabular and prismatic crystals	{120} indistnct	Occurs in joint planes and veins in schists and gnesses
Almandine $Fe_3Al_2Si_3O_{12}$	Red to black	4.0–4.3	7–7.5	Isometric;dodecahedra, trapezohedra, and combinations; also massive, granular	None	Occurs in metamorphic rocks; the most common granet
Zircon $ZrSiO_4$	Brown, grey	3.9–4.7	7.5	Tetragonal; prismatic crystals terminated by pyramids, and as rounded grains	{110} distinct; {111} indistinct	An accessory mineral of igneous and metamorphic rocks; a detrital mineral
Rutile TiO_2	Red-brown to blcak	4.2–4.5	6–6.5	Tetragonal; commonly as long prismatic crystals or massive	{110} distinct	Common accessory mineral of igneous and metamorphic rocks ; also in quartz veins
Cassiterite SnO_2	Brown to black	7.0	6.5	Tetragonal; prismatic crystals; massive, botryodial or concretionary	{100} distinct; {110} indistinct	Occurs in quartz veins and granite pegmatites; also detrital

IGNEOUS ROCKS

GRANITE CLAN

Plutonic

Adamelite
Ailsyte
Alaskite
Birkremite
Gneissoid Granite
Granite
Granite Porphyry
Granodiorite
Graphic granite
Greisen
Kalialaskite
Luxullianite
Northfieldite
Protogine
Unakite

Hypabyssal

Alsbachite
Aplite
Beresite
Brandergite
Falsite
Granophyre
Kalitordrillite
Leopardite
Lindoite
Markfieldite
Microadamellite
Paisanite
Pegmatite
Tollite
Topazite
Tsingtauite

Extrusive

Aporhyolite
Cantalite
Comendite
Cumbraite
Dacite
Khagiarite
Nevadite
Pantellerite
Perlite
Pumice
Rhyolite
Todrillite
Toscanite

SYENITE CLAN

Plutonic

Akerite
Beloeilite
bigwoodite
Canadite
Congressite
Covite
Deldoradite
Foyaite
Highwoodite
Hilairite
Husebyite
Juvite
Kalisyenite
Lakarpite
Larvikite
Laurdalite
Ledmorite
Litchfieldite
Lugarite
Lusitanite
Malignite
Monmouthite
Nordmarkite
Orthoclasite
Pulaskite
Shonkinite
Sommaite
Syenite
Tonsbergite
Trowlesworthite
Umptekite
Uncompahgrite

Hypabyssal

Albitite
Albitophyre
Allochetite
Borolanite
Cascadite
Durbachite
Elkhornite
Hedrumite
Heronite
Heumite
Kamperite
Minette
Muniongite
Rhomb porphyry
Solvsbergite
Sussexite
Tinguite
Ulrichite
Vogisite
Wennebergite

Extrusive

Arsoite
Blairmorite
Ciminite
Drakonite
Gibelite
Jumillite
Keratophyre
Madupite
Mugearite
Orendite
Selagite
Shackanite

DIORITE CLAN

Plutonic

Anchorite
Cavalorite
Diorite
Dungannonite
Glenmuirite
Gooderite
Hirnantite
Kentallenite
Laugenite
Oroite
Puglianite
Sebastianite
Tonalite

Hypabyssal

Akenobeite
Andesinite
Beringite
Camptonite
Cuselite
Helsinkite
Holyokeite
Kersantite
Krageroite
Kulaite
Leideite
Maenaite
Mondhaldeite
Plumasite
Timazite
Vaugnerite
Vintilite
Yatalite

Extrusive

Absaroikite
Andesite
Hungarite
Kaiwekite
Marloesite
Ordanchite
Ottajanite
Pollenite
Propylite
Santorinite
Skomerite
Volcanite
Weiselberigite

GABBRO CLAN

Plutonic

Allalinite
Anorthosite
Bojite
Corsite
Essexite
Gabbro
Labradite
Norite
Oligoclasite
Ossipite
Rougemontite
Routivarite
Theralite
Troctolite
Yamaskite

Hypabyssal

Aasby diabase
Alnoite
Ankaramite
Anorthitissite
Baldite
Crinanite
Devonite
Dolerite
Eucrite
Harrisite
Heptorite
Holmite
Kylite
Navite
Odinite
Soggendalite
Teschenite
Tholeiite

Extrusive

Alboranite
Anamesite
Arapahite
Atlanite
Basalt
Buchonite
Caltonite
Domite
Dorgalite
Macedonite
Mimosite
Nonesite
Schonfelsite
Shoshonite
Spilite
Subdutite
Tephrite

SILICA UNDER STURATED FELDSPATHOID ROCKS

Plutonic

Fasinite
Fergusite
Ijolite
Italite
Melteigite
Missourite

Hypabyssal

Arkite
Katzenbuckelite

Extrusive

Basanite
Kajanite
Leucitite
Leucitophyre
Nephelinite
Wyomingite

PYROCLASTIC IGNEOUS ROCKS

Agglomerate
Breccia (volcanic
Ignimbrite
Palagonite tuff
Sillar
Tuff

ULTRABASIC

Plutonic

Alexoite
Algarvite
Allivalite
Anabohitsite
Bahiate
Bebedourite
Biotitite
Bronzitfels
Davidite
Dunite
Harzburgite
Hornblendite
Hypersthenite
Jacupirangite
Kiirunavaarite
Madeirite
Montrealite
Peridotite
Perknite
Pyroxenite
Saxonite
Scyelite
Serpentinite
Websterite
Wohrlite
Kvellite
Lherzolite
Monchiquite
Newlandite
Ouachitite
Schriesheismite
Valbellite

Extrusive

Augitite
Batukite
Bielenite
Coppaelite
Picrite
Verite

Hypabyssal

Avezakite
Barashawite
Chromitite
Cortlandtite
Cromaltite
Dlopsldlte
Ilmenitite
Kimberlite

SEDIMENTARY ROCKS

CLASTIC

Rudaceous

Basal conglomerate
Breccia
Calcrete
Conglomerate
Coombe rock
Ferricrete
Molasse
Placer deposits
Pseudoconglomerate
Scree deposits

Arenaceous

Bone bed
China stone
Ganister
Greensand
Greywacke
Grit
Laterite
Lateritite
Placer deposits
Sandstone
Turbidite

Argillaceous

Adobe
Alphitite
Alum shale
Argillite
Bentonite
Blae
Blue mud
Boulder clay
Caliche
Clay
Fireclay
Fuller's earth
Loess
Marl
Mudstone
Ooze
Siderite mudstone
Terra rossa
Tillite
Tilloid
Turbidite
Varve clay
Varvite

CARBONATE ROCKS

Algal limestone
Anthraconite
Cementstone
Chalk
Cipolino
Coquina
Cornstone
Landscape marble
Limestone
Magnesite rock
Oolitic limestone
Pisolitic limestone
Pseudobreccia
Sideritic limestone
Tufa

FERRUGINOUS ROCKS

Black-band ironstone
Chert ironstone
Clay ironstone
Ironstone
Oolitic ironstone
Pisolitic ironstone

SILICEOUS ROCKS

Diatomite
Lydite
Malmstone

ORGANIC ROCKS

Anthracite
Asphalt
Bitumen
Bituminous coal
Boghead coal
Coal
Lignite
Oil shale

EVAPORITES

Evaporite

METAMORPHIC ROCKS

REGIONAL	CONTACT	DYNAMIC
High grade	***High grade***	Cataclasite
Amphibolite	Astite	Flaser gabbro
Eclogite	Buchite	Flaser gneiss
Gneiss	Calc-flinta	Flinty crush rock
Gondite	Catawberite	Kakirite
Migmatite	Charnockite	Mylonite
Ollenite	Cornubianite	Protomylonite
Prasinite	Enderbite	
Quartzite	Epidiorite	
	Epidosite	
Low grade	Hornfels	
Augen schist	Limurite	
Green schist	Porcellanite	
Itabirite	Quartzite	
Kinzigite		
Lavialite	***Low grade***	
Leptynite	Adinole	
Marble	Phyllite	
Mica schist	Skarn	
Murasakite	Slate	
Phyllite	Spotted slate	
Pinolite		
Schist		